DPH PHYSICS SERIES

ATOMIC AND NUCLEAR STRUCTURE

By

D.K. Jha

DISCOVERY PUBLISHING HOUSE
NEW DELHI-110002

Reprinted - 2019

First Published - 2004

ISBN: 978-81-7141-869-5

Atomic and Nuclear Structure

Published by:

DISCOVERY PUBLISHING HOUSE PVT. LTD.
4383/4B, Ansari Road, Darya Ganj
New Delhi-110 002 (India)
Phone: +91-11-23279245, 23253475; 43596065
E-mail: discoverybooksindia@gmail.com
discoverypublishinghouse@gmail.com
web: www.discoverypublishinggroup.com

Printed at:
Infinity Imaging Systems
Delhi

Preface

Atomic and nuclear structure has been chiefly governed by our adherence to the aim of covering the syllabi prescribed by the various Indian Universities for the students offering physics at the graduate and post graduate level. All along, it has been our endeavour to present the subject matter in a simple and lucid style and to arrange it in a sequential way so that the reader may not feel any discreteness in going through the book from beginning to the end.

The subject matter in such a simple way that the student should feel no difficulty to understand the article have been explain in details in a nice manner and all the example has been completely solved.

Author

Contents

1

Theory of Atomic Structure

1.1 J.J. THOMSON'S MODEL

Rutherford's experiment of radioactivity proved the facts that atom consists of positively and negatively charged particles. The question arose for the equilibrium of the atoms. Thomson suggested that the atom was spherical in shape. The whole mass of the atom was evenly distributed and the positive charge was distributed all over the mass. The electrons with negative charge were embedded within the atom (Fig. 1.1). The whole positive charge of the atom was equal to the total charge of the electrons. The atom was like a plum *pudding*. The electrons oscillate or vibrate about their mean positions.

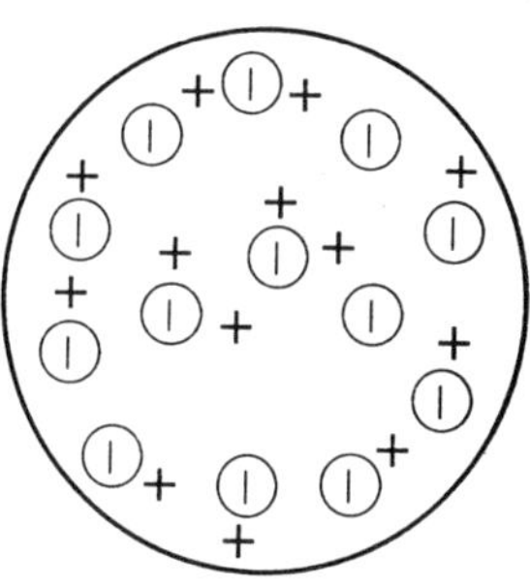

Fig. 1.1

The main feature's of Thomson Model were:

1. It gives a mechanism for the emission of the eletro-magnetic waves. The electro-magnetic waves were supposed to be emitted by the oscillating electrons in the atom.
2. It explained the emission of electrons by heating as thermions and in photo-electric effect.
3. It provided an elastic hard model required for the kinetic theory of gases.

1.2 RUTHERFORD'S EXPERIMENT SCATTERING OF α-PARTICLES AND RUTHERFORD MODEL OF THE ATOM

Rutherford's experiments are shown in Fig. 1.2. α-particles are emitted from the radioactive source A. Which incident on a thin gold

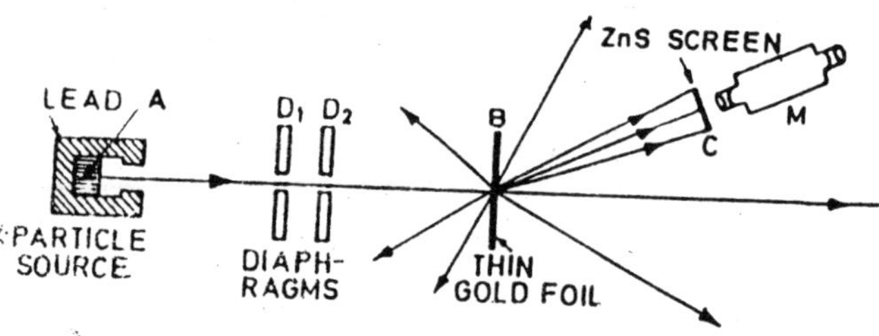

Fig. 1.2

foil B. While passing through the gold foil, the α-particles are scattered through different angles. The scattered α-particles in a particular direction are allowed to strike a screen C coated with zinc sulphide. When an α-particle is incident on zinc sulphide, it produces fluorescence and it is detected with the help of the microscope M. In some cases, the α-particles are scattered through large angles and some are even reversed.

Observation of Rutherford's experiment concluded that the atoms are consisted by a heavy positive charge portion in the atom. Rutherford said to it nucleus which is containing the positive fundamentals which particles and he concluded are in other space than the nucleus. The orientation of electrons was not clear to Rutherford how the electrons are arranged.

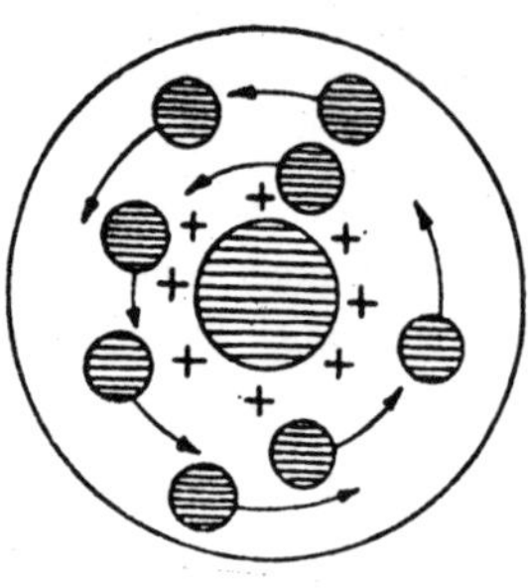

Fig. 1.3

1.3 DRAWBACK THE RUTHERFORD'S EXPERIMENT

After the Rutherford's Model the question of stability are that if there are negatively charged electrons are present in the atom. It must be continuously attracted towards the nucleus. An accelerated electrons emits electro-magnetic radiations. If must looses its energy and finally falls in the nucleus but such types of phenomenon does not exist in the atoms, so Rutherford's experiment was unable to explain the stability of the atom.

1.4 DERIVATION OF RUTHERFORD'S SCATTERING FORMULA

In deriving the scattering formula the following assumption are made:

(i) The wave aspect of the atom being neglected and considered that call is an between two particles, α-particles and nucleus.

(ii) The nucleus is considered to be so heavy that its impact may be disregarded.

(iii) The nucleus and the α-particle are considered as point charges, *i.e.,* mere centres of Coulombian force; thus, the dimensions of the interacting particles are not taken into account.

An α-particle moving along PO approach the relatively heavy nucleus, stationary at N (Fig. 1.4). Since both are positively charged, there is a force of repulsion between them, which being governed by the Coulomb's law of inverse squares, will increase enormously as the α-particle gets closer to the nucleus. Applying the properties of motion in central orbits to the α-particle thus repelled by the nucleus, it can be shown that the path of the α-particle will, in general, change from a straight line to a hyperbola PAP', one of whose foci is N and whose asymptotes PO give the initial and final directions of the α-particle.

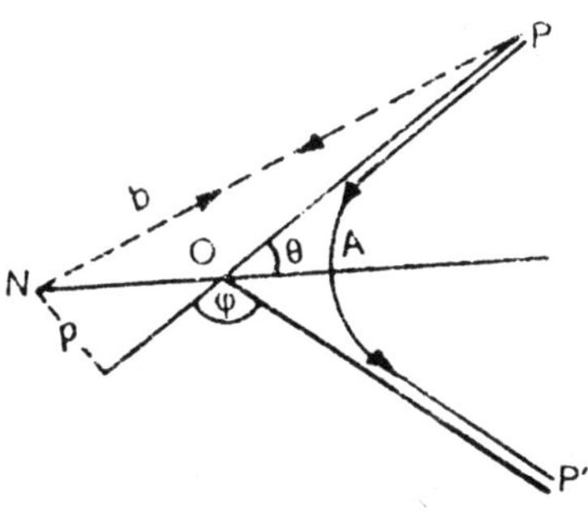

Fig. 1.4

Let the perpendicular distance of PO from N be p, *i.e.,* the shortest distance from the nucleus to the initial direction, which is called the *impact parameter*. Let m be mass of the α-particle, 2e its charge and υ_0 its initial velocity. Let Z be the atomic number of the element that scatters the α-particles, so that Ze is the charge on the nucleus. The angle of deviation or scattering of the α-particle is evidently

The magnitude of ϕ will depend on several factors, such as the charge of the nucleus Ze, the charge of the α-particle 2e, its mass m, its velocity υ_0 and impact parameter P.

To derive an expression for ϕ, let us consider the case of a central impact, *i.e.,* when the α-particle is directed straight towards N so that $p = 0$. On account of the repulsive force, the α-particle will be stopped at a certain distance b from the nucleus N and made to retrace its path, in which case ϕ is equal to 180°. This distance of closest approach b can be determined by using the principle of conservation of energy.

The electrostatic potential at a distance b due to the nucleus

$$= \frac{Ze}{4\pi \epsilon_0 b}$$

This acts on a charge 2e of the α-particle. Hence the potential energy of an α-particle when it is at the distance b from the nucleus

$$= \left(\frac{Ze}{4\pi \epsilon_0 b}\right) 2e = 2\frac{Ze^2}{4\pi \epsilon_0 b}$$

Since the α-particle is momentarily stopped at the distance b, its initial kinetic energy is completely changed into potential energy. Neglecting the loss of energy due to interaction with the peripheral electrons

$$\frac{1}{2} m\upsilon_0^2 = \frac{2Ze^2}{4\pi \epsilon_0 b}$$

or

$$b = \frac{Ze^2}{\pi \epsilon_0 m\upsilon_0^2} \qquad ...(1)$$

As it is not possible, in practice, to direct the α-particle exactly towards the nucleus, we must consider the case when $p \neq 0$. In such a case, the α-particle will be deflected through an angle ϕ, which is less than 180° and will travel along the hyperbolic path PAP'

Let V be the velocity of the α-particle at the vertex A. Using the principles of conservation of energy and of momentum and taking distance NA = d

$$\frac{1}{2} m\upsilon_0^2 = \frac{1}{2} m V^2 + \frac{2Ze^2}{4\pi \epsilon_0 d} \qquad ...(2)$$

$$m\upsilon_0 p = mVd \qquad ...(3)$$

Substituting from (1), the value of $2Ze^2$, *viz.*, $2\pi\in_0 bm\upsilon^2_0$ in equation (2), we get

$$\frac{1}{2}m\upsilon_0^2 = \frac{1}{2}m\ V^2 + \frac{2\pi \in_0 bm\upsilon_0^2}{4\pi \in_0 d}$$

$$\therefore \qquad V^2\ \upsilon_0^2\left(1-\frac{b}{d}\right) \text{ or } \frac{V^2}{\upsilon_0^2}\left(1-\frac{b}{d}\right)$$

From equation (3), we get

$$p^2 = \frac{V^2}{\upsilon_0^2}d^2 = d^2\left(1-\frac{b}{d}\right) = d(d-b)$$

Using the properties of the hyperbola, *viz.*,

$$\in = \frac{1}{\cos\theta}, \text{ where } \theta = \left(\frac{r-\phi}{2}\right), \text{ and On } = \in.OA.$$

$$\text{or} \qquad d = NO + OA = NO\left(\frac{1+1}{\in}\right) = NO(1+\cos\theta)$$

$$= \left(\frac{p}{\sin\theta}\right)(1+\cos\theta) = p\left\{\frac{(1+2\cos^2(\theta/2)-1}{(2\sin\theta/2\cos(\theta/2)}\right\}$$

$$= \frac{p.\cot\theta}{2}$$

$$\therefore \qquad p^2 = p\left(\frac{\cot\theta}{2}\right)\left[p\ \cot\left(\frac{\theta}{2}\right) - b\right]$$

$$\text{or} \qquad p = p\left[\cot^2\left(\frac{\theta}{2}\right)\right] - b\cot\frac{\theta}{2}$$

$$\therefore \qquad b = p\frac{[\cot^2\theta/2)-1]}{\cot\theta/2} = 2p\ \cot\theta = 2p\cot\left(\frac{\pi-\theta}{2}\right)$$

$$\text{or} \qquad b = \frac{2p\tan\phi}{2} = \frac{2Ze^2}{4\pi \in_0 b}$$

Substituting for b from equation (1)

$$\tan\frac{\phi}{2} = \frac{b}{2p}\ \frac{2Ze^2}{4\pi \in_0 m\upsilon_0^2 p} \qquad ...(4)$$

This relation shows that Z, m and υ_0 being kept constant, as the impact parameter p decreases from relatively high value upto the limit zero, ϕ increases from 0 to 180°. This means that when the α-particle passes for away from the nucleus (*i.e.*, p is large) the angle of scattering is very small. As the distance p diminishes, *i.e.*, as the α-particle passes closer and closer to the nucleus, the angle of scattering will become larger and larger and in the limiting case of central impact (p = 0), ϕ = 180°, *i.e.*, the α-particle will be forced to retrace its path after approaching the nucleus upto a distance b.

We shall now deal with the case realised in the actual experiment where a narrow beam of α-particles is incident normally on a thin foil of the scatterer, and the scattered particles are detected by means of scintillations produced by them on a fluorescent screen normal to the direction of view.

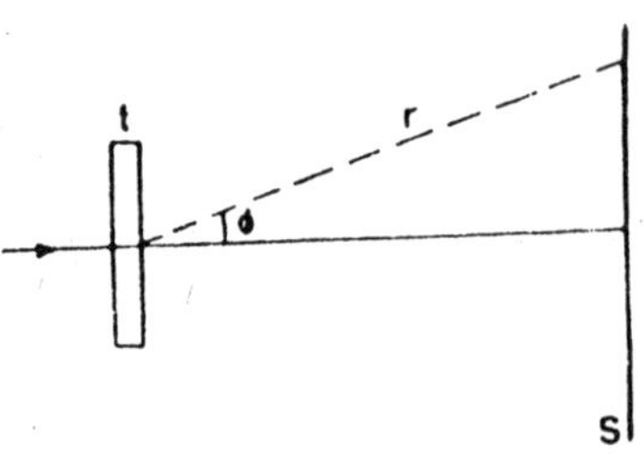

Fig. 1.5

Let n be the number of atoms per unit volume of the scatterer of thickness t. Let Q be the total number of a-particles that strike unit area of the scatterer. From simple probability considerations the number of α-particles N that are scattered through an angle ϕ and strike unit area of the fluorescent screen S at a distance r (Fig. 1.5) can be estimated as follows:

The scattered atoms is distributed, the probable number of α-particle within the distance of the impact parameter p of a nucleus is given by

$$\pi p^2 Qtn \qquad ...(5)$$

Hence the number of α-particles having an impact parameter between p and p + dp is

$$d(\pi p^2 Qtn) = 2\pi Qntpdp$$

After impact, these particles will be deflected through an angle between ϕ and $\phi + d\phi$. Thus the number of α-particles scattered between angles ϕ and $\phi + d\phi$ is

$$2\pi Qntpdp \qquad ...(6)$$

From equation (4),

$$p = \left(\frac{b}{2}\right)\cot\frac{\phi}{2} \text{ and } dp = \left(\frac{b}{2}\right)\left(-\frac{1}{2}\text{cosec}^2\frac{\phi}{2}\right)d\phi$$

Substituting these values in equation (6), the number of α-particles scattered between ϕ and $\phi + d\phi$ is

$$= -2\,\pi Q n t \times \left(\frac{b}{2}\right)\cot\frac{\phi}{2} \times \left(\frac{b}{4}\right)\text{cosec}^2\left(\frac{\phi}{2}\right)d\phi$$

$$= -\left(\frac{\pi}{4}\right)Q n t b^2 \cot\frac{\phi}{2}\,\text{cosec}^2\left(\frac{\pi}{2}\right)d\phi$$

All these α-particles will strike the screen within a circular annulus, of area $2\pi r \sin\phi \times r d\phi = 2\pi r^2 \sin\phi d\phi$. Hence the number N of α-particles striking unit area of the screen at an inclination ϕ from the incident direction is given by

$$N = \frac{(\pi/4)\,Q\,n t b^2 \cot\phi/2\,\text{cosec}^2\,(\phi/2)\,d\phi}{2\pi t^2 \sin\phi\, d\phi}$$

$$= \frac{Q\,n t b^2 \cot\phi/2\,\text{cosec}^2\,\phi/2}{16 r^2 \sin\phi/2 \cos\phi/2}$$

$$= \frac{Q\,n t b^2}{16 r^2 \sin^4\phi/2}$$

Substituting the value of b from eqn. (1),

$$N = \frac{Q\,n t Z^2 e^4}{16\pi^2 \epsilon_0^2 r^2 m^2 \upsilon_0^4 \sin^4\phi/2} \qquad ...(7)$$

According to the above expression, if Rutherford's concept of the nuclear atom is correct, N must be proportional to (i) $1/\sin^4\phi/2$, (ii) the thickness t of the scatterer, (iii) Z^2 the square of the atomic number of the and (iv) $\left(\frac{1}{m\upsilon_0^2}\right)^2$ *i.e.,* inversely to the square of the initial kinetic energy or to the fourth power of the initial velocity of the α-particles. These observations were verified by the Rutherford by the above experiment Rutherford found the size of nucleus of the of 10^{-14}m and size of the atom of the order of 10^{-14}m.

The experiments on the scattering of α-particles confirmed Rutherford's theory so well that the nuclear atom model was at once universally accepted Rut Rutherford himself soon realised that his model was not free from limitations, the chief among them arising from considerations of the distribution of the electrons outside the nucleus and the stability of the atom as a whole. For, it became obvious that in the nuclear atom, *equilibrium could not be secured by the operation of electrostatic forces* alone between the positively charged nucleus and the negatively charged electrons outside the nucleus. For instance, considering the case of an atom with two electrons, the nuclear charge is +2e. If the electrons are symmetrically placed at a distance r from the nucleus, the force of attraction between the nucleus and each of the electrons is $\dfrac{2e^2}{4\pi \epsilon_0 r^2}$ while the force of repulsion between the electrons is $\dfrac{e^2}{4\pi \epsilon_0 (4r^2)}$ Since the force of attraction is eight times greater than that of repulsion, the condition of stability is not satisfied and the electrons will fall into the nucleus, thus destroying the stable structure of the atom. In order to overcome this difficulty, Rutherford suggested that *electrons might be assumed to revolve round the nucleus*, like the planets round the sun, at such a speed that the mechanical centrifugal force would just balance the net excess of electrostatic attraction and in consequence stability of the atom could be secured.

But such an assumption brought in its wake a very serious difficulty from the point of view of the electromagnetic theory, according to which *a revolving electron should radiate energy continuously*. Now, this energy can only come from the atomic system, which will therefore steadily lose energy. As a result, the electron will approach the nucleus by a spiral path giving out radiation of constantly increasing frequency and finally fall into the nucleus. Thus, the orbital motion of the electron destroys the very purpose for which it was postulated *viz.*, the stability of the atom. Further, emission of radiation of constancy increasing frequency has no experimental support whatsoever, since elements are actually found to emit discrete spectral lines of definite frequencies. One is therefore forced to conclude that the Rutherford nuclear atom model with revolving electrons is defective or the classical electromagnetic theory fails in the present case. The dilemma was solved in 1913 by Niels Bohr, who admitting the failure of the classical theory applied with remarkable success the quantum theory to the Rutherford nuclear atom model with revolving electrons.

1.5 BOHR MODEL OF THE ATOM

According to the Bohr, negatively charged particle an revolving around the nucleus in a fixed orbit known as energy revolving around the nucleus there is no loss of energy. Loss or gain of energy takes place when electron jumps from one orbit to the there orbit.

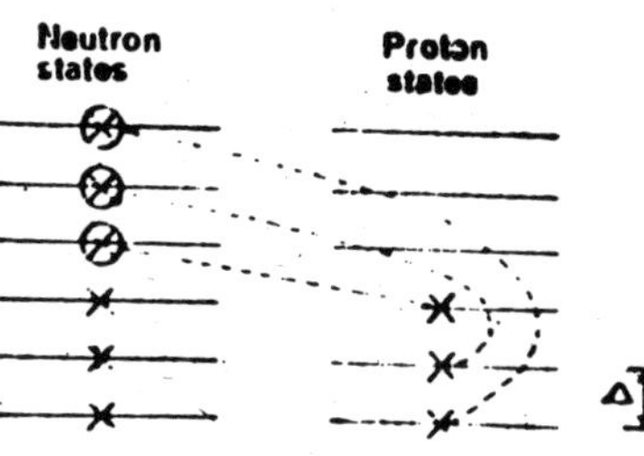

Fig. 1.6

1.6 BOHR'S THEORY OF THE HYDROGEN SPECTRUM

Rutherford's experiment on the scattering of α-particles led Bohr to the Theory of Atomic Structure that the atom consists of a positively charged nucleus at its centre. Moreover, Bohr applied the quantum theory of radiation as developed by Planck and Einstein to the Rutherford's Model. His theory is mainly based on the following postulates.

1. An Atom consists of a positively charged nucleus at the centre.
2. The negatively charged particles known as electrons move round the nucleus in various orbits known as stationary energy levels. The electrons cannot emit radiations when moving in their own stationary levels.
3. The Coulombian and Newtonian forces are applicable in the domain of the atom.
4. The electrons revolve round the nucleus in various circular orbits and the angular momentum $m\upsilon r = \frac{nh}{2\pi}$ where n (= 1, 2, 3. 4 etc.) is called the quantum number, and h is the Planck's constant.
5. When an electron jumps from a higher energy level to a lower energy level, it gives out electromagnetic radiations of a particular frequency.

$$W_{n2} - W_{n1} = h\nu$$

where W_{n2} is the energy of the n_2 energy level, W_{n1} is the energy of the n_1 energy level and v is the frequency of the radiation.

Bohr's postulates are the combinations of some of the ideas of classical physics and quantum; physics. While in some ways they agree with classical physics but in some aspects they are contradictory to it. *Here the energy levels are quantized.* It means that the electrons can move only in some particular orbits of *definite radii* as suggested by Rutherford's mode).

Consider an atom whose nucleus has a positive charge E and let an electron of negative charge e move round the nucleus is an orbit of radius r.

$$E = Ze, \quad \text{where } Z = 1 \text{ for hydrogen.}$$

The force of attraction between the nucleus and the electron

$$= \frac{Ee}{4\pi \epsilon_0 r^2} = \frac{Ze^2}{4\pi \epsilon_0 r^2}$$

As the electrons move along the circular orbit, the centripetal force $\left(= \frac{m\upsilon^2}{r}\right)$ is provided by the attractive force on the electron.

$$\therefore \quad \frac{m\upsilon^2}{r} = \frac{Ze^2}{4\pi \epsilon_0 r^2} \quad \text{...(i)}$$

$$\therefore \quad m\upsilon^2 = \frac{Ze^2}{4\pi \epsilon_0 r^2} \quad \text{...(ii)}$$

The angular momentum, $m\upsilon r = \frac{nh}{2\pi}$ ...(iii)

$$\therefore \quad \upsilon = \frac{nh}{2\pi mr} \quad \text{...(iv)}$$

Substituting the value of υ in equation (ii),

$$m\left(\frac{nh}{2\pi mr}\right)^2 = \frac{Ze^2}{4\pi \epsilon_0 r}$$

$$\therefore \quad r = \frac{\epsilon_0 n^2 h^2}{\pi mZe^2} \quad \text{...(v)}$$

Substituting this value of r in equation (iv)

$$\upsilon = \frac{Ze^2}{2 \in_0 nh} \qquad \text{...(vi)}$$

From equation (v), we find that $r \propto n^2$. The radii of the orbits are in the ratio of 1 : 4 : 9 : 16 : 25 etc. The radius of the second orbit is four times the radius of the first orbit. For hydrogen atom, the radius of the first orbit

$$r = \frac{\in_0 n^2 h^2}{\pi m Z e^2}, \qquad \text{Here } n = 1 \text{ and } Z = 1$$

Equation (vi) shows that the velocity of the electron decreases with the increase in the order of the orbit. The velocity of the electron in the first orbit $\upsilon = \dfrac{Ze^2}{2 \in_0 h}$

Substituting the values of n, h, Z and e

$$\upsilon = 2.2 \times 10^6 \text{ m/s}$$

Let us consider the energy of an electron in the orbit. The energy is partly potential and partly kinetic. From definition, the potential energy of an electron $= \displaystyle\int_{\infty}^{r} \frac{Ze^2}{4\pi \in_0 r^2} dx = \frac{-Ze^2}{4\pi \in_0 r}$

$\therefore$ Potential energy of the electron $= \dfrac{-Ze^2}{4 p \in_0 r}$

Kinetic energy $= \dfrac{1}{2} m\upsilon^2$

From equation (ii), $m\upsilon^2 = \dfrac{Ze^2}{4\pi \in_0 r}$

Kinetic energy $= \dfrac{Ze^2}{8\pi \in_0 r}$.

The potential energy is negative because energy must be given to the electron to bring it far away from the nucleus to zero energy level.

Total energy = potential energy + kinetic energy

$$= \frac{-Ze^2}{4\pi \in_0 r} + \frac{Ze^2}{8\pi \in_0 r} = -\frac{Ze^2}{8\pi \in_0 r} \qquad \text{...(vii)}$$

Substituting the value of r, we get

$$W_n = \frac{-me^4 Z^2}{8 \epsilon_0^2 n^2 h^2} \text{ joule} \quad \text{...(viii)}$$

Here $m = 9.1 \times 10^{-31}$ kg. $e = 1.6 \times 10^{-19}$ coulomb

$$\epsilon_0 = 8.85 \times 10^{-12} \frac{\text{coulomb}^2}{\text{newton} - \text{m}^2}$$

$$h = 6.624 \times 10^{-34} \text{ joule-second}$$

For an electron to be removed from its first orbit in a hydrogen atom, energy must be supplied. Therefore ionization potential for hydrogen,

$$\phi = \frac{me^4 Z^2}{8 \epsilon_0^2 n^2 h^2}$$

For hydrogen, $Z = 1$, $n = 1$

$$\therefore \quad \phi = \frac{me^4}{8 \epsilon_0^2 n^2 h^2} \text{ joules} \quad \text{...(ix)}$$

$$\phi = \frac{me^4}{8 \epsilon_0^2 h^2 \times 1.6 \times 10^{-19}} \text{ electrons volts} \quad \text{...(x)}$$

When an electron jumps from energy level n_2 to energy level n_1

$$W_{n2} - W_{n1} = \frac{-me^4 Z^2}{8 \epsilon_0^2 n_2^2 h^2} + \frac{me^4 Z^2}{8 \epsilon_0^2 n_1^2 h^2}$$

$$W_{n2} - W_{n1} = \frac{me^4 Z^2}{8 \epsilon_0^2 h^2} \left(\frac{1}{n_1^2} - \frac{1}{n_2^2} \right)$$

But $\quad W_{n2} - W_{n1} = hv$

$$hv = \frac{me^4 Z^2}{8 \epsilon_0^2 h^2} \left(\frac{1}{n_1^2} - \frac{1}{n_2^2} \right)$$

and

$$v = \frac{me^4 Z^2}{8 \epsilon_0^2 h^3} \left(\frac{1}{n_1^2} - \frac{1}{n_2^2} \right)$$

In the case of hydrogen atom $Z = 1$

$$\therefore \qquad v = \frac{me^4}{8\epsilon_0^2 h^3}\left(\frac{1}{n_1^2} - \frac{1}{n_2^2}\right)$$

But $$c = v\lambda, \quad \therefore \; v = \frac{c}{\lambda}$$

$$c = \frac{me^4}{8\epsilon_0^2 h^3}\left(\frac{1}{n_1^2} - \frac{1}{n_2^2}\right)$$

$$\frac{1}{\lambda} = \frac{me^4}{8\epsilon_0^2 ch^3}\left(\frac{1}{n_1^2} - \frac{1}{n_2^2}\right)$$

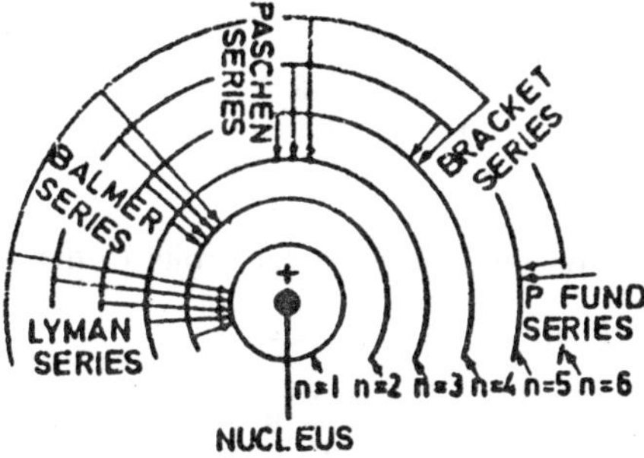

Fig. 1.7 : Spectral Series of Hydrogen.

$\frac{1}{\lambda} = \bar{v}$ called the wave number

$$\bar{v} = \frac{me^4}{8\epsilon_0^2 ch^3}\left(\frac{1}{n_1^2} - \frac{1}{n_2^2}\right)$$

$\frac{me^4}{8\epsilon_0^2 ch^3}$ = R and is known as Rydberg constant.

Substituting the values, R = 109678 cm^{-1}

= 1.096×10^7 per metre

$$\therefore \qquad \bar{v} = R\left(\frac{1}{n_1^2} - \frac{1}{n_2^2}\right) \qquad \text{...(xi)}$$

1.7 SPECTRAL SERIES OF HYDROGEN ATOM

1. Lyman Series

When an electron jumps from the outer orbits to the first orbit, the spectral lines are in the ultraviolet region (Fig. 1.7).

Here $n_1 = 1, n_2 = 2, 3, 4, 5.......$

$$\bar{v}_1 = R\left(\frac{1}{1} - \frac{1}{2^2}\right) = \frac{3}{4}R$$

$$\bar{v}_2 = R\left(\frac{1}{1} - \frac{1}{3^2}\right) = \frac{8}{9}R$$

$$\bar{v}_3 = R\left(\frac{1}{1} - \frac{1}{4^2}\right) = \frac{15}{16}R$$

2. Balmer Series

When are electron jumps from outer orbits to the second orbit,

$n_1 = 2, n_2 = 3, 4, 5.....$ etc.

$$\therefore \quad \bar{v}_1 = R\left(\frac{1}{2^2} - \frac{1}{3^2}\right) = \frac{5}{36}R$$

$$\bar{v}_2 = R\left(\frac{1}{2^2} - \frac{1}{4^2}\right) = \frac{3}{16}R$$

This series lies in the visible region of the spectrum.

3. Paschen Series

When an electron jumps from the outer orbits to the third orbit,

$n_1 = 3, n_2 = 4, 5, 6,$ etc.,

$$\therefore \quad \bar{v}_1 = R\left(\frac{1}{3^2} - \frac{1}{4^2}\right) = \frac{9}{144}R$$

$$\bar{v}_2 = R\left(\frac{1}{3_2} - \frac{1}{5_2}\right) = \frac{16}{225}R$$

4. Bracket Series

When an electron jumps from outer orbits to the fourth orbit.

$n_1 = 4, n_2 = 5, 6, 7,$ etc.,

$$\therefore \qquad \bar{v}_1 = R\left(\frac{1}{4^2} - \frac{1}{5^2}\right) = \frac{9}{400}R$$

$$\bar{v}_2 = R\left(\frac{1}{4^2} - \frac{1}{6^2}\right) = \frac{5}{144}R$$

5. Pfund Series

When an electron Jumps from outer orbits to the fifth orbit

$$n_1 = 5,\ n_2 = 6,\ 7,\ 8,\ \text{etc.},$$

$$\therefore \qquad \bar{v}_1 = R\left(\frac{1}{5^2} - \frac{1}{6^2}\right) = \frac{11}{900}R$$

$$\bar{v}_2 = R\left(\frac{1}{5^2} - \frac{1}{7^2}\right) = \frac{24}{1225}R.$$

The last three series are in the infra-red region. When the Bohr's Theory was suggested, only the Balmer and Paschen series for hydrogen atom were known. The other series as suggested theoretically according to equation (xi) were verified experimentally also. The Lyman Series (1916), the Bracket Series (1922), and Pfund Series (1929) studied experimentally were found to show the same wave numbers as suggested theoretically. When an electron absorbs energy, it moves from lower energy level to higher level as in the absorption spectrum. When an electron jumps from higher energy level to the lower energy level, it emits radiation as in the emission spectrum.

1.8 ENERGY LEVELS OF HYDROGEN ATOM

In the case of Hydrogen Atom, the work function for the electron in the first orbit is given by

$$\phi = \frac{me^4}{8\epsilon_0^2 h^3}$$

This value of ϕ = 13.60 eV.

The energy of electron in the first orbit U_1 = 13.60 eV.

Similarly, for the second, third etc., orbits, the energy of the electron will be

$$U_2 = \frac{13.6}{2^2} = -3.4\,\text{eV}$$

$$U_3 = \frac{13.6}{3^2} = -1.511\,\text{eV}$$

$$U_4 = \frac{13.6}{4^2} = -0.85\text{eV}$$

$$U_5 = \frac{13.6}{5^2} = -0.544\text{eV}$$

... ...

... ...

The energy level diagram is shown in Fig. 1.8.

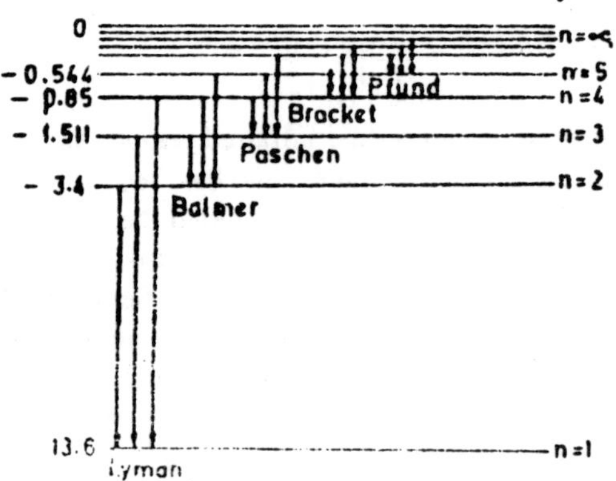

Fig. 1.8

The energy is inversely proportional to the square of the number of the orbit and is negative. The lowest energy state (n = 1) is called the ground state. With increase in the order of the orbit, the energy of the orbit increases. If n = ∞, the energy is zero. In all other orbit, the energy is negative. Due to this reason the electron is bound to the nucleus. The different spectral series are shown in the Fig. 1.8.

1.9 RESONANCE, EXCITATION AND IONIZATION POTENTIALS

Resonance Potential

The minimum potential required to provide energy to the electron in the ground state to the first excited state *i.e.*, from n = 1 to n = 2 is called resonance potential.

The energy of electron in the ground state of hydrogen atom is −13.6 eV and first excited state (n = 2) is −3.4 eV. Therefore, the energy required to move the electron from the ground state to the first excited state is (−3.4) eV − (−13.6) eV = 10.2 eV

Therefore, the resonance potential for hydrogen is 10.2 V.

Excitation Potential. The state n > 1 are called excited states. The energy required to move the electron to the first, second, third excited states is given by

$$E_1 = 10.2 \text{ eV}$$

$$E_2 = -\left(\frac{13.6}{3^2}\right) - (-13.6)$$

$$= -1.51 + 13.60 = 12.09 \text{ eV}$$

$$E_3 = -\left[\frac{13.6}{4^2}\right] - (-13.6)$$

$$= -0.85 + 13.60 = 12.75 \text{ eV.}$$

Therefore, the successive excitation potentials are, 10.2V, 12.09V 12.75V and so on.

The excitation potential is the potential required to provide energy to raise; the electron from the ground state to the state n > 1 *i.e.*, n = 2, 3, 4...

Ionization potential is the minimum potential required that provides energy to bring the electron from the ground state out of the atom for hydrogen atom, ionization potential = 13.6 V the energy to the electron in the atom can be provided by electron emitted from a hot filament and accelerated though a creation potential V electron will move with a velocity υ'

$$\text{Where } \upsilon = \left(\frac{2\text{eV}}{\text{m}}\right)^{1/2}$$

(i) If the energy of the striking electron is just equal to energy required by the electron in atom to move from the ground state to the first state the electron will move to first excited state and striking electron gives whole of its energy to the electron in the atom.

(ii) If the energy of the electron is just equal to or more than the energy required by the electron to come out of the atom, the electron in the atom absorbs energy and comes out of the atom.

Screening Effect

The energy for one electron atom in various orbits such as hydrogen or He^{+}(ion) is given by

$$E_n = \left(\frac{-13.6Z^2}{n^2}\right)eV$$

For multi-electron atoms, the nuclear charge Ze is largely cancelled or shielded by the negative charge of inner care electrons. Hence, the outer electrons interact with a net of the order of electronic charge. The expression (i) for allowed energies with Z is replaced by Z_{eff}

$$E_n = \left(\frac{-13.6Z_{eff}^2}{n^2}\right)eV$$

for k shell of large atoms, Z_{eff} is Z – 1 for higher energy states Z_{eff} decreases gradually from Z – 1 to 1

1.10 BOHR'S CORRESPONDENCE PRINCIPLE

Classical mechanics deals with the macroscopic bodies where as Quantum physics deals with the microscopic object. Bohr's Correspondence Principle is the combination of classical physics and Quantum Physics

According to correspondence principle a system in a higher quantum number higher excitation is cavenened by the laws of classical mechanics. Classical mechanics classical mechanics is not applicable for large systems. For systems at very low temperature or in very low state of quantum numbers, lows of quantum mechanics are applicable the example are He^{3+} He^{2+}, super conductors etc., correspondence principle can be illustrated as follows.

1. In quantum mechanics, one of the important features is the uncertainty principle. The uncertainty in determining the position of a particle is given by $h = \dfrac{h}{m\upsilon}$

   For a particle having high momentum λ is small and its position can be accurately determined and it be haves as a particle similar to the behaviour of particles in classical mechanics.

If the momentum is very small λ is very large and wave characteristics are exhibited. This explains the dual behaviour of various particles.

2. In classical thermodynamics, Maxwell-Boltzmann's statistics are applicable for gases having low density and high temperature. However, in quantum mechanics, three different types of statistics are applicable to particles of half integral spin. Bose Einstein statistics is applicable for particles having integral spin. The two statistics behave differently at low temperature *i.e.,* for systems of low excitation. At higher temperatures, however, both the statistics merge into the classical Maxwell-Boltzmann statistics.

1.11 QUANTIZATION OF ANGULAR MOMENTUM *(Wilson-Sommerfeld Quantization Rule)*

According to Bohr's theory the stationary energy levels of the hydrogen atom can be explained with the assumption that the angular momentum of a system has only discrete values in term of simple integral multiples of a universal constant. Wilson and Sommerfeld Quantization Rule successfully explains the various predicated energy levels of periodic systems.

A periodic system having N degrees of freedom can be described with the help of:

(i) N coordinates q, and

(ii) N canonically conjugate momenta p_i

Here $i = 1, 2, 3.............N$

Therefore, the phase integral of such a periodic system is represented by

$$\int_i = \oint p_i dq_i$$

The integration is taken over one period of the variable q_i.

According to Wilson-Sommerfeld Quantization rule, *only those orbits or stationary states are permissible in which the value of each phase integral is an integral multiple of universal Plank's constant.*

$$\therefore \quad \oint p_i dq_i = n_i\, h \qquad ...(ii)$$

Here $n = 1, 2, 3...$ etc.

This rule helped in correlating the experimental results with theoretical considerations.

1.12 ELLIPTICAL-ORBITS—WILSON SOMMERFELD QUANTIZATION THEORY

Bohr's Theory for the spectrum of hydrogen atom was developed with the assumption that the electrons move around the nucleus in circular orbits. From the analogy of Kepler's Laws of planetary motion and motion of planets in elliptical orbits, Bohr also considered the possibility of an electron moving around the nucleus in an elliptical orbit.

The study of the hydrogen spectrum with spectrographs of high resolving power showed tine structure of the spectral lines. The spectral lines as predicted by Bohr's Theory are to be simple but it is found that each line is complex and consists of a small number of lines close together. These give rise to the so-called fine structure of spectral lines.

Wilson and Sommerfeld applied a generalized quantum condition for the elliptical orbits. Wilson and sommerfeld generalized that quantum condition is given by

$$\oint p.dq = nh$$

Here q is the position coordinate of the electron which varies periodically as the electron traverses the elliptical path and p is the corresponding value of the momentum of the electron, n is the principle quantum number and h is the Plank's constant.

Consider an electron in its elliptical orbit with polar coordinates r and θ (Fig. 1.9). The nucleus is at one of its foci. Here θ is azimuthal angle and r is the radial distance. For the electron there are two degrees of freedom corresponding to the two coordinates r and θ. According to Wilson-Sommerfeld theory each of the two degrees of freedom are *individually quantized.* This dual requirement is expressed as

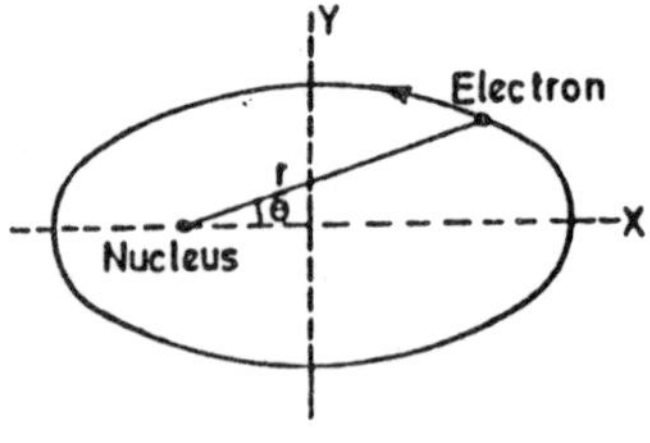

Fig. 1.9

$$\oint_0^{2\pi} p_\theta . dq = kh$$

$$\oint_0^{2\pi} p_r dr = n_r h$$

Here k is the azimuthal quantum number and n_r is the radial quantum number, k and n_r are both integers. Also

$$n = (k + n_r)$$

P_θ and P_r represent angular momentum and radial momentum respectively

$$p_\theta = mr^2 \left(\frac{d\theta}{dt}\right)$$

and $$p_r = m\left(\frac{dr}{dt}\right)$$

Here m is the mass of the electron.

The force experienced by the electron is due to electrostatic attraction between the positively charged nucleus and the electron. This force acts along the radius vector at any instant. Consequently there is no force (acceleration) at right angles to the radius vector. Therefore the transverse component of acceleration is always zero *i.e.,*

$$\therefore \quad \frac{d}{dt}\left[\frac{d\theta}{dt}\right] \text{ is zero}$$

or $$\frac{1}{r^2}\frac{d}{dt}\left[r^2 \frac{d\theta}{dt}\right] \text{ is zero}$$

It means that $\frac{r^2 \, d\theta}{dt}$ is a constant.

Hence

$p_\theta \; mr^2 \frac{d\theta}{dt}$ is also a constant and equal to p

$$\therefore \quad p_\theta = p$$

$$\therefore \quad \oint_0^{2\pi} p_\theta \, d\theta = \int_0^{2\pi} p \, d\theta = 2\pi p = kh$$

$$\therefore \quad p = \frac{kh}{2\pi} \qquad \text{...(i)}$$

But $$p = m\upsilon r$$

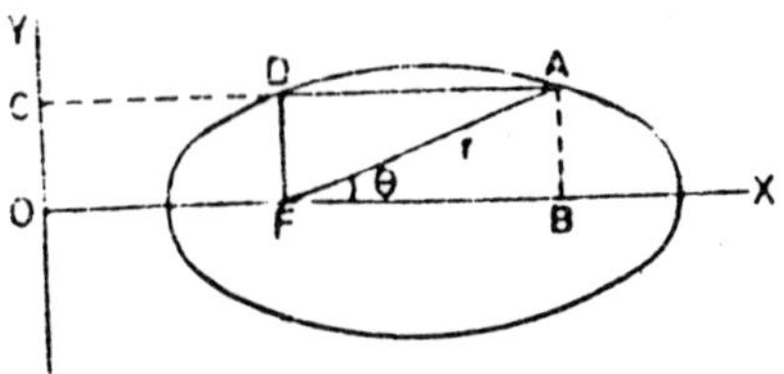

Fig. 1.10

$\therefore$ $$m\upsilon r = \frac{kh}{2\pi} \qquad \text{...(ii)}$$

To evaluate the integral containing radial quantum number, consider the electron at A in its elliptical path. The nucleus is at F. Here OY is the directrix and FD is the semi-latus rectum of length l. AB is perpendicular to OX at B (Fig. 1.10).

$$FA = \in (AC)$$

$$r = \in (AC) = \in (OB)$$

$$r = \in (OF + FB)$$

$$r = \in\left(\frac{l}{\in} + r\cos\theta\right)$$

$$r = l + \in r \cos\theta$$

or $$\frac{l}{r} = (1 - \in \cos\theta)$$

or $$r = l\,(1 - \in \cos\theta)^{-1}$$

Differentiating r with respect to θ.

$$\frac{dr}{d\theta} = \frac{-l \in \sin\theta}{(1-\in\cos\theta)^2}$$

$$\frac{1}{r}\left(\frac{dr}{d\theta}\right) = \left(\frac{l}{r}\right)\left[\frac{-\in\sin\theta}{(1-\in\cos\theta)^2}\right]$$

But $$\frac{l}{r} = (1 - \in\cos\theta)$$

$$\therefore \qquad \frac{l}{r}\left(\frac{dr}{d\theta}\right) = \frac{-\in\sin\theta}{1-\in\cos\theta}$$

$$p_r\ m\left(\frac{dr}{dt}\right) = m\left(\frac{dr}{d\theta}\right)\left(\frac{d\theta}{dt}\right) \qquad ...(iii)$$

But $$p = mr^2\left(\frac{d\theta}{dt}\right) \qquad ...(iv)$$

$$\therefore \qquad p_r = \left(\frac{p}{r^2}\right)\frac{dr}{d\theta} \qquad ...(v)$$

$$dr = \left(\frac{dr}{d\theta}\right)d\theta \qquad ...(vi)$$

From equations (v) and (vi)

$$p_r\,dr = \left(\frac{p}{r^2}\right)\left(\frac{dr}{d\theta}\right)^2 d\theta$$

or $$p_r\,dr = p\left(\frac{1}{r}\frac{dr}{d\theta}\right)^2 d\theta \qquad ...(vii)$$

Substituting the value of $\frac{1}{r}\frac{dr}{d\theta}$ in equation (viii)

$$p_r\,dr = p\left[\frac{\in^2\sin^2\theta}{(1-\in\cos\theta)^2}\right]d\theta$$

$$\therefore \qquad \oint_0^{2\pi} p_r dr = \oint_0^{2\pi} p\left[\frac{\in^2\sin^2\theta}{(1-\in\cos\theta)^2}\right]d\theta$$

But $$\oint_0^{2\pi} p_r dr = n_r h$$

$$\therefore \qquad \oint_0^{2\pi} p_r dr = \oint_0^{2\pi}\left[\frac{\in^2\sin^2\theta}{(1-\in\cos\theta)^2}\right]d\theta = n_r h$$

But $$p = \frac{kh}{2\pi}$$

$$\frac{kh}{2\pi}\oint_0^{2\pi}\left[\frac{\in^2\sin^2\theta}{(1-\in\cos\theta)^2}\right]d\theta = n_r h$$

$$\frac{1}{2\pi}\oint_0^{2\pi}\left[\frac{\epsilon^2 \sin^2\theta}{(1-\epsilon\cos\theta)^2}\right]d\theta = \left(\frac{n_r}{k}\right) \qquad ...(viii)$$

Take integral

$$l = \oint_0^{2\pi}\left(\frac{\epsilon^2 \sin^2\theta}{(1-\epsilon\cos\theta)^2}\right)d\theta$$

The integral can be integrated by parts.

Take $\quad \epsilon \sin\theta = u$

and $\quad \dfrac{1}{1-\epsilon\cos\theta} = \upsilon$

$$du = \epsilon \cos\theta \, d\theta$$

and $\quad d\upsilon \dfrac{\epsilon \sin\theta}{(1-\epsilon\cos\theta)^2} = d\theta$

$\therefore \quad \int u \, d\upsilon = u\upsilon - \int \upsilon \, du$

$\therefore \quad l = \left[\dfrac{\epsilon \sin\theta}{1-\epsilon\cos\theta}\right]_0^{2\pi} - \displaystyle\int_0^{2\pi}\left[\frac{\epsilon\cos\theta}{(1-\epsilon\cos\theta)}\right]d\theta$

$\therefore \quad l = \displaystyle\int_0^{2\pi}\left[\frac{\epsilon\cos\theta}{1-\epsilon\cos\theta}\right]d\theta$

$$l = \int_0^{2\pi}\left[\left(\frac{1}{1-\epsilon\cos\theta}\right) - 1\right]d\theta$$

$$= \frac{2\pi}{(1-\epsilon^2)^{1/2}} - 2\pi$$

Substituting the values in equation (viii)

$$\frac{1}{(1-\epsilon_0^2)^{1/2}} - 1 = \frac{n_r}{k}$$

or $\quad 1 - \epsilon^2 = \dfrac{k^2}{(k-n_r)^2}$

But $\quad k + n_r = n$

$$\therefore \qquad 1 - \epsilon^2 = \frac{k^2}{n^2} \qquad \text{...(ix)}$$

$$\text{or} \qquad \epsilon = \sqrt{1 - \left(\frac{k}{n}\right)^2}$$

$$\epsilon = \sqrt{1 - \left(\frac{k}{k + n_r}\right)^2}$$

If a and b are the semi-major and semi-minor axes of the ellipse, then

$$1 - \epsilon^2 = \frac{b^2}{a^2} \qquad \text{...(x)}$$

From equation (ix) and (x)

$$\text{or} \qquad \frac{b^2}{a^2} = \frac{k^2}{n^2}$$

$$\text{or} \qquad \frac{b}{a} = \frac{k}{n} \qquad \text{...(xi)}$$

The total energy of the electron on the elliptical orbit is given by

$$E = E_k + E_p$$

Here E_k is the kinetic energy and E_p is the potential energy of the electron.

$$E_k = \frac{1}{2} m\upsilon^2 = \frac{m}{2}\left(\frac{ds}{dt}\right)^2$$

$$ds^2 = dr^2 + (r\, d\theta)^2$$

$$\therefore \qquad E_k = \frac{1}{2} m\left[\left(\frac{dr}{dt}\right)^2 + \left(r\frac{d\theta}{dt}\right)^2\right] \qquad \text{...(xii)}$$

$$\text{But} \qquad p_r = m\left(\frac{dr}{dt}\right)$$

$$\text{and} \qquad p = mr^2\left(\frac{d\theta}{dt}\right)$$

$$E_k = \frac{1}{2m}\left[p_r^2 + \frac{p^2}{r^2}\right]$$

Substituting the value of

$$p_r = \frac{p}{r^2}\left(\frac{dr}{d\theta}\right) \text{ in this equation we get}$$

$$E_k = \frac{1}{2m}\left[\left(\frac{p^2}{r^4}\right)\left(\frac{dr}{d\theta}\right)^2 + \frac{p^2}{r^2}\right]$$

$$E_k = \frac{p^2}{2mr^2}\left[\left(\frac{1}{r}\frac{dr}{d\theta}\right)^2 + 1\right]$$

The potential energy of the electron is given by

$$E_p = -\frac{Ze^2}{4\pi \epsilon_0 r}$$

Total energy

$$E = E_k + E_p$$

$$E = \frac{p^2}{2mr^2}\left[\left(\frac{1}{r}\frac{dr}{d\theta}\right)^2 + 1\right] - \frac{ze^2}{4\pi \epsilon_0 r}$$

$$\therefore \quad \left(\frac{1}{r}\frac{dr}{d\theta}\right)^2 = \frac{2mEr^2}{p^2} + \frac{2Ze^2mr}{(4\pi \epsilon_0)p^2} - 1 \qquad \text{...(xiii)}$$

But $\quad \frac{1}{r}\left(\frac{dr}{d\theta}\right) = \frac{-\epsilon \sin\theta}{(1 - \epsilon\cos\theta)}$

$$\therefore \quad \left(\frac{1}{r}\frac{dr}{d\theta}\right)^2 = \frac{\epsilon^2 \sin\theta}{(1 - \epsilon\cos\theta)^2}$$

But $\quad \frac{l}{r} = (1 - \epsilon \cos\theta)$

$$\therefore \quad \left(\frac{1}{r}\frac{dr}{d\theta}\right)^2 = (\epsilon^2 \sin^2\theta)\,\frac{r^2}{l^2}$$

$$\frac{l}{r} = (1 - \in \cos\theta)$$

$$\in \cos\theta = \left(1 - \frac{l}{r}\right) = \left(\frac{r - l}{r}\right)$$

$$\cos\theta = \left(\frac{r - l}{\in r}\right)$$

$$\sin^2\theta = \left[1 - \left(\frac{r - l}{\in r}\right)^2\right]$$

$$\therefore \qquad \left(\frac{1}{r}\frac{dr}{d\theta}\right)^2 = \frac{\in^2 r^2}{l^2}\left[1 - \left(\frac{r - l}{\in r}\right)^2\right]$$

$$= \frac{\in^2 r^2}{l^2}\left[1 - \frac{1}{\in^2}\left(1 - \frac{l}{r}\right)^2\right]$$

$$= \frac{\in^2 r^2}{l^2}\left[1 - \frac{1}{\in^2}\left(1 + \frac{l^2}{r^2} - \frac{2l}{r}\right)\right]$$

$$= \frac{\in^2 r^2}{l^2} - \frac{r^2}{l^2} - 1 + \frac{2r}{l}$$

$$= \frac{r^2(\in^2 - l)}{l^2} + \frac{2r}{l} - 1 \qquad \text{...(xiv)}$$

From equation (xiii) and (xiv) we get

$$\frac{2mr^2 E}{p^2} + \frac{2mr Ze^2}{(4\pi \in_0)p^2} - 1 = \frac{r^2(\in^2 - 1)}{l^2} + \frac{2r}{l} - 1 \qquad \text{...(xv)}$$

Equating the coefficients of r^2 and r we get,

$$\frac{2mE}{p^2} + \frac{1}{l^2}(\in^2 - 1) \qquad \text{...(xvi)}$$

and

$$\frac{mZe^2}{(4\pi \in_0)p^2} = \frac{1}{l} \qquad \text{...(xvii)}$$

Dividing equation (xvi) by (xvii) we get

$$\frac{2E(4\pi \in_0)}{Ze^2} = \frac{1}{l}\left(\in^2 - 1\right)$$

$$E = \frac{1}{4\pi \in_0}\left[\frac{Ze^2}{2l}\right](\in^2 - 1)$$

$$E = -\left(\frac{Ze^2}{4\pi \in_0}\right)\left(\frac{1}{2l}\right)(1 - \in^2) \qquad \text{...(xviii)}$$

But $$\frac{1 - \in^2}{l} = \frac{1}{a}$$

$\therefore$ $$E = -\frac{Ze^2}{(4\pi \in_0)2a} \qquad \text{...(xix)}$$

Here a is the semi major axis.

Thus the total energy of an electron in an elliptical orbit depends upon the semi major axis a of the ellipse. It is just equal to the energy of an electron moving in a circular orbit of radius a.

To eliminate l, square equation (xvii) and divide it by (xvi)

$$\frac{m^2Z^2e^4}{(4\pi \in_0)^2 p^4} \times \frac{p^2}{2mE} = \frac{1}{l^2} \times \frac{l^2}{(\in^2 - 1)}$$

$$E = \frac{mZ^2e^4\ (\in^2 - 1)}{(4\pi \in_0)^2\ (2p^2)}$$

$$E = \frac{mZ^2e^4\ (1 - \in^2)}{(4\pi \in_0)^2\ (2p^2)} \qquad \text{...(xx)}$$

As $$P = \frac{kh}{2\pi} \text{ and } (1 - \in^2) = \frac{k^2}{n^2}$$

$\therefore$ $$E = \frac{mZ^2e^4\,(k^2/n^2)}{(4\pi \in_0)^2\,(2k^2h^2/4\pi^2)}$$

$$E = \frac{2\pi^2 mZ^2e^4}{(4\pi \in_0)^2\, n^2h^2} \qquad \text{...(xxi)}$$

$$E = \frac{mZ^2e^4}{8 \in_0^2 n^2h^2} \qquad \text{...(xxii)}$$

$$\therefore \qquad E \propto \frac{1}{n^2}$$

i.e., the total energy of the electron is inversely proportional to the square of the principal quantum number.

Now, consider the case when the principal quantum number n = 1. Also n = k + n,

It means when $k = 1, n_r = 0$

when $k = 0, n_r = 1$

From the relation $\frac{b}{a} = \frac{k}{n} = \frac{k}{(k + n_r)}$

when $k = 1$, and $n = 1$, $\frac{b}{a} = 1$

It is a case of circular orbit.

When $k = 0$, and $n_r = 1$, then $b = 0$.

It means the ellipse is reduced to a straight line. Therefore in this case the electron should pass through the nucleus twice in each period. This is not possible.

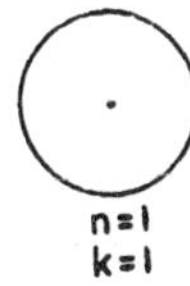

Fig. 1.11

Hence for n = 1, the only possible case is $k = 1$ and $n_r = 0$,

The orbit will be circular (Fig. 1.11)

Take the case when n = 2

Here, for $k = 1, n_r = 1, k = 2, n_r = 0$

when $k = 1$, $\frac{b}{a} = \frac{1}{2}$, or $b = \frac{a}{2}$

when $k = 2$, $\frac{b}{a} = 1$, or $b = a$

It means that the electron has two possible orbits (i) an ellipse with eccentricity

$$\epsilon = \left[1 - \frac{b^2}{a^2}\right]^{1/2}$$

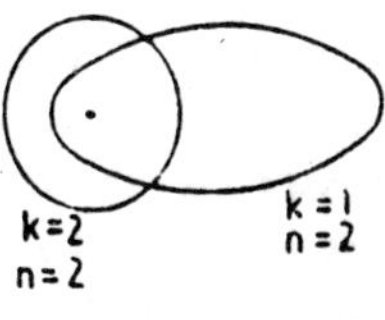

Fig. 1.12

$$= \frac{\sqrt{3}}{2}$$

(ii) a circle of radius a (Fig. 1.12)

Take the case when n = 3

Here, for $k = 1,\ n_r = 2$ and $\frac{b}{a} = \frac{1}{3}$

for $k = 2,\ n_r = 1$ and $\frac{b}{a} = \frac{2}{3}$

for $k = 3,\ n_r = 0$ and $\frac{b}{a} = 1$

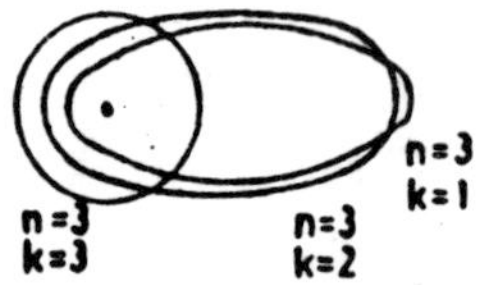

Fig. 1.13

For the first two cases the orbits are elliptical. In the third case the orbit is circular (Fig. 1.13).

1.13 EFFECT OF VARIABLE MASS OF THE ELECTRON

The energy of an electron in terms of the principal quantum number does not introduce new energy levels. Hence, there is no possibility for new energy transitions apart from that predicted by Bohr's theory. The fine structure of the spectrum can be explained by taking into account the relativistic mass of the electron instead of the constant rest mass. The relativistic mass of the electron is given by

$$m = \frac{m_0}{\left[1 - \frac{\upsilon^2}{c^2}\right]^{1/2}}$$

Bohr and Sommerfeld took into account the relativistic mass of the electron. Sommerfeld in his treatment of the elliptical orbit had shown that the path of an electron was not simply an ellipse but an ellipse which processes *i.e.*, the major axis of the ellipse slowly turns about the focus (nucleus) in the plane of the ellipse. The path of the electron orbit is called Rosette orbit (Fig. 1.14).

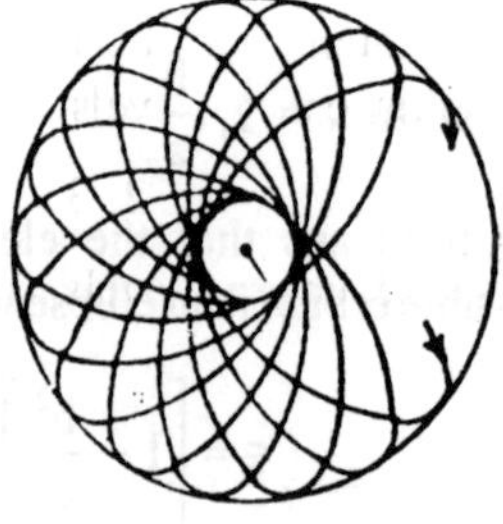

Fig. 1.14

For the Rosette orbit the equation for the path of the electron is taken as

$$\frac{l}{r} = 1 - \in \cos(\phi\theta)$$

where $$\phi = \left[\frac{1 - Z^2e^4}{c^2p^2}\right]^{1/2}$$

Therefore, the value of $\phi\,\theta$ increases by 2π before the value of r is repeated. Thus, the value of θ increases by $\frac{2\pi}{\phi}$ during the time the major axis of the ellipse has processed by $\left(\frac{2\pi}{\phi} - 2\pi\right)$. Hence, the energy expression for the electron including the relativistic correction is given by

$$E = -\left(\frac{mZ^2e^4}{8\in_0^2 n^2h^2}\right)\left[1 + \frac{Z^2\alpha^4}{n^2}\left(\frac{n}{k} - \frac{3}{4}\right)\right] \qquad \text{...(xxiii)}$$

Here $$\alpha = \frac{2\pi e^2}{(4\pi \in_0)ch}$$

$$\alpha = \frac{2 \times 3.14 \times (1.6 \times 10^{-19})^2 \times 9 \times 10^9}{3 \times 10^8 \times 6.624 \times 10^{-34}}$$

$$\alpha = \frac{1}{137} \text{ (approximately)}$$

α is called the fine structure constant.

In equation (xxiii) the correction term depends on the azimuthal quantum number k *i.e.,* for any principal quantum number n, there are n different energy levels associated with values of

$$k = 1, 2, 3 \text{ etc.. upto } n.$$

These energy levels explain the fine structure in the hydrogen spectrum.

1.14 CORRECTION FOR THE FINITE MASS OF THE NUCLEUS

In the simple theory developed above it was assumed that the mass of the nucleus was so large compared with the mass m of the electron that the nucleus remained fixed at the centre of the circular orbits. This is rigorously true only if the mass of the nucleus is infinite. In reality, this is not the case. For instance, in the hydrogen atom the nuclear mass is only about 2,000 times that of the nucleus. On account of its finite mass the will also revolve. But the electron and the nucleus will evidently rotate about their common centre of gravity O (Fig. 1.15), the former in the larger circle of radius r_1, while the latter in the smaller one of radius r_2. A simple theorem of centre of gravity gives the relation $Mr_2 = mr_1$. As before, if a represents the distance between the nucleus and the electron, we get

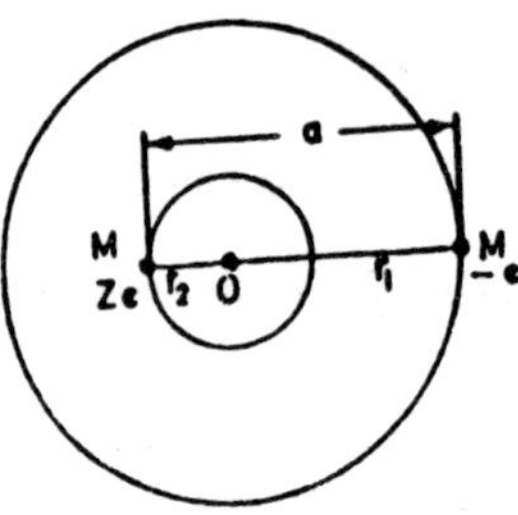

Fig. 1.15

$$r_1 = a - r_2 = a - \frac{mr_1}{M}$$

$$\left(\frac{1+m}{M}\right) r_1 = a; \qquad \therefore \; r_1 = \frac{aM}{(M+m)} \qquad ...(1)$$

and

$$r_2 = \left[\frac{mr_1}{M}\right] = \left[\frac{am}{(M+m)}\right] \qquad ...(2)$$

Since the two masses are revolving in circular orbits, we shall have two equations as conditions of stability, *viz.*

$$\frac{Ee}{4\pi \epsilon_0 a^2} = \frac{M\upsilon^2}{r_1} \text{ for the electron}$$

$$\frac{Ee}{4\pi \epsilon_0 a^2} \; \frac{M\upsilon^2}{r_2} \text{ for the nucleus,}$$

where the velocity of the electron $\upsilon = \omega r$, and the velocity of the nucleus $V = \omega r_2$, here ω being the common angular velocity of the system about O.

The total kinetic energy is made up of two parts, that of the electron and that of the nucleus.

$$\therefore \text{Total K.E.} \quad = \frac{1}{2} m\upsilon^2 + \frac{1}{2} MV^2$$

$$= \frac{1}{2} m\omega^2 r_1^2 + \frac{1}{2} M\omega^2 r_2^2$$

Substituting the values of r_1 and r_2 from equations (1) and (2) and simplifying we get

$$\text{Total K.E.} = \frac{1}{2}\left[\frac{mM}{M+m}\right]\omega^2 a^2$$

$$= \frac{1}{2}\mu\omega^2 a^2$$

where $$\mu = \frac{mM}{M+m} = \frac{m}{(1+m/M)} \quad ...(3)$$

usually called the "reduced mass".

The expression for the kinetic energy in the present case where the motion of the nucleus is taken into account differs from that in which the nucleus is supposed to be at rest in that the reduced mass μ replaces the mass m of the electron.

Without going further into the entire deduction we may write the final equation for the wave numbers of spectral lines of the atom, in which the nuclear mass is not infinite:

$$\bar{v} = \frac{mE^2e^2}{8\epsilon_0^2 ch^3}\left(\frac{M}{M+m}\right)\left(\frac{1}{n_1^2} - \frac{1}{n_2^2}\right) \quad ...(4)$$

We see that the factor multiplying the quantity within the bracket, which factor gives the Rydberg constant, depends upon the *ratio of the mass of the electron to that of the nucleus*. For, m is replaced by mM/(M + m) = m/(1 + m/M). If M is infinite, this expression is reduced to m. But actually the ratio m/M is not the same for different atoms. Therefore, the Rydberg constant will vary from element to element, though the variation will be small, since the correction factor, 1/(1 + m/M), is small. Including the atomic number Z by the relation E = Ze, equation (4) becomes

$$\bar{v} = \frac{me^2}{8\upsilon_0^2 ch^3}.Z^2\left[\frac{M}{(M+m)}\right]\left(\frac{1}{n_1^2} - \frac{1}{n_2^2}\right) \quad ...(5)$$

The Rydberg constant for any element is given by

$$R_z = \frac{me^4}{8\epsilon_0^2 xh^3}\left[\frac{M_z}{M_z + m}\right]$$

$$= R_\infty \left[\frac{1}{1 + m/M_z}\right] \qquad ...(6)$$

where $R_\infty \dfrac{me^4}{8\epsilon_0^2 ch^3}$ and M_z the mass of the nucleus of the element of atomic number Z

If Mz = ∞, the expression reduces to

$$R_H = R_\infty = \left[\frac{1}{(1 + m/M_H)}\right]$$

The constant R_∞ can be computed by substituting the values of the quantities involved ; but they are not known with spectroscopic precision. It is therefore preferably estimated from the spectroscopically observed value of the Rydberg constant for hydrogen R_H = 109,67,770 per metre and the known value of m/M_H = 1/1840,

$$R_H = R_\infty = \frac{1}{(1 + m/M_H)}$$

The value of R_∞, thus found is 109,73,740 per metre

Equation (5) can be written as

$$\bar{v} = Z^2 R_z\left(\frac{1}{n_1^2} - \frac{1}{n_2^2}\right) \qquad ...(7)$$

which expresses the modification to be made when the motion of the nucleus is included, and holds good for all elements.

The above relation has been experimentally confirmed in the following three cases:

1. Spectrum of the Singly Ionised Helium

The helium atom (Z = 2) has a mass four times that of hydrogen, a two-fold positive charge and normally two peripheral electrons. If it is singly ionised by the removal of one of the electrons, the residual atom, usually represented by He+, will have a hydrogen-like structure, consisting of a positive nucleus about which a single electron rotates.

It is to be expected therefore that the spectrum of singly ionised helium should be similar to that of hydrogen, but there will be differences due to the charge of the former being two instead of one and its nuclear mass being nearly four times than that of the latter.

Considering first the effect of the increased nuclear charge and neglecting the difference in nuclear mass, the ionised helium should give series of spectral lines represented by the equation

$$\bar{\nu} = 4R_H\left(\frac{1}{n_1^2} - \frac{1}{n_2^2}\right)$$

This is identical in form with that of hydrogen; but the wave numbers of the He+ lines would be four times as large as those of the lines in the corresponding series of hydrogen, since $4R_H$ replaces R_H.

Such series have been actually observed in the *spark* spectrum of helium and are known as the Fowler and Picketing series, the former corresponding, to $n_1 = 3$, while the latter to $n_1 = 4$. This not only verifies Bohr's theory but also provides an irrefutable proof for the nuclear charge of helium being double than that of hydrogen.

Careful examination of the spectra of the two atoms, however, showed that the coincidence was not quite exact. Taking for instance, the Pickering series, the second line in it is given by $\bar{\nu} = 4R_H\left(\frac{1}{4^2} - \frac{1}{6^2}\right)$ $= 4R_H\left(\frac{1}{2^2} - \frac{1}{3^2}\right)$ which is the wave number of the first (H_α) line of the Balmer series. Accurate spectroscopic measurements showed that the two lines do not coincide but the second Pickering line is shifted toward the shorter wavelength side of the H_α line.

Theory accounts for this discrepancy by considering the difference in nuclear masses in the two cases, which results in a slightly different value of the Rydberg constant for the two atoms, R_{He} being greater than R_H.

From the general expression for the Rydberg constant (equation 6) we get

$$R_H = R_\infty \{M_H/(M_H + m)\}$$

$$R_{He} = R_\infty \{M_{He}/M_{He} + m\}$$

for corresponding lines in the two spectra

$$\frac{R_{He}}{R_H} = \frac{M_{He}(M_H + m)}{M_H(M_{He} + m)}$$

Since the mass of the helium nucleus is nearly four times that of the hydrogen nucleus.

$$\frac{R_{He}}{R_H} = \frac{4M_H(M_H + m)}{M_H(4M_H + m)}$$

$$= \frac{4(M_H + m)}{4M_H + m}$$

Since the numerator is slightly greater than the denominator, $R_{He} > R_H$, in consequence, a line in the helium spectrum will have a slightly greater wave number and hence shorter wavelength than the corresponding line in the hydrogen spectrum, as experimentally observed.

The values of the Rydberg constant estimated from the hydrogen and helium spectra are

$$R_H = 109{,}67{,}770 \text{ m}^{-1}; \; R_{He} = 109{,}72{,}240 \text{ m}^{-1}$$

Other hydrogen-like ions whose spectra have been observed are Li^{++} and Be^{+++} and the values of the Rydberg constants obtained are R_{Li} = 109,72,890, R_{Be} = 109,73,080. Hence, the Rydberg constant is least for hydrogen and increases with the increase of nuclear mass, approaching the universal limit R = 109,73,740 which corresponds to a nucleus of infinite mass.

2. Determination of the Ratio of the Mass of the Electron to That of the Proton : m/M_H

Taking the relation

$$\frac{R_{He}}{R_H} = \frac{4(M_H + m)}{4M_H + m}$$

$$\frac{R_{He} - R_H}{R_H - \frac{1}{4}R_{He}} = \frac{4M_H + 4m - 4M_H - m}{4M_H + m - M_H - m}$$

$$= \frac{3m}{3M_H} = \frac{m}{M_H}$$

Substituting the values for R_{He} and R_H obtained from spectroscopic data, m/M_H can be calculated and is found to the 1/1840, which is in excellent agreement with the value obtained by other methods.

3. Discovery of Deuterium or Heavy Hydrogen of Mass Number Two

According to theory, for atom of the same value of Z there should be lines of slightly different wave numbers if their nuclei have different masses. Hence, in the case of hydrogen, if the isotope of mass number 2 existed, there should be isotopic components of the Balmer lines on the short wavelength side. Calculation showed that the separation between the isotopic components would be 1.793, 1.326, 1.185 and 1.119 Å for the first four lines H_α , H_β , H_γ and H_δ respectively. We have already described the very interesting circumstances that led to the, discovery of deuterium. We shall here summarise the researches of Prof. Urey and his collaborators, who succeeded in 1932, to prove the existence of deuterium. They used a twenty-one foot concave grating, capable of producing a dispersion of 1.3 Å per mm and photographed the lines of the Balmer series. Using first ordinary hydrogen in the discharge tube, they found that H_α and H_β had weak satellites separated by exactly the calculated distance. In itself, the appearance of a very weak line near each Balmer line would not have been quite convincing ; but Urey and his colleagues made their proof definite by preparing three samples of hydrogen in which the proportion of H^2, if it existed, had been raised by fractional distillation, and showing that the intensities of the lines in the positions calculated for H^2 were in agreement with the proportion of the heavy isotope believed to be in the sample. Thus, the third sample which was prepared so as to be considerably richer in H^2 than the others gave more intense lines at the required positions. With this last sample they obtained 1.791, 1.313, 1.176 and 1.088 Å for the separation of the isotopic components of the first four Balmer lines. Their experiments also showed that H^2 was present in ordinary hydrogen in the proportion of 1 in 5,000.

1.15 ZEEMAN EFFECT

A spectral line emitted by the excited atoms is split up into a doublet or a triplet when the emitting atoms are placed in a magnetic field. This effect of the splitting of a spectral line under the action of a magnetic field is known as *Normal Zeeman Effect.*

To produce Zeeman Effect, the source of light such as a sodium lamp or a mercury arc or gas discharge in a Geissler tube is placed between the poles of a powerful electro-magnet (Fig. 1.16). The light coming from the source is examined by means of a spectroscope of high

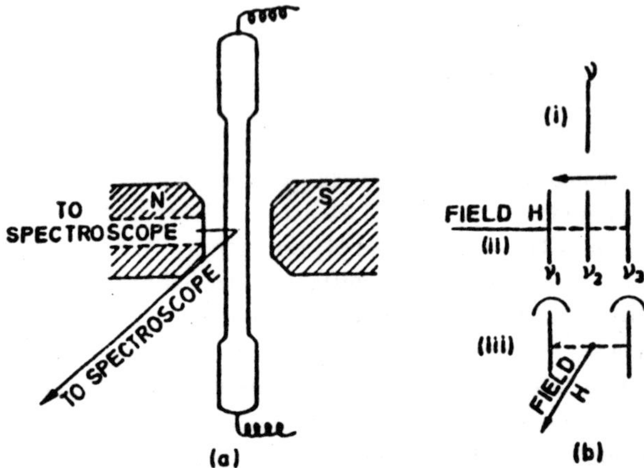

Fig. 1.16 : Normal Zeeman Effect (i) No field (ii) Perpendicular to the field (iii) Parallel to the field

resolving power. In order to view the light parallel to the magnetic field, a hole is drilled in one of the pole-pieces along the axis of the magnet.

When no magnetic field is applied, the spectroscope is focussed on one of the lines in the spectrum of the source of light. When a magnetic field is applied, it is observed:

1. That when the light is viewed in a direction perpendicular to the magnetic field, three component lines are observed. One of the lines is in the same position as the original line and the other two lines are on the two sides of the original line. The outer two lines, when observed by means of a nicol prism as an analyser, are polarized at right angles to the undisplaced line. This effect is known as *Transverse Zeeman Effect.*
2. When the light is viewed in a direction parallel to the direction of the field, there is no line in the position of the original line, only two outer lines are present. These lines are found to be circularly polarized in opposite directions. The effect is known as *Longitudinal Zeeman Effect.*

The normal Zeeman Effect is explained on Lorentz electron theory Consider an electron moving in a circular orbit of radius r with a velocity υ (Fig 1.17). The centripetal force is, $F = \frac{m\upsilon^2}{r}$. If an external magnetic field is applied, an additional force acts which is directed perpendicular to the direction of motion of the electron. This force is also perpendicular to the direction of the magnetic field and is along the radius. When/this force acts inwards along the radius, the velocity of the electron increases. When this force acts outwards along the radius, the velocity of the electron decreases. Suppose, this force due to the magnetic field = F_1 and let the velocity of the electron be increased to υ_1 by the application of the magnetic field. Then, $F_1 = Be\upsilon_1$. Suppose, this force is directed towards the centre, total force along the radius

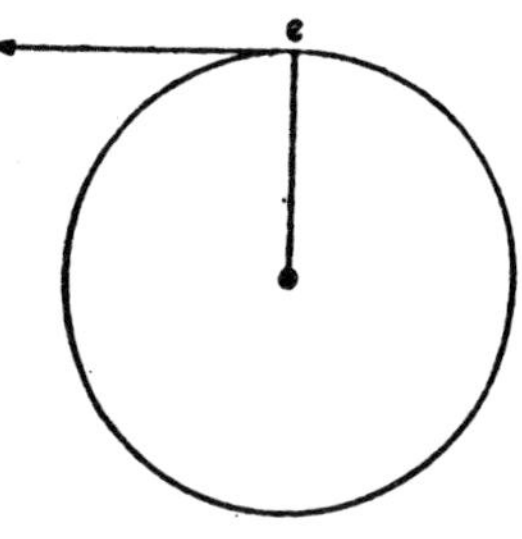

Fig. 1.17

$$= F + F_1 = \frac{m\upsilon^2}{r} + Be\upsilon_1$$

$$\text{Total force} = \frac{m\upsilon_1^2}{r}$$

$$\therefore \quad \frac{m\upsilon_1^2}{r} = \frac{m\upsilon^2}{r} + Be\upsilon_1 \quad \text{...(i)}$$

But, $\upsilon = r\omega$ and $\upsilon_1 = r\omega_1$

where ω and ω_1, are the respective angular velocities.

From equation (i)

$$\frac{mr^2\omega_1^2}{r} = \frac{mr^2\omega^2}{r} + Ber\omega_1$$

$$mr\omega_1^2 - mr\omega^2 = Ber\omega_1$$

$$\omega_1^2 - \omega^2 = \frac{eB\omega_1}{m}$$

$$(\omega_1 + \omega)(\omega_1 + \omega) = \frac{eB\omega_1}{m}$$

$$(\omega_1 + \omega) = \frac{eB\omega_1}{m(\omega_1 + \omega)}$$

$(\omega_1 + \omega)$ is approximately equal to $2\omega_1$

$$\therefore \qquad \omega_1 - \omega = \frac{eB\omega_1}{2m\omega_1} \frac{eB}{2m}$$

$$\therefore \qquad \omega_1 = \omega + \frac{eB}{2m}$$

If v_1 and v are the frequencies, then, $\omega_1 = 2\pi v_1$, $\omega = 2\pi v$

$$2\pi v_1 = 2\pi v + \frac{eB}{2m} \qquad \text{...(iii)}$$

$$v_1 = v + \frac{eB}{4\pi m}$$

When the electron moves in the opposite direction, the magnetic field produces a force in the opposite direction and the velocity decreases to υ_2. In that case

$$\text{Total force} = F - F_2 = \frac{m\upsilon^2}{r} + Be\upsilon_2$$

$$\therefore \qquad \frac{m\upsilon_2^2}{r} = \frac{m\upsilon^2}{r} - Be\upsilon_2$$

But $\qquad \upsilon_2 = r\omega_2$ and $\upsilon = r\omega$

$$\therefore \qquad \frac{mr^2\omega_1^2}{r} = \frac{mr^2\omega^2}{r} + Ber\omega_2$$

$$mr\omega_2^2 - mr\omega^2 = Ber\omega_2$$

$$\omega_2^2 - \omega^2 = \frac{eB\omega_2}{m}$$

$$(\omega_2 - \omega) = -\frac{eB\omega_2}{m(\omega_2 + \omega)}$$

$$\omega_2 - \omega = -\frac{eB}{2m} \qquad (\therefore \omega_2 + 2\omega_2 \text{ approx.})$$

$$2\pi v_2 - 2\pi v = -\frac{eB}{2m}$$

$$\therefore \qquad v_2 = v - \frac{eB}{4\pi m} \qquad \text{....(iv)}$$

From (iii) and (iv) we get

$$\Delta v \text{ (in each case)} = \frac{eB}{4\pi m} \qquad \text{...(v)}$$

This shows the two lines are displaced equally on the two sides of the original line,

$$v_1 - v_2 = \frac{eB}{2\pi m} \quad ...(vi)$$

In general $$\Delta v = \frac{eB}{4\pi m} \quad ...(vii)$$

Alternative Treatment: Consider an electron moving in a circular orbit of radius r with velocity υ.

The centripetal force $F \frac{m\upsilon^2}{r} = mr\omega^2$. If an external field B is applied, the change in the force,

$$dF = \pm Be\upsilon = \pm Ber\omega \quad ...(i)$$

$$F = mr\omega^2 \quad ...(ii)$$

Differentiating, $dF = 2mr\omega d\omega$

Equating (i) and (ii)

$$\pm Ber\omega = 2mr\omega d\omega$$

$$\frac{e}{m} = \pm \frac{2d\omega}{B}$$

Also $$\omega = 2\pi v = \frac{2\pi c}{\lambda} \quad ...(iii)$$

Differentiating $d\omega = \frac{-2\pi c}{\lambda^2} d\lambda$

Substituting the value of $d\omega$ in equation (iii)

$$\frac{e}{m} = \pm \frac{4\pi c d\lambda}{\lambda^2 B} \quad ...(iv)$$

$$d\lambda = \pm \frac{eB\lambda^2}{4\pi mc} \quad ...(v)$$

$$\therefore \quad \lambda_1 = \lambda + \frac{eB\lambda^2}{4\pi mc}$$

$$\lambda_2 = \lambda - \frac{eB\lambda^2}{4\pi mc}$$

$$\lambda_1 - \lambda_2 = \frac{eH\lambda^2}{2\pi mc} \quad ...(vi)$$

$$v = \frac{c}{\lambda}$$

Differentiating, $dv = -\frac{cd\lambda}{\lambda^2}$

Substituting the value of $d\lambda$ from equation (v)

$$dv = \pm \frac{eB}{4\pi m}$$

$$v_1 = v + \frac{eB}{4\pi m}$$

$$v_2 = v - \frac{eB}{4\pi m}$$

$$v_1 - v_2 = \frac{eB}{2\pi m}.$$

The quantity $\frac{eB}{2\pi m}$ is known as *Normal Zeeman* separation, knowing the values of v and B, $\frac{e}{m}$ can measurement of *Zeeman Effect* = 1.757×10^{11}e/kg and if is in agreement with value of $\frac{e}{m}$ of the electron obtained from Thomson's experiment. This experiment established that the electron in the form is responsible for the emission of special lines.

1.16 MOLECULAR SPECTRA

The spectra emitted by atoms as a result of energy changes in their electronic system, which we have considered so far, are known as atomic spectra. There is another class of spectra called the band or molecular spectra, emitted by molecules, *i.e.,* when the emitting element is in the molecular state. The name 'band spectra' was given to these at a time when spectroscopes of low resolving power alone were available so that the component lines of the spectra were indistinguishable and appeared merged together into strips of shaded intensity, giving the impression of a continuous spectrum but divided into several bands. The fact that spectra of this type are actually due to molecules is readily established by the disappearance of the bands when the emitting substance is heated to temperatures at which the molecule dissociates into atoms.

With high resolving power instruments band spectra disclose a threefold structure : (i) Each band is composed of a *large number of lines* which are arranged with great regularity often stretching in both directions from a gap in the band and crowding together at the band-head, (ii) several bands are found to follow one another in a regular sequence, constituting a *group of bands*, (iii) a close and regular arrangement of different groups of bands forming a *band system*. Several such band systems are comprised in a complete molecular spectrum; sometimes adjacent bands overlap and obscure the regularity of arrangement of lines, which makes the analysis of band spectra very difficult.

Many substances are known to produce band spectra in different spectral regions, *viz*., in the *far infra-red* (about 150Å to 30μ, 1μ = 10,000Å). The *near infra-red* (about 5μ to 1μ and *visible and ultra-violet region* 700Å to 1000Å. These three systems of bands according to the accepted ideas of the mechanism of their production, are called *pure rotation bands, rotation-vibration bands and electronic bands respectively.*

1.17 EXPERIMENTAL STUDY

The *rotation bands* in the far infra-red were first noticed in the absorption spectra of the hydrogen halides, by Czerny, who found in the case of HCI a series of seven absorption maxima in the region 120μ to 44μ and was able to show that the observed absorption lines could be represented by the relation :

$$\bar{v} = 20.84m - 0.001814m^3$$

where $m = 4, 6, 7, 8, 9, 10, 11.$

He obtained similar relations in the case of HBr and HI. The above relation shows that the group of lines that constitutes the rotation bands are separated by an approximately constant frequency difference, since in assume successive integral values, the missing integers corresponding to lines not observed. The second term in m^3 indicates that the frequency interval between the successive lines is not quite constant, but decreases with higher wave numbers.

Water and ammonia also show complicated absorption spectra in this region. Till of late, it has not been easy to observe the pure rotation bands in many cases owing to experimental difficulties. But with the recent introduction of the microwave spectrometer, it has now become comparatively easy to observe and analyse these bands and interesting results have been obtained.

The *rotation-vibration* bands in the near infra-red have also been observed in the absorption spectra of the hydrogen halides. The general structure of a typical band of this sort (HCI) is shown in Fig 1.18. The

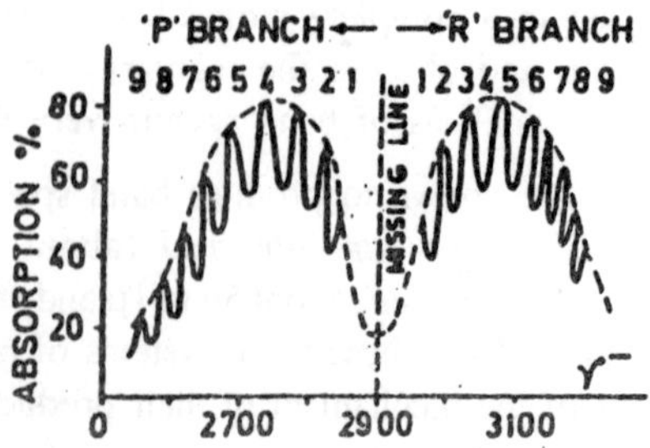

Fig. 1.18 : Rotation-vibration band of HCI.

form is seen to be doubled—the so-called *double band*, in the sense that a curve drawn through the intensity maxima of the individual lines shows two maxima, as indicated by the dotted line. There is a gap between the two bands—the *missing central line*. This feature of a double band with a central gap distinguishes the rotation vibration spectra from the pure rotation spectra which have only a single band with no gap.

The wave numbers of the individual line maxima are well represented by the relation:

$$\bar{v} = 2886.2 + 20.5379m - 0.30318m^2 - 0.001814m^3.$$

The first term indicates the region in which this type of spectra appears (2886 cm^{-1}), *i.e.,* near infra-red; the second term] gives the approximate distance between the lines (20.5 cm^{-1}), which is seen to be nearly the same as the distance between the lines of the pure rotation spectra (20.8 cm^{-1}). The other two terms are correction factors which indicate that the frequency interval between the line is not constant, but decreases with higher members. The lines which are on the higher wave number side of the central missing line constitute the so-called R branch, while those on the lower wave number side the P branch.

Rotation-vibration bands are found also with substances other than the hydrogen halides, but their structure is often more complex with no gap in the centre of the band. Thus the absorption spectrum of CO_2 shows many rotation-vibration bands from 1.46μ. to 15μ that of water vapour in the region 0.69 μ to 6.26μ. These bands together with the rotational lines of water vapour are responsible for the marked absorption of the earth's atmosphere in the infra-red.

The electronic bands occur in the visible and ultra-violet regions both in emission and absorption spectra. Probably the best known case of this type is the cyanogen (CN) bands in the red and violet regions, which appear in the spectrum of an electric arc between carbon electrodes. The cyanogen bands in the violet region obtained by Graaff are reproduced here (Fig. 1.19). In the complete electronic band system there are several groups of bands, each group containing in turn, several partial bands,

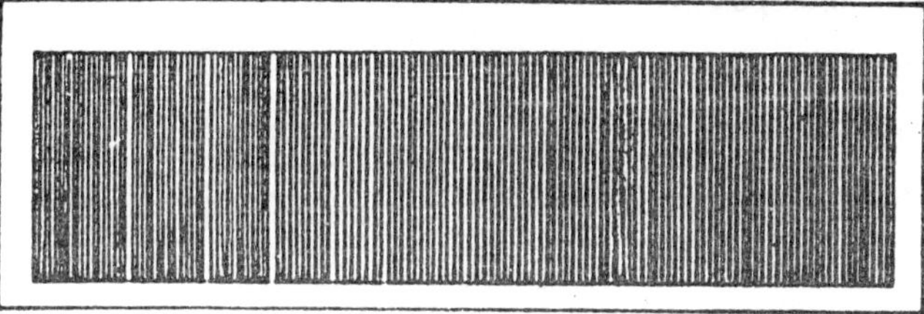

Fig. 1.19 : Cyanogen bands in the violet region.

sharply defined, bright on the long wavelength side, called the *band-head* and fading away towards shorter wavelength. Under greater resolution, the partial bands are found to be composed of a large number of lines of regular structure, but crowded together at the band-head. Among these lines it has been possible to distinguish three branches known as the P, Q and R branches, the Q branch being absent in many bands. The distance between the partial bands is of the order of magnitude of the vibrational frequency of the molecules, while the spacing of the lines in the bands is that of rotational frequency. The empirical relation discovered by Deslandres in connection with these bands is in the form:

$$\bar{v} = A + Bm + Cm^2$$

where $\bar{v}$ is the wave number of a line in a partial band, A and wave number of the band-head, B and C constants characteristic of the band m = 1, 2, 3, etc.

1.18 THEORETICAL EXPLANATION

As molecular spectra are more complex than atomic spectra, they naturally involve a more complicated mechanism for their production. Bohr's theory can, however, be used to explain, in an elementary manner, the chief features of the three types of band spectra detailed above. Considering the simple case of a diatomic molecule such as HCI, CN etc., it is supposed to be constituted with two atoms separated by a distance which is large in comparison with the atomic dimensions. The molecule as a whole can rotate about an axis passing through the centre of gravity of the system; there can also be relative vibratory motion of the two atoms, inside the molecule along the line joining the two atoms; finally the orbital motions of the electrons associated with the molecule must also be taken into account. Corresponding to these three types of motions, the total internal energy E of the molecule is assumed to be made up of three parts:

(a) rotational energy E_r,

(b) vibrational energy E_υ and

(c) electronic energy E_e. Hence

$$E = E_r + E_\upsilon + E_e$$

From the spectral regions in which the three types of bands occur it is readily seen that

$$E_e > E_\upsilon + E_r$$

Assuming that the same quantum laws used by Bohr to explain atomic spectra hold good also in the case of molecular, spectra, the three components of the molecular energy can have only discrete values, characterised by quantum numbers say m_r, m_υ, m_e respectively, changes in which play the same part in band spectra as the quantum number n in atomic spectra. Further, energy changes which give rise to the spectra are governed by Bohr's frequency condition

$$hv = W_1 - W_2$$

The *pure rotation bands* in the far infra-red region arise due to the rotation of the molecules. Analytical treatment on the lines indicated above leads to the relation

$$v_r = \frac{h}{8\pi^2 I}(2m_r + 1) \qquad ...(1)$$

where v_r is the frequency of a spectral line in the rotation bands, h Planck's constant, I the moment of inertia of the molecule about the axis of rotation, m_r = 0, 1, 2 etc.

This equation (1) predicts a spectrum of equally spaced lines with frequencies equal to integral multiples of a fixed quantity $\frac{h}{8\pi^2 I}$ and a constant frequency interval $\frac{h}{4\pi^2 I}$ in-agreement with experimental findings, the correlation term in the empirical formula being accounted for by the change in with the variation in the velocity of rotation.

The *rotation-vibration bands* in the near infra-red are caused by the simultaneous rotation and vibration of the molecules. In this case, the total energy of the molecule is partly rotational and partly vibrational. Hence any given change of state of the molecule involves a simultaneous-change of both types of energy, *i.e.*, the energy hv of an emitted radiation comes partly from the vibrational energy change and partly from rotational energy change. The frequency v of the individual lines in the rotation-vibration band is given by

$$v = v_\upsilon + v_r$$

where v_υ is the contribution to the frequency from vibration and v_r from rotation. It can be shown that $v_\upsilon = v_0(m_{\upsilon 1} - m_{\upsilon 2})$ where v_0 is the fundamental (natural) frequency of vibration, $m_{\upsilon 1}$ and $m_{\upsilon 2}$ are the vibrational quantum numbers of the two states between which transition takes place. Substituting for v_r from equation (1), we get

$$v = v_0(m_{\upsilon 1} - m_{\upsilon 2}) + \frac{h}{8\pi^2 I}(2m_r + 1) \qquad ...(2)$$

Taking the relation $v_\upsilon = v_0\,(m_{\upsilon 1} - m_{\upsilon 2})$ we see, first of all, that the vibrational frequency v_υ is an integral multiple of a fundamental value since $m_{\upsilon 1}$ and $m_{\upsilon 2}$ are integers. Secondly, considering the normal case of absorption in which the molecule in the lower state ($m_{\upsilon 2} = 0$) is raised to an upper state ($m_{\upsilon 1} = 1$, it is seen that $v_0 = \tau_0$. This means that the position of the vibrational bands is determined by the mechanical vibrational frequency of the molecule, whose order of magnitude is such that the bands corresponding to the transition $m_{\upsilon 2} = 0 \rightarrow m_{\upsilon 1} = 1$ will be found between 1000 and $4 \times 10^5 m^{-1}$ for light molecules; this range is in the near infra-red where all the rotation-vibration bands are actually observed.

Since the vibrational energy is much larger than the rotational energy ($E_\upsilon > E_r$), every change in the former will be accompanied by a number of smaller changes in the latter. That is to say, for a given change in m_υ there will be many possible changes in m_r. In other words, for a given value of v_v, υ_r has a series of values for the several integral values of m_r Hence in the rotation-vibration band there will be a series of equidistant lines with a frequency interval of $\frac{h}{4\pi^2 I}$. The Theory of Atomic Structure is in good agreement with experimental observation.

As in the case of pure rotational lines, here also the frequency interval between the lines is not constant but decreases as we pass to higher members, as indicated by the correction terms in the empirical formula. The explanation for this is to be sought for, along the same lines as in the case of the pure rotational lines, *viz.*, due to the alteration of the moment of inertia of the molecule. The mean moment of inertia of a vibrating molecule must be expected to differ from that of the molecule not vibrating. Moreover, the centrifugal force which comes into play during rotation must alter the strength of the linkage between the atoms and so must affect the vibrations. There must be therefore an integration between rotation and vibration. This can, in fact, be calculated and the observed variations in the spacing of the lines can be accounted for.

In relation (2), even if we put m, = 0 in the second terms, this term does not vanish, but is equal to $\frac{h}{8\pi^2 I}$. Hence it is impossible to have v equal to v_υ alone, which means that the molecule cannot alter its vibrational energy without a simultaneous alteration of its rotational energy. This is the reason why the central line corresponding to $v = v_\upsilon$ is missing. Changes in rotational energy for which m, increases by unity give rise to an array of lines extending towards higher frequencies—the positive R branch, while those for which m_r decreases by unity give rise to another set of lines extending towards lower frequencies—the negative branch. Thus the theory proposed gives a satisfactory explanation for all the observed facts about rotation-vibration spectra.

The *electronic band spectra* in the visible and ultra-violet regions are produced by the electronic transitions on which the molecular vibrations and rotations, are superposed. Although, it is the most complex of band spectra, its explanation becomes relatively easy after what has been said about the other two types. Here the total energy of the molecule in a given quantum state is made up of all the three components, *viz.*, electronic vibrational and rotational energies. Transitions between two such energy states, for which

$$hv = (E_{e_1} + E_{\upsilon_1} + E_{r_1}) - (E_{e_2} + E_{\upsilon_2} + E_{r_2})$$

are those which give rise to the electronic band spectrum. The frequency v of the individual lines in it is given by

$$v = \frac{E_{e_1} - E_{e_2}}{h} + \frac{E_{\upsilon_1} - E_{\upsilon_2}}{h} + \frac{E_{r_1} - E_{r_2}}{h}$$

$$= v_e = v_\upsilon + v_r$$

where v_e represents the contribution to the frequency due to the change in electronic energy, v_υ the contribution due to the change in vibrational energy and v_r due to rotational energy changes.

Since the electronic energy is of greater magnitude that the vibrational and rotational energies ($E_e > E_\upsilon > E_r$) the position of the spectrum is determined by the change in E, which as we know from atomic spectra is in the visible and ultra-violet regions.

Further, every single change in E_e will be accompanied by a number of changes in E_υ according to the many possible values of ($m\upsilon_1 - m\upsilon_2$), which accounts for the existence of a number of partial bands in a group, as well as different groups in a band system. Each of the Vibrational transitions will, in turn, be accompanied by a variety of rotational changes, as m_r assumes different integral values, giving rise to the rotational lines in each band. Thus, just as the pure rotation spectrum can be superposed upon a vibrational frequency, so also the whole rotation-vibration spectrum can be superposed on an optical 'frequency due to an electronic transition. This process produces a band system. The manifold of all such band systems connected with different electronic transitions in a molecule represents a complete electronic band spectrum.

The electronic configuration has a resultant angular momentum, which combines vectorially with the rotation of the molecule as a whole and permits transitions in the vibrational stale without alteration of the rotational state, unlike in the case of rotation-vibration bands. This means the transition corresponding to $m_r = 0$ is not forbidden so that the central line will not be missing in the electronic bands.

The effect on the rotational and vibrational states due to alterations in the electronic configuration is very much greater than the interaction between vibration and rotation in the rotation-vibration bands. In consequence the molecule has to be considered to have different strengths of linkage and different moments of inertia in the initial and final states of a transition. This fact can be shown to lead to the existence of the three branches P, Q and R in the rotational lines of the electronic bands.

Band spectra research has now developed as a science of considerable dimensions on account of its importance in the study of molecular structure, isotropic constitution of elements, Raman effect, etc. For the interpretation of the extensive experimental data, ingenious mathematical devices, such as a group theory, consideration of symmetry, etc., have been called into service.

1.19 RAMAN EFFECT

Raman Effect may be said to be the *optical analogue* of Compton effect, since it also belongs to the category of *incoherent scattering*, which finds an adequate interpretation only on the basis of the quantum theory principles. Hence the inclusion of Raman effect among the applications of Bohr's theory may be justified. Moreover, as Raman effect is closely connected with molecular spectra the study of the former naturally follows the latter.

Discovery

In 1928, Sir C.V. Raman was studying the scattering of light by liquids with the intention of reproducing such natural phenomena as the *blue of the sea and of the sky*. Already, in 1871, Lord Rayleigh had given a correct explanation of the blue colour of the sky. He attributed the phenomenon to the scattering of light by the individual air molecules or groups of them found in abundance in the earth's atmosphere. He showed also *that the coefficient of scattering of light in any medium is inversely proportional to the fourth power of the wavelength of light.* Form this it follows that the intensity of the scattered light must increase rapidly as the wavelength decreases. Hence when sunlight passes through the atmosphere, the intensity of the blue colour in the scattered light will be very high compared with most of the other components of white light, since its wavelength is smaller than those of most of the other visible colours. This is the reason why the sky appears blue. It is to be noted-that this type of scattering, known as *Rayleigh scattering*, is also called *coherent scattering*, since, though there is a change in the intensity of the scattered light, there is no change in the spectral character, *i.e.,* in the wavelength.

In his experiments, Raman found that when a beam of monochromatic light was passed through organic liquids such as benzene, toluene, etc., the scattered light contained other frequencies in addition to that of the incident light. Although, such an effect had been predicted on theoretical grounds as early as 1923 by Smekal, Raman was the first to observe it experimentally. The arrangement used by him was simple in design. A round-bottomed glass flask was filled with dust-free benzene and the liquid was strongly illumined by the 4358 line from a mercury arc, suitably filtered and concentrated by a lens. The scattered light was examined by means of a spectroscope placed transversely, *i.e.,* in a direction at right angles to that of the incident radiation. In the spectrum of the scattered light, number of new lines were observed of both sides of the main line. Those on the low frequency side were numerous and more intense than those on the high frequency side; the spacing between the lines was symmetrical about the main line; most of them were strongly polarised. The observed spectrum is now generally referred to as *Raman spectrum* and the lines; as *Raman lines*; more specifically those on the low frequency side as Stokes' lines and those on the high side as *anti-Stokes' lines*.

Raman, having thus observed a set of a new discrete frequencies in the scattering process, contrary to the expectation of classical theory, was able to establish that they constituted a new phenomenon distinct from both the simple Rayleigh or coherent scattering and the more complex Stokes' or fluorescent scattering. He argued that in Rayleigh scattering no frequency change was caused, but only some of the already existing frequencies were selected, whereas in the phenomenon observed new lines of different frequencies appeared even when a single frequency (monochromatic light) was scattered.

Nor could the observed effect be considered similar to fluorescence, although there was a certain amount of superficial resemblance between the two, in so far as with a given exciting line new lines made their appearance in both. For, in the first place, the frequencies in the fluorescent spectrum were always less than the incident frequency, while the Raman lines had frequencies both higher (anti-Stockes' lines) and lower (Stokes' lines) than that of the incident line. Secondly, the frequencies of the fluorescent lines were really independent of that of the exciting line, provided the latter was able to produce fluorescence, but in the case of the Raman lines, their frequencies were directly related to that of the incident light, since if the latter was varied the former changed at the same rate. Thirdly, while the frequencies of the fluorescent; lines were fixed by the nature of the scatterer, it was the frequency shifts of the Raman lines (known also as *Roman frequencies*) that were determined by the scatterer rather than the frequencies themselves. Fourthly, the lines observed by Raman were strongly polarised unlike the lines in the fluorescent spectrum. Finally, the most important fact that confirmed

Raman's discovery as a new effect was that the Raman frequencies, were either actual infra-red frequencies in the absorption spectrum of the scatterer or differences in such frequencies which was not the case in fluorescent scattering. This fact proved further that the observed effect was molecular phenomenon. Hence the modified frequencies observed in the scattering process was a new type of secondary radiation, to which the name *Raman Effect* was given and for which Raman was honoured by the scientific world, in 1930, by conferring on him the Nobel Prize.

Almost simultaneously with Raman, Landsberg and Mendelstam in Russia, studying the light scattered by certain crystals discovered the effect in the case of solids. It was soon found that many liquids, vapours, gases and transparent solids exhibited the effect which thus proved to be a general phenomenon.

Experimental Study

Raman effect has been extensively investigated by a large number of workers. The *general principle* used in these researches is to illumine the substance under test with an intense monochromatic source of light and photograph the scattered radiation by means of a spectrograph arranged in a transverse direction. But the technical details vary according to the nature of the substance studied, *i.e.,* solid, liquid or gas, the chief aim being to obtain the best and quick results.

Apparatus

The original simple arrangement of Raman was not quite efficient and required very long exposures, about 100 hours and more, to obtain good records of the Raman spectrum. Hence improvements were made as regards the container of the substance, the source of radiation, filters, spectrograph, etc.

The apparatus shown in Fig. 1.20 is the one, first designed by Wood and now ordinarily used in the study of Raman effect in *liquids*.

The container C of the liquid to be investigated, called the *Raman tube*, consists of a glass tube of about 1 or 2 cm in diameter, and 10 to 15 cm long, one end of which is drawn out into the shape of a horn and blackened outside to provide a suitable background, the other end being closed with an optically plane glass plate constituting the window

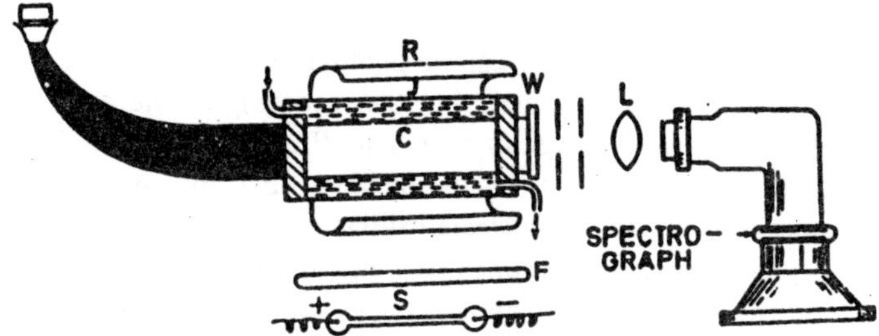

Fig. 1.20 : Apparatus for the study of Raman effect.

through which the scattered light emerges. The container is surrounded by a water jacket in which cold water is circulated to prevent overheating of the liquid in the Raman tube due to the proximity of the exciting arc.

An ideal *source* S would be light from a helium discharge tube filtered by nickel oxide glass, giving a strictly monochromatic line of wavelength 3888 Å. But on account of the many technical difficulties involved in the construction and manipulation of this source, it is not widely used. The source ordinarily employed is the mercury arc, the next best available, from which it is possible to get single wavelengths by the use of suitable filters. Thus, for instance, to obtain the ,4358 Å line, slightly acidulated quinine sulphate solution contained in a *novial glass* vessel is used as a filter, which cuts off all the other lines except the 4358Å. To get the 4046Å line, a solution of iodine in carbon tetrachloride contained in a novial glass cell is found to be a very satisfactory filter. The filter solution may be arranged either to surround the Raman tube or in front of the arc. The mercury arc is placed as close to the Raman tube as possible, which results; in a large intensity of the incident light. A semi-cylindrical aluminium reflector R, enhances the intensity of illumination still further.

The chief features of a *spectrograph*, suited for the study ,of Raman spectra, are:

1. large light gathering power,

2. special prisms of high resolving power, and
3. a short focus camera, A lens L in front of the window W directs the scattered radiation upon the slit of the spectrograph, which is carefully aligned along the axis of the Raman tube and screened from the direct rays of the arc. The more intense Raman lines of a liquid such as CCl_2 can be photographed in about an hour with a small spectrograph, but the recording of the complete spectrum may require up to ten to fifteen hours, depending largely on the intensity of the incident light, the speed of the spectrograph and the intrinsic brilliance of the Raman lines. It may be noted that instruments of high resolving power such as gratings are not used with advantage, on account of their poor luminosity which necessitates long exposures.

Certain modifications in the apparatus are necessary for the excitation of Raman effect in solids and gases. In the case of solids 'which are available as *large* and *transparent blocks*, like gypsum, quartz etc., a container is not required and the light from the arc can be directly focussed on the substance with a large condensing lens. With solids which are in the form of *loose crystal powders*, Raman effect can be obtained as was first shown by Baer and Menzies, by reflecting light from the crystal surfaces. But special precaution has to be taken to avoid the masking effect due to large amount of direct light coming out of the container by repeated reflection at the crystal faces and entering the spectrograph. To achieve this, two techniques have been used : *one* the use of complementary filters as developed by Ananthakrishnan at the Raman laboratory and the *other* a special type of spectrograph with two parts, each part having a prism and two lenses with a common slit in between the two, devised by Billroth, Kohlrausch and Reitz in Germany. Both give very good results with crystal powders and the latter even with very small quantities of the substance.

In the case of *gases*, the intensity of the light scattered is normally very weak, but the difficulty has been overcome by intense illumination of the gas under high pressure and the use of a spectrograph of large light-gathering power. Wood employed a very long tube of HCl gas and obtained its Raman spectrum at atmospheric pressure using a specially made mercury arc which was placed in contact with the gas tube, hollow cylindrical reflectors enclosing both of them. The illumination produced in this way was very intense as the light from the arc was reflected back and forth between the walls of the reflectors. Rosetti was the first to develop the technique of exciting Raman effect in gases under high pressure, which shortened considerably the time of exposure. He used a thick-walled quartz tube 20 cm long and 2.2 cm internal diameter, which could stand pressures of 10 to 15 atmospheres. With the 2537Å line of the mercury arc he was able to obtain the Raman spectra of several gases under pressure. Bhagavantam has constructed a Raman tube for gases which can stand pressures up to 50 atmospheres. It is made of transparent silica and enclosed in an outer steel tube for protection. He has been able to obtain with his arrangement good spectrographs of Raman spectra of gases in about 40 or 50 hours time of exposure using the 3650Å, 4046Å and 4358 A mercury lines.

1.20 RESULTS OF RAMAN EFFECT

Raman effect has been observed and studied in a large number of liquids, organic and inorganic, many vapours and gases and some solids both crystalline and amorphous. An idea of the type of Raman spectra produced by solids, liquids and gases may be formed by inspecting the photographs with benzene, carbon tetrachloride (liquids), acetylene (gas), calcite and diamond (solids) [Fig. 1.21]. It is seen that a number of new lines and bands, exhibiting a variety of characters of intensity and fine structure are recorded on either side of the exciting radiation. Each line in the incident spectrum, if of sufficient intensity, gives rise to its own set of lines or bands. We shall now outline the main results obtained from the study of such photographic records.

Raman Effect in Liquids

About a hundred liquids, so far examined, show the effect in an unmistakable manner. With benzene, the frequency shifts of the Raman lines correspond to an infra-red wavelength of 3.27μ in which region benzene exhibits a strong band in its absorption spectrum. A clear examination, however, of the infra-red absorption spectrum of benzene shows that none of the Raman lines are represented in it. An interesting feature with CCI_4 is the triad of Stokes' and anti-Stokes, lines equally spaced on either side of the exciting 4358Å line. With well purified water, two broad bands are found instead of sharp lines at about $\lambda = 3\mu$. Solutions of salts in water give Raman spectra characteristic of the salts and water.

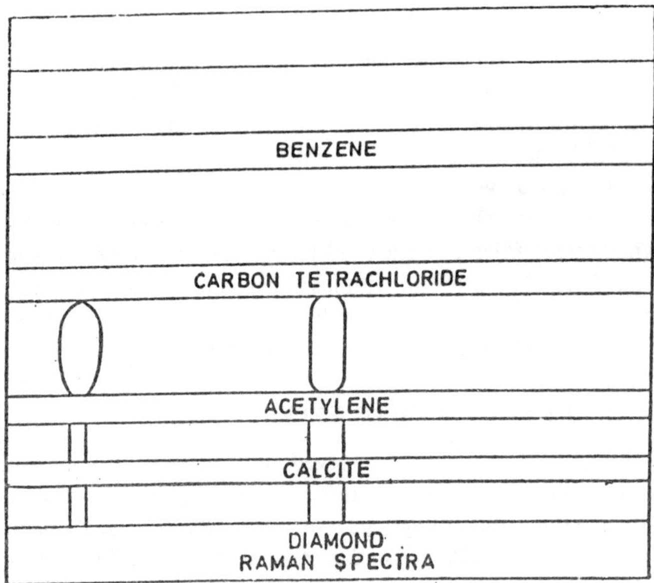

Fig. 1.21 : Raman spectra of different substances.

Raman Effect in Gases

Several gases such as HCl, CO, CO_2, NO, N_2O, CS_2, hydrogen, oxygen, nitrogen, ammonia, etc., have been studied. The frequency shift of the Raman line of HCl gas corresponds to 3.46μ, which is almost exactly the wavelength of the central missing line of the infra-red absorption band of HCl. CO gives a Raman line whose frequency shift is equal to the frequency of its infra-red bands, while CO_2 gives a line whose shift is equal to the frequency difference of two of its infra-red bands. Oxygen, hydrogen and nitrogen give a pattern of equally-spaced lines. The intensities of the individual lines alternate in the case of nitrogen and hydrogen, while with oxygen the alternate lines are absent. These peculiarities in intensity of alternate lines-have led to very significant Theory of Atomic Structures as regards molecular and nuclear structure.

Raman Effect in Solids

The Raman lines obtained with crystals are becoming diffuse with rise of temperature. In calcite ($CaCO_3$) two lines which are nearest the exciting line have been identified with the oscillations of the crystal lattice. Diamond exhibits a strong and sharp line of a comparatively large frequency shift, which has to be ascribed to, a lattice oscillation.

Intensity of Roman Lines

The intensity of a Raman line when expressed as a function of the parent line is usually a few hundredths in liquids and few thousands in gases. In spite of the practical difficulties in measuring the intensities of Raman lines results of fundamental importance have been obtained. The variations in intensity as we pass from line to line and from substance to substance are of great significance in the study of molecular structure and chemical constitution. The Stokes lines are always more intense than the corresponding anti-Stokes lines; these, however, grow more intense as the temperature is raised. All the Raman lines move inward towards the parent line with increase of temperature.

Polarisation of Raman Lines

Just as the Raman lines vary greatly in their intensities, so also their states of polarisation. The fact that different lines are differently polarised is probably connected with their relative intensity.

The experimental arrangement used for determining the polarisation of Raman lines is essentially the same as that described earlier, but with the following modifications. The light from the source is concentrated by means of a condenser into the substance contained in the Raman tube. A suitably oriented double image prism whose function is to separate the vertical and horizontal components in the scattered light is placed in front of the slit of the spectrograph, so that two images, one above the other are formed on the slit, which are simultaneously photographed.

The state of polarisation of a Raman line is measured by a quantity known as the *depolarisation factor* which is simply, the ratio of the intensities of the horizontal and vertical components when the incident light is vertically polarised. The ratio is readily obtained from the traces photographed as described above, by one of the usual methods employed for comparing the intensities of two beams of the same wavelength. In order to get fairly accurate values of the depolarisation factor, the following precautions should be taken:

(i) Crystalline quartz should not be used for condenser, spectrographs or windows, since its optical activity complicates the phenomenon of polarisation;

(ii) the window of the Raman tube through which the scattered light emerges should be stain free and plane; and

(iii) errors arising from oblique refraction at the prism surfaces, want of transversality in the incident beam and slit width should be eliminated.

Cabannes found that the *Raman Lines* in crystals, such as quartz, are differently polarised, the intensity and depolarisation of the lines depending upon the orientation of the crystal. Menzies has investigated the polarisation of the Raman lines in liquids, such as CCI_4, in directions perpendicular and obliquely forward to the incident beam and has shown that many of the observed facts could be accounted for by considering the initial and final directions of vibrations in the molecule involved to be parallel in the case of polarised lines, perpendicular in that of unpolarised lines and at an oblique angle for partially polarised lines. The following are some of the important results obtained:

(a) The depolarisation factor varies from 0 to 0.86 for the vibrational Raman lines, while it has a constant value of 0.86 for the rotational lines.

(b) With circularly polarised incident light, part of the Raman line is circularly polarised in the reverse direction and part circularly polarised in the same direction as that of the incident light. Highly depolarised rotational lines exhibit reverse circular polarisation.

(c) Sharp and strong lines are ordinarily characterised by low depolarisation factors, while diffused and weak ones by high depolarisation factors.

(d) Corresponding line in molecules having similar structures have nearly the same depolarisation factors. This is to be expected as the polarisation of a Raman line is decided by the symmetry of the oscillation.

Nature of Raman Effect

From the many and varied experimental data, it is clear that Raman effect is a molecular phenomenon. In the case of free molecules, as in gases, three different kinds of Raman Effect can therefore be distinguished, *viz.*, rotational, vibrational and electronic. A mixed "rotation-vibration" effect can also take place under certain circumstances. Solids can exhibit another type, in which the crystal lattice as a whole takes the place of the molecule. Most of the observed Raman lines and bands with moderate or large frequency shifts are due to the vibrational effect, since pure rotational Raman lines lie very close to the parent line and are often masked by it except in the case of certain light gases. The vibrational Raman lines are due to the various normal modes of vibration of the molecule or the crystal lattice. The lines arising from oscillating crystal lattice are characteristic only of the solid state and are not present in liquids or gases. Raman lines due to the electronic effect are rarely observed, except in the single case of NO obtained by Rosetti.

The rotational and vibrational. Raman spectra resemble closely the far infra-red and near infra-red absorption spectra of molecules respectively. Further, in many cases the Raman frequencies are equal either to the actual infra-red frequencies of the absorption spectrum of the scatterer or to the differences of such frequencies. But there seems to be little or no correlation between the relative intensities of Raman lines and the intensities of the corresponding infra-red bands. In addition, not all Raman lines have their corresponding infra-red bands, nor all infra-red bands their corresponding Raman lines. Hence it appears that although the ultimate result as far as the molecule is concerned, is the same in both the Raman and band spectra, yet the mechanism of production and the intervening laws are quite different in the two cases.

Theoretical Explanation

A simple and satisfactory explanation based on the quantum theory was put forward by Prof. Smekal, in 1923, which accounts for the essential features of the Raman effect. According to the quantum theory, a beam of monochromatic light of frequency υ_0 has its energy distributed in quanta, each of energy $h\upsilon_0$. When such a light quantum or photon hits a molecule of the scatterer, three things might happen. The molecule might merely deviate the photon without absorbing its energy, which would result in the appearance of the *unmodified line* in the scattered beam. The molecule might, on the other hand, absorb part of the energy of the incident photon, giving rise to the *modified Stokes' line*, whose frequency would evidently be less than that of the incident radiation. It may also happen that the molecule itself, being in an excited state, imparts some of its intrinsic energy to the incident photon and this would produce the anti-Stokes' line of frequency greater than that of the incident radiation. The mechanism of Raman scattering can therefore be analytically expressed as follows.

Treating the phenomenon as a kind of collision between the photon and the molecule and applying the principle of conservation of energy we may write

$$E_p + \frac{1}{2}m\upsilon^2 + hv_0 = E_q + \frac{1}{2}m\upsilon'^2 + hv' \qquad ...(1)$$

where E_p and E_q are the intrinsic energies of the molecule before and after collision respectively, m the mass of the molecule, υ and υ' its velocities before and after impact, v_0 and v' the frequencies of the incident and scattered photon.

As the collision does not cause any appreciable change of temperature, we may assume that the kinetic energy of the molecule remains practically unaltered in the process and hence simplifying the above relation as

$$E_p + hv_0 = E_q hv'$$

$$h(v' - v_0) = E_p - E_q$$

or

$$v' = v_0 + \frac{E_p - E_q}{h} \qquad ...(2)$$

In this expression, if $E_p = E_q$, $v' = v_0$ which accounts for the unmodified line; if $E_p < E_q$, $v' < v_0$, which refers to the Stokes' line; and finally if $E_p > E_q$, $v' > v_0$, which corresponds to the anti-Stokes' line.

Considering the molecule involved in the process, since the same property of possessing energy in quanta appears in molecules and atoms, we may apply quantum principles to the change in the intrinsic energy of the molecule and write

$$E_p - E_q = nhv_m \qquad ...(3)$$

where n = 1, 2, 3, etc. and υ_m is the characteristic frequency of the molecule.

Hence equation (2) reduces, in the simplest case of n = 1, to:

$$\upsilon' = v_0 + v_m \qquad ...(4)$$

The frequency of scattered radiation *i.e.,* of Stoke's line is

$$\upsilon_s = v_0 - v_m \qquad ...(i)$$

The frequency of scattered radiation *i.e.,* of Anti-Stoke's line is

$$\upsilon_{as} = v_0 + v_m \qquad ...(ii)$$

This expression shows that the difference in frequency ($v_0 - v'$) between the incident and scattered lines in the Raman Effect corresponds to the frequency υ_m characteristic of the molecule, which thus accounts for the experimentally observed fact that the Raman lines are symmetrically situated on either side of the parent line at intervals corresponding to the characteristic frequency of the molecule and hence to the infra-red absorption lines of the scatterer.

1.21 PRACTICAL IMPORTANCE OF RAMAN EFFECT

Over and above the great theoretical interest attached to it as further confirmation of the quantum theory, the Raman Effect is of immense practical importance on account of its many useful applications in physics and chemistry. The universal nature of the phenomenon, the relative simplicity of experimental technique, the ease with which the effect can be controlled, making it appear as a part of the visible spectrum, which can be chosen at will (its position depending only on the choice of the incident frequency) and the complementary character of the spectrum obtained with reference to infra-red absorption (the former containing lines which are not found in the latter) make the Raman effect a very convenient and powerful tool of research in problems concerning the intimate structure of matter, chiefly as regards the constitution of molecules in gaseous, liquid and solid states. As an illustration, we shall now briefly state how the Raman effect is utilised in the study of molecular structure.

1.22 RAMAN EFFECT AND MOLECULAR CONSTITUTION

Raman spectra are, in general, determined by those factors which affect most the nature of vibration *viz. The number* of atoms in the molecule, the masses of the atoms and the strength of the chemical bonds between the atoms.

Taking a *diatomic molecule* its natural frequency of vibration is given by $\upsilon_0 = \left(\frac{1}{2\pi}\right)\sqrt{\frac{F}{\mu}}$, where F is restoring force per unit displacement and μ the resultant mass. This means that:

(a) there is only one-vibration frequency in diatomic molecules,

(b) a molecule containing light atoms should have a higher frequency than the one containing heavier atoms, and

(c) a molecule in which the force binding the atoms is large should have higher characteristic frequency than the one in which the force is weak. This force depends on the nature and strength of the inter-atomic bonds. Thus a diatomic molecule with a double bond should have a higher frequency than the one containing only a single bond. The Raman lines are also expected to appear with high intensity when the bond is of the covalent type (homopolar or electrically non-polar molecules) while the reverse is the case when the bond is of the electrovalent type (polar or heteropolar molecules). The reason for this is to be found in the fact that the Raman lines essentially depend upon the symmetry of the molecules and the extent to which the polarisability is affected by the oscillations. In covalent molecules the binding electrons remain common to the nuclei, so that the polarisability of the molecule is considerably modified by nuclear oscillations and this vibration in polariability gives use to Raman lines. The electrovalent molecules the binding electrons definitely change over from one nucleus to the other in the formation of the molecule so that the polarisability of the molecule is little affected by nuclear oscillation, which means no Raman lines will occur.

Considering *polyatomic molecules* which are constituted by more than two atoms, the Raman spectra will naturally be much more complex. First of all more than one characteristic frequency is to be expected *e.g.*, triatomic molecules will, in general, have three such frequencies, Secondly, the arrangement of atoms, such as symmetrical or unsymmetrical, linear or non-linear symmetry, etc. is an additional factor by which the intensity of the Raman lines is essentially determined. Hence, from the number and intensity of the observed lines in the Raman effect, in conjunction with infra-red data, it is possible to draw important Theory of Atomic Structures about molecular structure. Theory leads to the following important rule, *known as the rule of mutual exclusion*. For molecules with a centre of symmetry transitions that are allowed in the infra-red are forbidden in the Raman spectrum and vice-versa. The rule implies that for molecules without a centre of symmetry there are transitions that can occur both in the infra-red and the Raman Effect. It does not, however, imply that all transitions that are forbidden in one must occur in the other. For, some transitions may be forbidden in both.

The following examples may serve to illustrate these remarks:

Diatomic Molecules

Some of the diatomic molecules studied are H_2 D_2, N_2, O_2, HCl, HBr and HI. The first four are homonuclear molecules *i.e.,* constituted with identical atoms, while the last three heteronuclear molecules, *i.e.,* made up of non-identical atoms. In all these cases there is at only one vibration frequency and its value obtained from Raman spectra is given below. The restoring force per unit displacement F in each case, calculated with the help of the relation $\upsilon_0 = \left(\frac{1}{2\pi}\right)\sqrt{\frac{F}{\mu}}$, is also given.

	H_2	D_2	N_2	O_2	HCl	HBr	HI
$v_0(cm^{-1})$	4156	2993	2331	1556	2880	2558	2233
$F \times 10^{-5}$	5.1	5.3	22.4	11.4	4.8	3.8	2.9

It is readily seen from such a tabulation that:

1. the heavier the molecule the lower is the vibration frequency, and
2. the values of F, which is a measure of the binding strength, may be roughly classified in the ratio of 3 : 2 : 1, thereby indicating the existence of three different types of bonds triple, double and single, between the atoms in the molecule.

Triatomic Molecules

Among the several triatomic molecules analysed, we shall, consider here only three, *viz.*, CO_2, N_2O and H_2O as typical cases illustrating some special points of interest in molecular structure.

Carbon dioxide (CO_2) is one of the most frequently and thoroughly studied molecules, It has two very strong bands in the infra-red absorption spectrum at 668 and 2349cm^{-1}, while only one strong band in its Raman spectrum at 1389 cm^{-1}. As none of these occurs both in the Raman and infra-red spectrum, it follows from the rule of mutual exclusion that the molecule of CO_2 must have a centre of symmetry. For triatomic; molecules this implies that the molecule is linear and symmetric. Hence, the molecular structure of CO_2 is O-C-O. This symmetric structure is confirmed by the rotational Raman spectrum of CO_2, which consists only of alternate (odd) levels, like that of O_2. The three bands stated above, two in the infra-red spectrum and one in the Raman spectrum represent the three fundamental frequencies of CO_2. Carbon disulphide (CS_2) also belongs to this class.

Nitrous oxide (N_2O). This molecule has the same number of electrons as CO_2 and one might therefore expect it to have a linear symmetric structure. But analysis of the infra-red and Raman spectra of N_2O shows clearly that this molecule although, linear, is not symmetrical. The three fundamental frequencies of N_2O are 2224, 1285 and 589cm^{-1}. All the three appear in the infra-red absorption and two of them, *viz.*, 2224 and 1285, appear in the Raman spectrum; 589 has, however, not been recorded, due to weak intensity. At any rate, the two Roman lines appear also in the infra-red, which argues to the absence of the centre of symmetry in the molecule. For, it was a centre of symmetry, only one fundamental should appear in the Raman spectrum and this fundamental should not appear in the infra-red, according to the rule of mutual exclusion. Hence the molecule has the unsymmetrical structure N-N-O. This Theory of Atomic Structure is confirmed by the rotational Raman spectrum of N_2O, which consists of both (odd and even) sets of lines without any alteration in the intensity. Other molecules of this type are HCN, ClCN, BRCN, ICN, etc. Water H_2O the observation of a strong pure rotation—Vibration spectrum lead.

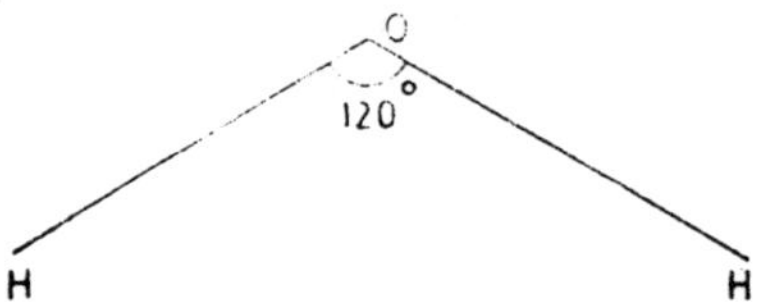

Fig. 1.22

unambiguously to the Theory of Atomic Structure that H_2O, though symmetrical, is hot linear but bent (Fig. 1.22). According to theory, ail triatomic molecules of bent symmetrical structure should give rise to three Raman lines all of are very frequencies. There is also evidence to show that the bend in the system is 120°. In the Raman spectrum, two frequencies have been recorded at 1665 and 3605cm^{-1} which correspond roughly with the infra-red frequencies. There is some difficulty in recording the Raman lines of H_2O, which are very weak, due evidently to the fact that the movements of the light hydrogen atoms do not cause appreciable variation in the polarisability tensor. Water gives some other extra bands which have to be ascribed to polymerised molecules $(H_2O)_2$, $(H_2O)_3$, D_2O, S_2O and H_2S have similar bent symmetrical structures.

Thus, the study of Raman spectra of different substances enables one to classify them according to their molecular structure.

1.23 VECTOR ATOM MODEL

The vector atom model is an extension of the Rutherford, Bohr. Sommerfeld atom model on new lines, devided not only to remove the limitations of the previous model's but also to cover new fields of experimental observations. The original simple theory of Bohr was absolutely in capable of explaining the fine structure of spectral lines even in the simplest hydrogen atom Sommerfelds modifications through giving a theoretical justification of the splitting of the individual spectral lines of hydrogen in to fine structure components met with only a partial success. Hence these older atom models were quite inadequate to tackle more complex atoms. Soon other new experimental data, such as the anomalous Zeeman Effect, Stark Effect, etc., brought out this insufficiency of the older theories with still greater force. New ideas were therefore introduced partly by analogy, partly by empirical methods in the interpretation of more complex spectral phenomena and their relation to atomic structure, which finally, resulted in what is known as the *vector atom model.* Among the physicists who have contributed to the accomplishment of this really monumental work, special mention must be made of Bohr, Sommerfeld, Uhlenbeck, Goudsmit, Pauli Stern and Gerlach.

Description of the Vector Atom Model

The two essential features that characterise the vector atom model and differentiate it from the other models are the conception of quantisation of direction or *spatial quantisation and spinning electron.*

Spatial Quantisation

We have seen the Bohr-Sommerfeld orbits are quantised as regards their magnitude, *i.e.,* their size and form. But quantum theory demands more that this, *viz.* quantisation also of direction of orientation of the orbits in space, *i.e.,* it selects from the continuous manifold of all possible positions of the orbits in space, permitted by classical ideas, a discrete number conformable to quantum conditions. The introduction of such a spatial quantisation makes the orbits vector quantities. To quantise spatially we need, of course, a certain preferential direction with respect to which the orbits may have their discrete orientations. Such a direction of reference may be obtained by an external field of force, but whose strength tends to zero so that it may disturb the orbits to the minimum. To determine the quantised orientations relative to the field direction, we are guided by the fact that *the projections of the orbits on the field direction must themselves be quantised.* This new idea of spatial quantisation was first proposed and worked out by Sommerfeld in connection with the explanation of Zeeman effect, *i.e.,* the splitting of spectral lines under the influence of an applied magnetic field.

1.24 SPINNING ELECTRON

In order to explain satisfactorily some of the intricate spectral phenomena such as the fine or multiple structure of lines, Zeeman Effect, both normal and anomalous, etc., Uhlenbeck and Goudsmit put forward in 1925, the hypothesis of the spinning electron. According to it, the electron revolves not only in an orbit round the nucleus but also about an axis of its own, somewhat like our earth, which not only revolves round the sun but also rotates on an axis of its own.

These two new factors, spatial quantisation and spinning electron brought in profound modifications in the atom model. The electron in the atom will be endowed with two angular momenta, one due to orbital motion and the other due to spin. In consequence the total angular momentum of the atom will be no longer due simply to the orbital motion of the electron but also due to the spin of the electron. According to the quantum theory, the spin motion, like the orbital motion, is to be quantised, which will therefore introduce a new quantum number, known as the *spin quantum number*, in addition to the orbital quantum number. Further, as both the orbital and spin motions are to be quantised not only in magnitude but also in direction according to spatial quantisation, they are to be considered as quantised vectors. It is this special feature which is a valuable addition to the atom model.

Since the rotation of a charged body gives rise to a magnetic moment, the electron will be endowed with two magnetic moments, one due to orbital motion and the other due to spin, so that the magnetic moment of the atom will be the resultant of two magnetic moments, the orbital and spin magnetic moments, which are also quantised vectors for reasons given above.

As the different components that determine the state of the atom such as the orbital and spin motions are all quantised vectors, the atom model built on such considerations is aptly called the vector atom model to which vector laws apply.

1.25 QUANTUM NUMBERS ASSOCIATED WITH THE VECTOR ATOM MODEL

In this model, to each of the component parts is assigned; quantum number, the numerical value of which may be conveniently thought of as the length of the vector which represents it. In vector analysis, angular momentum is represented by a straight line whose direction is parallel to the axis of rotation and whose length is proportional to the magnitude of the momentum. The quantum numbers associated with each of the electrons in a given atom are the following:

(i) *A total quantum number n*, identical with the one used in Bohr-Sommerfeld Theory. It can take any integral values l, 2, 3, etc.

(ii) *An orbital quantum number l*, may take any integral value between 0 and (n – 1) inclusively. Thus, if n = 4, *l* can have values 1, 2, 3. This quantum number is similar to the azimuthal quantum number n_φ of Sommerfeld's Theory. The orbital angular momentum p_l of the electron is given by $l(h/2\pi)$, as it is quantised. By convention for which $l = 0$ is called an S electron, if $l = 1$, p electron, $l = 2$, d electron, $l = 3$, electron and so on.

(iii) *A spin quantum number s*, the magnitude of which is always 1/2. The half-integral value, as it contradicts the integral multiple rule of quantisation, appeared to raise a serious difficulty against the 'spinning electron' theory. But an answer was readily found in the fact that the use of half-integers for 's' consistently leads to results which are in complete agreement with experimental facts. The spin angular momentum *p sub s* is given by $s(h/2\pi)$ with s = 1/2.

(iv) *A total angular quantum number j*, which refers to the resultant angular momentum of the electron due to both orbital and spin motions, is the numerical value of vector sum of *l* and s. Hence $j = l + s$ and the value of j is evidently half-integer, since one of its components, s, is always equal to half. It is usually expressed as j = $l \pm 1/2$, plus sign when s is parallel to l, and minus sign when s is anti-parallel. The total angular momentum of the electron p is given by j(h/2π).

If the atom is placed in a magnetic field so that a preferential axis in space is thereby determined due to the fact that a field exerts a directive influence on both the orbital and spin motion, which constitute elementary magnets, three more quantum numbers will be associated with the electron.

(v) *A magnetic quantum number* m_l, which is the numerical value of the projection of the orbital vector l on the field direction m, is an integer and may have only any of the (2 + 1) values from $-l$ to $+l$ including zero. For, according to spatial quantisation, the projection of l in the field direction must itself be quantised. Hence l can be inclined to the field direction only at such discrete values that its projection m_l may also be an integer. For instance, if the vector l is inclined to the field direction at an angle θ (Fig. 1.23), the projection $m_l = l \cos \theta$. Now since m_l has to be an integer and cos 6 never exceed unity, the permitted values of m_l are from $+l$ to $-l$ at unit intervals *viz.* l, $(l-1)$, $(l-2)$...1, 0, -1... $-(l-2)$, $-(l-1)$, l. Using the laws of arithmetical progression it is seen that the total number of possible value of m_l is $(2l+1)$.

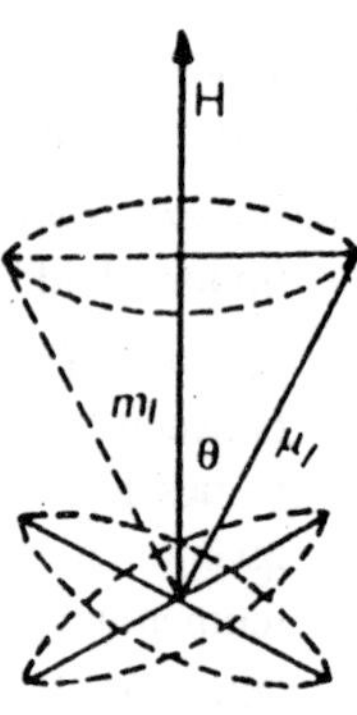

Fig. 1.23

Conversely, the permitted orientation of the l vector relative to the field direction is also $(2l+1)$. Thus, for instance, if $l = 1$, the permitted orientations of l are three, for which m_l has the values + 1, 0, –1. If $l = 2$, the permitted orientations of l are five, for which m, = +2, +1, 0, –1, –2, as shown in Fig. 1.20.

(vi) *A magnitude spin quantum number* m_l which is the numerical value of the projection of the spin vector s on the field direction. By analogy with the orbital vector, the spin vector can have only (2s + 1) permitted orientations and in consequence ms can have any of the (2s + 1) values from – s to + s, excluding zero, however, since s is always equal to 1/2 and never zero. This means that m, can have only two values + 1/2 and −1/2.

(vii) *A magnetic total angular momentum quantum number m_j*, which is the numerical value of the projection of the total angular momentum vector j in the field direction. Since we are dealing with a single electron, J can have only half-integral value, since j = $l \pm 1/2$ and in consequence m must have only half-integral values. The permitted orientations of j are (2j + 1) and hence, m_j can have only (2j + 1) values from +j to −J, zero excluded.

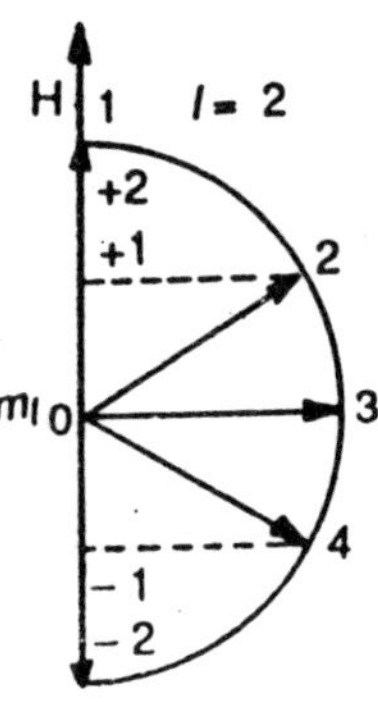

Fig. 1.24

It may be noted that the quantum numbers mi and m, come into play only in very strong magnetic fields where the coupling between l and s is broken while m_j is effective in ordinary fields.

Having thus defined the state of each of the electrons in a given atom, the state of the atom is determined as follows: The resultant quantised vectors representing the atom as a whole can be obtained by:

(a) suitably combining the vectors of the individual electrons in the atom, and

(b) applying the rule of spatial quantisation. It is to be noted while small letters as l, s, j and s, p, d, f, etc., are used to designate the state of the electron, capital letters as L, S, J and S, P, D, F etc. are used to describe the state of the atom.

1.26 COUPLING SCHEMES

There are several ways in which the different vectors of the electrons may combine to give the vectors representing the atom as a whole. The method of combination depends on the interaction or "coupling" between the component vectors, since the orbital and spin motions of the electron produce magnetic fields and thereby result in mutual perturbation. It is usual to distinguish two types of combination, known as the Russell-Saunders or L-S coupling and j-j coupling.

The **L-S coupling** is the one which occurs most frequently and hence is known as the normal coupling. In this type the several spin vectors of the electrons combine to form a resultant vector **S**; the several orbital vectors of the electrons likewise go to form a resultant Vector **L**; then the **S** and **L** combine to make the vector J, which represents the total angular momentum of the atom. The process may be symbolically represented as

$$(s_1 + s_2 + s_3 + \ldots\ldots) + (l_1 + l_2 + l_3 + \ldots\ldots) = \mathbf{S} + \mathbf{L} = \mathbf{J}.$$

This sort of coupling is the most natural one when the, interaction between the individual spins on the one hand and the individual orbital momenta on the other are very strong.

The principles that govern this type of coupling are:

(i) All the three vectors **L, S** and J are quantised.

(ii) **L** is always an integer, including zero, *i.e.,* **L** may be 0, 1, 2, 3, etc.

(iii) S is an integer or half-integer depending on the number of electrons involved and the direction of the spin vectors. Thus, for distance, for an atom containing a *single electron* S can have only the value 1/2; for a two electron system S may be either

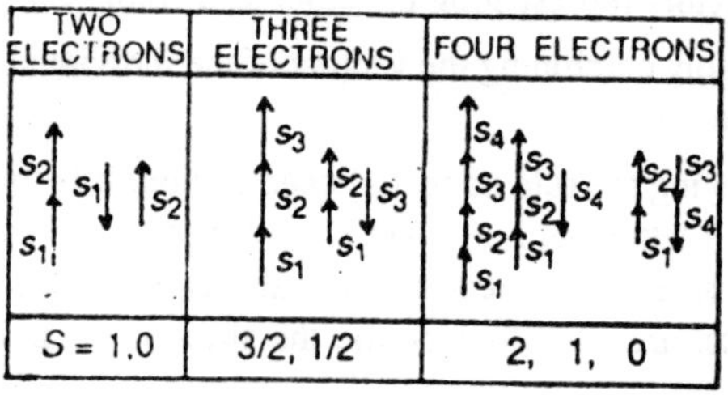

TWO ELECTRONS	THREE ELECTRONS	FOUR ELECTRONS
s_2 s_1; s_1 s_2	s_3 s_2 s_1; s_2 s_3 s_1	s_4 s_3 s_2 s_1; s_3 s_2 s_4 s_1; s_2 s_3 s_4 s_1
S = 1,0	3/2, 1/2	2, 1, 0

Fig. 1.25

1 or 0 depending on whether the spin vectors are parallel or antiparallel. For the *three electron* system S is 3/2 or 1/2 and for a *four electron* system S may be 2, 1, 0 due 10 the same reason as readily seen from Fig. 1.25.

(iv) Hence **J**, the vector sum of **L** and **S** must be an integer (0, 1, 2, 3, etc.) if **S** is an integer, *i.e.,* for *odd electron* systems and *half integer* (1/2, 3/2, 5/3, etc.) if **S** is half-integer, *i.e.,* for *even* *electron system*. It can be proved that, in general, the possible number of values which **J** can assume are:

$$(2S + 1) \text{ if } L \geq S$$

and $$(2L + 1) \text{ if } L < S.$$

The quantity (2S + 1) is known as the multiplicity of the **L** state. It gives the permitted values of **J** for a given value of **L**. In particular, if **L** = 0, J can have only one value, *viz.* **J** = **S. J** *is always positive, never negative*, since it represents the total angular momentum of atom.

The j-j Coupling

Under certain circumstances, the interaction between the spin and orbital vectors in each electron may be stronger than that between either the spin vectors or the orbital vectors of the different electrons. If this is the case, each electron is considered separately and its contribution to the total angular momentum of the atom is obtained by combining first its individual spin and orbital vector using the relation $j = l + s$. The vector sum of all the individual j vectors of the different electrons thus got gives the total angular momentum **J** of the atom. This sort of coupling may be symbolically represented as

$$[(s_1 + l_1) + (s_2 + l_2) + ...] = (j_1 + j_2 + ...) = \mathbf{J}$$

These two types of coupling are limiting cases, between which a whole range of intermediate types may occur, which makes the problem very difficult of treatment. But for most known cases, the L-S coupling is effective.

1.27 APPLICATION OF SPATIAL QUANTISATION

Applying the principle of spatial quantisation to the resultant vectors representing the atom, obtained by the coupling schemes described above, the number of permitted orientations of **L, S** and **J** with respect to a preferential field direction are (2**L** + 1), (2**S** + 1) and (2**J** + 1) respectively. The corresponding magnetic quantum numbers.

$$M_L = \Sigma m_l\,,\ M_s = \Sigma m_s \text{ and } Mj = \Sigma m_j$$

can only have (2L + 1), (2S +1) and (2J + 1) values respectively. It is to be noted that in the one electron system, *i.e.,* an atom with a single effective electron, the state of the atom as a whole is identical with the state of the electron, so that **L** = l, **S** = s and **J** = j.

1.28 APPLICATIONS OF THE VECTOR MODEL

The great advance made by the vector atom model lies in the fact that it offers an adequate and rational explanation of several complex atomic and spectral phenomena, which cannot be satisfactorily explained on the basis of the older models. In order to be able to use effectively the somewhat artificial structure of the vector model described above, certain other important rules, such as Pauli's Exclusion Principle, selection, intensity and interval rules and Lande's splitting factor 'g' are required. With these additional helps, the electronic structure in atoms, the fine structure of spectral lines, the Zeeman Effect with anomalous characteristics, the complicated Stark effect, etc. have been interpreted very satisfactorily. Most of these topics are beyond the scope of this book. We shall, however, deal with Pauli's exclusion principle and use it in conjunction with the vector model' to elucidate the electronic structure in atom, which is of immediate interest to us. Then we shall briefly state the important selection, intensity and interval rules and employ them in the interpretation of the fine structure of spectral lines.

1.29 PAULI'S EXCLUSION PRINCIPLE

Pauli's Exclusion Principle, sometimes termed also as the equivalence principle states that every completely defined quantum state in an atom can be occupied by only one electron. In other words, it is impossible for two electrons in an atom to be identical as regards all their quantum numbers, *i.e.*, in such a case one of the two will be excluded from entering into the constitution of the atom. Hence the name "exclusion principle".

The principle may be stated in yet another way: two systems of quantum numbers which are deducible from each other by interchange of two electrons represent only one state. Thus, understood, the principle enunciates the indistinguishability of electrons which have identical quantum numbers. Hence the name "equivalence principle"

It is clear that the principle defines a certain minimum individuality of the electrons in the atom. Among the several quantum numbers associated with the electron, four, *viz.*, n, I, m_l and w, or n, l, j and m_j, are strictly required to specify completely the state of any particular electron, the former set being used in the presence of a strong magnetic field which breaks the coupling between l and s in the electron, whereas the latter in the rest of the cases.

The principle was first introduced by Pauli to explain certain experimental facts, such as:

(a) the occurrence or non-occurrence of spectral lines in optical and X-ray spectra it was found that for the missing energy states of the atoms all the four quantum numbers of the electron agreed, and

(b) the complete scheme of the successive formation of the atoms as arranged in the periodic table, discovered by Bohr, Stoner and others-such an arrangement required the association of the above-stated four quantum numbers with each electron.

Naturally then, the principle finds its chief use in the elucidation of electronic structure and atomic spectra. It has been realised later that the principle is much more fundamental and universal, holding good for the totality of electrons in any arbitrary molecule, nay, even for the comprehensive system of conduction electrons that belong to a metal bulk state on the one hand and for the constituent particles of the nucleus in the ultramicroscopic state on the other.

No theoretical proof of the principle can be given as yet and for the present it must be regarded as something empirical added to and regulating the vector atom model.

1.30 ELECTRONIC STRUCTURE IN ATOMS

Admitting that the number of electrons in the normal atom increases steadily as one passes from light to heavy elements, one is interested to know whether the several electrons in the given atom are distributed at random or in a regular manner.

Since, in general, a system is most stable when its energy is minimum, and as the energy of the atom is smaller, the lower the value of the total quantum number n designating the electronic orbit, it would seem that all the electrons should tend to crowd into the lower orbit of least energy. But investigations based on X-ray spectra and the periodic classification of elements have proved the contrary *viz.,* the electrons, instead of all crowding into the lowest orbit, are arranged in different orbits with a certain regularity.

The study of X-ray spectra of the elements strongly suggests that the *electrons in the atom are arranged in different groups or shells* round the nucleus, all the electrons having the same total quantum number forming one shell. Thus the innermost shell, known as the **K** shell, consists of electrons whose total quantum number n = 1; the next shell, called the **L** shell, contains electrons for which n = 2 and so on. The complete scheme, as far as it is required in building up the electronic structure in the atoms of all the elements is:

Shell	→	K	L	M	N	O	P	Q
n	→	1	2	3	4	5	6	7

For the conception and nomenclature of these electronic shells, we are indebted to Lewis, Langmuir, Bohr and Stoner.

According to the conception of orbits, as there may be several orbits having the same total quantum number, all the orbits having the same value of n are said to form one shell.

The **periodic table of elements**, with its clear manifestation of periodic variations in the properties of the elements, confirm, in its turn, such a grouping of electrons in different shells. The periodic table is an arrangement of the different elements found in nature in a scheme, based on their chemical properties and atomic weights, first drawn up by Mendeleeff in 1871 and then gradually perfected, chiefly by the study of X-ray spectra of the elements. Moseley's work on the characteristic X-ray lines, for instance, settled in an unambiguous manner that the determining factor in the arrangement is not the atomic weight but the atomic number which assigns the numerical position of the element in the classification, as 1, 2, 3..., up to 92.

This table of elements presents a complex scheme, with **seven** horizontal rows called **periods**, in which the chemical and physical properties of the elements vary gradually as a periodic function of the atomic number, and eight vertical columns, called **groups**, each of which contains elements manifesting similar properties. The number of elements in the different periods is not the same; the first consists of only two elements, the second and third **eight** elements each, the fourth and fifth eighteen elements each, the sixth thirty-two and the seventh sixteen elements. As the numerical position of the elements is determined by the atomic number, it gives also the number of electrons in their respective atoms.

Such a periodic classification of elements readily suggests the idea of the *formation, at regular intervals, of a new shell*, which, as it develops in the course of the period, passes through the same characteristic phases, determined by the number and distribution of its component electrons. There is also a further implication that a new shell which marks the beginning of a new period *will*, in general, be formed *only when the inner shells are completed* with their due quota of electrons. It must, however, be noted that there occur cases where a new outer shell is started before the inner one is filled.

To determine the *number of electrons required to complete each one of the shells*, we have a clue in the periodic occurrence of monotonic elements chemically inactive, hence known as inert gasses, *viz.*, helium, neon, argon, krypton, zenon and radon (classified under group 0 in the table). Considering their atomic numbers, *viz.*, 2, 10, 18, 36, 54, and 86 respectively, the total number of electrons in their respective atoms are also 2,10, 18, 36, 54 and 86. Rydberg made a suggestion that these numbers could be arranged in a simple numerical series as follows:

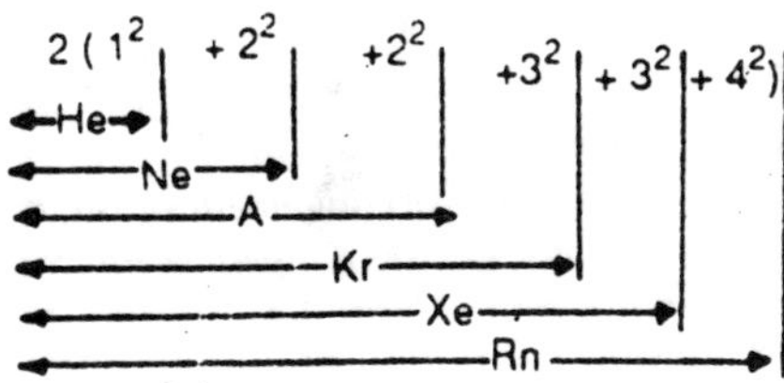

Fig. 1.26

Such an arrangement indicates *that electrons in these inert gases can be distributed in shells*, that the helium atom, for instance, has a single shell of two electrons, neon has two shells, the first containing two electrons and the second eight electrons, argon has three shells, the first one containing two electrons, the second eight, the third eight also, krypton has four shells, the first three containing the same number of electrons as argon, but the fourth eighteen electrons and so on. The factor two outside the bracket suggests some sort of symmetry, pairing of electrons.

Further investigations showed that Rydberg's series had to be modified as regards the order of the numbers within the bracket. Thus, in the case of Kr the arrangement should be $2(1^2 + 2^2 + 3^2 + 2^2)$, the outermost shell containing eight electrons like A, but the third eighteen electrons. Similarly, for Xe the order should be $2\ (1^2 + 2^2 + 3^2 + 2^2)$

and for Rn $2(1^2 + 2^2 + 3^2 + 4^2 + 3^2 + 2^2)$. This *would indicate that the number of electrons in the shells increases up to a certain limit and then decreases.*

The chemical inertness of these elements shows that they atoms have very stable structure, with a peculiar permanent nature for their electrons. This, in turn, suggests the *general law that* 2, 10, 18, 36, etc. *electrons tend to make a very stable combination in the atom.* On the other hand, the next element beyond any one of these inert gases has one more electron than necessary to form a stable structure, *e.g.,* Li (3), Na (11), K (19), Rb (37), Cs (55), the metals of the alkali group, whose properties are strikingly similar. The second element beyond an inert gas has two superfluous electrons, *e.g.,* Be (4), Mg (12), Ca (20), Sr (38), Ba (56), Ra (88), the alkaline earths, also with similar properties. Going in the other direction, the next element before inert gas lacks one electron to make a stable structure, *e.g.,* F (9), Cl (17), Br (35), I (53), forming the halogen, group, again with similar properties. The second next before the inert gases, *viz.* 0 (8), S (16). Se (34), Te (52) are likewise similar. Considerations of this kind indicate *a very fundamental principle in electronic structure, viz., the chemical and physical properties of elements are determined by the number and arrangement of the electrons in the outermost shell rather than by the total number of electrons in their respective atoms.*

Conversely, by studying the chemical and physical properties of elements one may obtain information as to the number of electrons in the outermost shell of their respective atoms. Then paying attention to the elements with respect to the nearest inert gas, it is possible to determine the whole arrangement of the electrons in the different shells. Thus for instance, in the case of Na the arrangement is two electrons in the first K shell, eight electrons in the second L shell and one in the third M shell as yet incomplete, represented ordinarily by the abbreviation (2, 8, 1). Similarly, the arrangement in Mg is (2, 8, 2) in F (2, 7) and in 0 (2, 6). Reasoning of this kind determine also the number of electrons required to complete the different shells. The K shell is complete or closed with 2 electrons, the N shell with 32 electrons, the 0 shell with 18, the P shell with 12 and the Q shell with 2.

Thus, the shell character of the electronic structure in atoms, which manifests itself in the periodic variation of the chemical and physical properties of elements and their consequent classification into a periodic system, was discovered by a close scrutiny of empirical data.

Pauli's Exclusion Principle gives a rational theory; of such an empirically deduced electronic shell structure of atoms, showing why the electrons are grouped in shells, why a certain definite number of them is required to complete a shell and why the number and distribution of electrons outside closed shells are mainly responsible for the chemical and physical properties of atom, thus, leading up to a *logical construction of the periodic table.*

Considering the **K** shell, for the electrons in it n = 1, l = 0; hence m_l = 0. Thus, three of the four quantum numbers that characterise an electron are fixed. The fourth, *viz.* m_s can assume only two values +1/2 and –1/2. Hence the maximum number of electrons in **K** shell can, be *only two*, since a greater number would mean that two more should necessarily be identical as regards all their four specific quantum numbers, which is against Pauli's exclusion principle. *The shell is therefore completed or closed with 2 electrons.*

For the electrons in the next **L** shell n = 2, l = 0 and 1. In the case n = 2, l = 0, and m = 0 and m ± 1/2; hence there can be only two electrons, according to Pauli's Principle. In the other case n = 2, l = 1, m_l can have (2l + 1) values, *i.e.*, three values 1, 0, –1; with each of these may be associated the two values for the four quantum members characterising an electron, which are not identical; hence in this case six electrons can co-exist. *The L shell with two sub-groups (n = 2, l = 0 and n = 2, l = 1) is therefore completed when it contains 2 + 6 = 8 electrons.*

The electrons in the third M shell have n = 3, and l = 0, 1, 2. Three sub-groups can therefore be distinguished in this shell, *viz.* first n = 3, l = 0, second n = 3, l = 1, and third n = 3, l = 2. The first and second sub-groups are completed by two and six electrons respectively for reasons given above. The third is completed with 2(2l + 1) electrons *i.e.*, ten electrons, since l = 2. Hence, *the total number of electrons required to complete the M shell is 18.*

Arguing in a similar manner, it can be shown that fourth **N** shell has four groups, since n = 4 and l = 0, 1, 2, 3, the first three are completed by 2, 6 and 10 electrons respectively, while the fourth by 2(2l + 1) = 2 (2 × 3 + 1) = 14 electrons. This brings up the total to *32 electrons to complete the n shell.*

From this discussion the following general Theory of Atomic Structure also can be drawn:

(i) *In the electronic configuration of an atom there can be only $2n^2$ electrons with the same total quantum number n.*

(ii) *In the shell there are n sub-groups and in each of these sub-groups designated by the orbital quantum number l, there can be only 2(2l + 1) electrons.*

Thus, Pauli's Principle in conjuction with the vector model supplies the reason why electrons of an atom do not all occupy the most stable shell closest to the nucleus but dispose themselves in different shells according to certain regulations.

A direct consequence of Pauli's Principle is that every closed shell or sub-shell is balanced with respect to angular momenta both as regards orbital and spin motions **and hence contributes nothing to the total angular momentum** of the atom as a whole. The proof of this is essentially contained in the statement that a closed shell or orbital quantum number l is built up of $2(2l + 1)$ electrons. Since the (7.1 + 1) different values of m_l and the two different values of w, are just used up in this way by the $2(2l + 1)$ electrons, there is nothing more of m_l or m_s which can contribute to the atom as a whole. In other words, for each closed sub-group as well as for the whole closed shell, $M_l = \Sigma m_l = 0$ and $M_s = \Sigma m_s = 0$. Thus, for instance, taking the L shell, for the first closed sub-group ($n = 2, l = 0$), w; = 0 and the vector sum of the two values of m_s is equal to zero, so that $\Sigma m_s = 0$. For the second closed sub-group ($n = 2, l = 1$), the vector sum of the three m_l values, *viz.* +1, 0, –1 is evidently zero, so that $\Sigma m_l = 0$; similarly the vector sum of the two :n, values is also zero.

If, therefore, M_L and M_s are zero for a closed- sub-group or shell, L = 0, S = 0 and J = L + S = Q. A closed sub-shell or shell cannot, in consequence, contribute anything to the atom as a whole. From this it follows that the total angular momentum of the atom is determined only by the electron outside closed shells. *This fact explains why the chemical and physical properties of elements depend only on the electronic configuration outside closed shells, i.e.,* on the electrons in the outermost incomplete shells, which are therefore known as the valency or optical electrons.

Having thus fixed the maximum number of electrons in each shell and sub-shell, it is possible to build up elements genetically and show that they can be arranged in a periodic system, starting from hydrogen and ending at uranium. The general procedure is to picture the nucleus of atomic number Z, initially stripped of all its electrons, as capturing one by one of the Z electrons necessary to make a neutral normal atom. The quantum numbers chosen for each added electron are such as to place the electron in a vacant shell, subjected to Pauli's principle and to minimum energy, the condition required for maximum stability. In deciding which shell of sub-shell the next captured electron will permanently occupy, spectroscopic and critical potential data are used.

When normal atoms of different elements are built like this, it is found that their chemical and spectroscopic behaviour are determined by the electron configuration of the outermost shell, as is to be expected from Pauli's Principle. Elements with outermost shells of similar electronic structure possess in a large measure equivalent chemical a id physical properties. This explains the occurrence of periods in the classification of elements. Thus the alkalis, to which belong Li, Na, K, Rb and Cs are characterised by a single electron in the outermost shell, the alkaline earths, Mg, Sr, Ba, and Ra, have two electrons outside closed shells, the halogens, F, Cl, Br, and I, lack one electron to make up a closed shell; the inert gases, Ne, A, Kr, Xe and Rn have the outermost shell closed with eight electrons.

Pauli's exclusion principle leads, therefore, to a periodic system agreeing closely with the one built on experimental data. It further represents an *ideal* system, much simpler than the real one of chemists and going deeper into the internal structure of the atom, which are of interest to the chemist and the spectroscopist in the optical *i.e.,* visible region, while the ideal system describing the interior of the atom is a great asset to the X-ray spectroscopist.

It is to be noted that the *ideal periodic system given by Pauli's exclusion principle does not however, completely agree with the real one.* The two coincide up to argon, but then with potassium and calcium, the N shell begins to form before the completion of the preceding M shell; likewise the 0 shell is initiated irregularly with Rb(37), while it has to begin with Hf(72): Another discrepancy is that in the real system the 0 shell has only 18 electrons although 50 electrons would be required according to the ideal system to complete it, the P shell only 12 electrons, while 72 would be necessary to fill it completely and the Q shell 2 electrons alone when 98 would be its maximum quota. These differences between the real and ideal systems are accounted for as follows : The minimum energy condition may not always be such that one shell must be completed before an electron settles in the next. Further, since the deviations of the real system form the ideal one concern only the less tightly bound electrons of the outermost shells, if we regard as representatives of the elements not only the normal neutral atoms but also the ionised ones, the theoretical rule is found to be exactly verified.

1.31 SOME EXAMPLES OF ELECTRON CONFIGURATIONS

In order to illustrate the theory proposed above, the electronic configurations in some of the elements will now be considered.

Hydrogen (Z = 1) has only one electron which will naturally occupy the first K shell for which n = 1, in order to give least energy to the system. According to Pauli's Principle, there can be two electrons in the **K** shell, or the single electron can exist in two states with $m_s = +1/2$ and $m_s = -1/2$. On account of the single electron which cannot complete the **K** shell, one may expect the atomic hydrogen to be very active chemically, combining its single electron with another hydrogen atom to form a stable helium-like pair, the molecule of hydrogen, which is known to be comparatively inactive. The symbolic representation of the electron configuration in the hydrogen atom is 1_S, 1 referring to the total quantum number of the electron and s the type of electron, since $l = 0$ in this case. This is confirmed by the fact that hydrogen is *monovalent and electropositive.*

Helium (Z = 2) has two electrons, both of which, according to Pauli's Principle, can occupy the first **K** Shell. Both spectroscopic and ionisation potential data show that to be really the case. Further, the two electrons complete the shell, which renders the helium atom an *inert* monotonic gas, completing the first period of the periodic table of elements. The normal helium atom is therefore represented by $1s^2$, *i.e.,* two electrons for which n = 1 and $l = 0$, the total number being affixed at the top of s and the rectangular enclosure indicating that the electrons are interlocked in a closed shell. The spark spectrum of helium resembles the arc spectrum of hydrogen. This is what we should exact from theory as the singly ionised. He atom found in the spark spectrum has a hydrogen-like structure, consisting of a positive nucleus about which a single electron rotates; but the frequencies of the corresponding lines in the helium spectrum are higher than in hydrogen, because the stronger nuclear charge in He causes all energy levels lie much lower.

Lithium (Z = 3) has three electrons, two of which occupy the **K** shell and the third the next **L** shell, in order to keep the energy of the system at a minimum. The third electron in the incomplete **L** shell is the valency electron. Hence the neutral normal atom of Li is represented by $1s^2$ 2s. *Optical* spectrum of Li shows that the lowest orbit is such that this third electron is really 1s and not 2p or any other. This valency 2s electron is much more free, *i.e.,* much more loosely bound than the $1s^2$ electrons, as is seen from the fact that Li is strongly *monovalent and electro-positive. The arc* spectrum of Li is predominantly that to be expected from a single electron system of the alkali type. The spark spectrum, on the other hand, which is ascribed to singly ionised Li is similar to the arc spectrum of He, for the same reason as that given above for the singly ionised helium.

Sodium (Z = 11) has a total number of 11 electrons. According to· Pauli's principle, 10 of them will be interlocked in the first two **K** and **L** shells, completed by 2 and 8 electrons respectively, so that the remaining single electron will have to occupy the third **M** shell. The two electrons in the **K** shell are represented by $1s^2$. Of the eight electrons in the **L** shell, will be in the first sub-group (l = 0) and hence represented by $2s^2$ the remaining six in the second sub-group (l = 1) and hence represented by 2p. The last electron in the **M** shell is the valency electron which is found to be as electron from energy considerations and spectroscopic data; it is therefore represented by 3s. The electron configuration of the normal sodium atom is represented by

$$\boxed{1s^2\ 2s^2\ 2p^6}\ 3s$$

The arrangement receives confirmation from the fact that sodium is *monovalent* and *electropositive*,

Magnesium (Z = 12) has 12 electrons of which 10 occupy the first two shells, while the remaining two will have to go to the third **M** shell. These are the valency electrons, making **Mg** *divalent.* A study of the arc and spark spectra show that the two valency electrons, are,} electrons. Hence the normal **Mg** atom is represented, by

$$\boxed{1s^2\ 2s^2\ 2p^6}\ 3s^2$$

Aluminium (Z = 13). Leaving aside the 10 electrons interlocked in the first two shells, the remaining 3 occupy the third **M** shell. Hence, Al is *trivalent.* Spectroscopic study and energy consideration show that the eleventh and twelfth electrons are of the s type, while the thirteenth is a p electron. The symbolic representation of Al atom is

$$\boxed{1s^2\ 2s^2\ 2p^6\ 3s^2}\ 3p$$

Copper (Z = 29). Of the 29 electrons, 2 will occupy the first **K** shell, 8 the two sub groups of the second L shell and 18 the third **M** shell, distributed as 2, 6 and 10 in the three sub-groups. What is left over is only one electron, the valency electron which occupies the fourth **N** shell. Hence copper should be *monovalent* and its electronic configuration is

$$\boxed{1s^2\ 2s^2\ 2p^6\ 3s^2\ 3p^6\ 3d^{10}}\ 4s$$

The arc spectrum of copper should therefore resemble that of the alkalis like Li, Na, etc. It is to be noted however, that the last of the electrons, which completes the 3d sub-shell is rather lightly bound and comes off easily so that copper frequently a valency of two as an alternative to the expected valency of one, as readily seen from the existence of both cupric oxide (CuO) and cupric oxide (Cu_2O).

Thallium (Z = 81). The total number of electrons is therefore 81. The normal atom of Tl is represented applying Pauli's Principle and paying due attention to energy considerations, by

$$\underset{K}{\underline{1s^2}}\quad \underset{L}{\underline{2s^2 2p^6}}\quad \underset{M}{\underline{2s^2 3p^6 d^{10}}}\quad \underset{M}{\underline{4s^2 4p^6 4d^{10} 4f^{14}}}\quad \underset{O}{\underline{5s^2 5p^6 sd^{10}}}\quad \underset{P}{\underline{6s^2 6p}}$$

Hence it has the first four shells complete. But before the fifth O shell is completed, the sixth P shell is started. Of the three electrons in the last shell, two are s electrons and one p electron. Hence Tl is *trivalent*. These few examples will serve to demonstrate the methods and some of the fundamental principles employed in ascertaining the distribution of electrons in atoms in the different elements. The same procedure is followed in the case of any element in the periodic table.

1.32 FINE STRUCTURE OF SPECTRAL LINES

The Selection Rules

It is experimentally found that all the possible combinations of permitted energy states of an atom do not actually appear as spectral lines. Selection rules are certain principles, which give the reason for such a state of affairs.

The first selection rule devised in connection with the older atom model referred to the azimuthal quantum number $n\,\phi$ or k and it was $\Delta k = \pm 1$ as we have seen. This was formulated on an empirical basis, but was later justified theoretically by Rubinowicz by applying the theorem of an emission of electromagnetic radiation by an atom. This simple rule, however, was soon found inadequate.

For the vector atom model, three selection rules have been devised, one for L, another for J and third for S.

(a) *The selection rule for L*: Most of the observed spectral lines are due to transitions between states, in which a single electron jumps from one orbit to another, and in such cases, the selection rule is $\Delta L = \pm 1$, *i.e.*, only those lines are observed for which the value of L changes by ± 1.

(b) *The selection rule for J*: Spectral lines arise only when transitions lake place between states for which $\Delta J = \pm 1$ or 0, $0 \rightarrow 0$ being, however, excluded.

(c) *The selection rule for S* is given by $\Delta S = 0$, which means that states with different S (hence different multiplicities) do not combine with one another. Theory and experiment, however, show that this selection rule is adhered to less and less strictly as the atomic number increases. Hence, it is only an approximate rule holding good in the case of light atoms.

Note: In the presence of a magnetic field, the orbital magnetic quantum number, m_L either does not change or changes by ± 1 *i.e.*, $m_L = 0$ or ± 1. The spin magnetic quantum m, remains unchanged, *i.e.*, $\Delta m_s = 0$. Inconsequence

$$\Delta m_J = 0 \text{ or } \pm 1$$

These selection rules were first introduced on a purely empirical basis in the study of optical and X-ray spectra and Zeeman Effect. Later, they were obtained by a rigorously deductive method in wave mechanics and hence they now rest on a permanent theoretical basis.

These rules furnish invaluable help in the allocation of observed spectral series to the proper quantum number. With their aid, energy level diagrams can be constructed both for the natural complex multiple lines and for the Zeeman Effect of such lines.

Transitions which contravene these rules are sometimes observed, but the intensities of such "*forbidden lines*", as they are called are usually weak in comparison with the normal lines.

The intensity rules have been devised to supplement the selection rules, in order to predict also the intensity of the lines that occur. These were originally postulated on an empirical basis in the study of optical and X-ray spectra. Later, a full theoretical derivation was given on a wave mechanical basis. The intensity rules are :

(a) *Those transitions are strong, giving rise to intense lines, in which L and J change in the same sense: the transitions are the weaker the more the change in direction of L and J is different.*

(b) *A transition in the decreasing sense (L → L – 1) is, ceteris paribus, stronger than a transition in the increasing sense (L → L + 1).*

(c) The case of oppositely directed transitions does not occur, in general, either in X-Ray spectra or in subject spectra; because it would lead to a final state, in which (J – L) would be two units greater than in the initial state, which is forbidden.

Hence we may distinguish the following cases:

$\Delta L = \pm 1, \Delta J = -1$: most intense line (a)

$\Delta L = -1, \Delta J = 0$: less intense (a)

$\Delta L = +1, \Delta J = +1$: weaker (b)

$\Delta L = +1, \Delta J = 0$: weakest (a) and (b)

$$\left.\begin{array}{l}\Delta L = -1, \Delta J = +1 \\ \Delta L = +1, \Delta J = -1\end{array}\right\} \text{no line (c).}$$

Note: It can be shown both theoretically and experimentally that the total intensity of all the lines coming to or starting from any given J term of a multiple is proportional to (2J + 1).

The Interval Rule

Lande discovered a rule regarding the interval in frequency between the different levels constituting a multiple. It is called the *Lande interval rule* and states that the frequency interval between two levels with total angular momenta (J + l) and J respectively is proportional to (J + l).

There are, however, many deviations from this interval rule, chiefly in cases when the coupling scheme follows neither the L-S type, nor the j – j type, but is intermediate between these extreme cases.

Spectral Terms

Since a spectral line arises due to the transition of the atom from one energy state to another, in the interpretation of the fine structure of a spectral line, one must necessarily start with the determination of the different possible energy states. Then by the use of adequate selection rules the states between which transitions actually take place are picked out. Such transitions should account for the observed fine structure.

The different energy states of the atom between which transitions can take place have come to be known by the special name in spectroscopic study, *viz., spectral terms*. The problem of the fine structure is thus essentially a problem of spectral terms. When the latter is solved, the former is readily understood. Hence in the application of the vector atom model to the fine structure phenomenon, it must, first of all be seen how the model can be used to fix up the spectra terms involved.

For the sake of clearness, the fine structure in optical and X-ray spectra will be considered separately, as the notations used in the two cases are different.

1.33 OPTICAL SPECTRA

Spectral Terms

In the classification of the spectral terms by the use of the vector atom model, it is convenient to divide atoms into two main categories, *viz. one-electron system and many-electron system*.

The atoms belonging to the first class have *only one valence or optical electron, e.g.*, hydrogen and hydrogen-like atoms, such as the alkalis. Even if the atom of this system actually contains several electrons, all of these, except one, are interlocked in closed shells, which, in consequence, do not contribute to the total angular momentum of the atom, according to Pauli's Principle. Hence, the single free electron in the outermost shell is alone effective in determining the spectral properties of the atom.

The atoms of the second class have *more than one valence or optical electron, i.e.*, several free electrons not interlocked in closed shells. They will, therefore, contribute their share to the total angular momentum of the atom and hence become effective in fixing the spectral properties, *e.g.*, the alkaline earths, belonging to the two electron system, aluminium, scandium, etc., belonging to the three electron system, titanium to the four electron system and so on.

Considering the one-electron system, since the state of the atom a whole is determined by that of a single free electron, the values of *l*, s and j referring to the electron are equal to those of L, S, J, defining the whole atom. Hence S = s = ± 1/2 and the multiplicity of state given by r = (2S + 1) is two. This system **will therefore give rise only to doublet terms with the exception of the ground terms** which corresponds to the normal state of the atom, *i.e.*, to L = 0. For, in present case, the possible values of J are only two *viz.*, J = (L + l/2) (and L – l/2). Now, if L = 0, J = ± 1/2; but since the net angular momentum of the atom is always a positive quantity, the only value of J in this case is J = + 1/2, which means a *singlet*. If L = 1, J = 1 ±1/2. *i.e.*, 3/2, or 1/2 (*doublet*). For L = 2, J = 2 ± 1/2, *i.e.*, 5/2 or 3/2 (*doublet again*) and so on.

In the **many electron system**, on the other hand, the state of the atom will be determined by the states of several free electrons. In consequence, the vectors L, S and J of these electrons. Hence the value of S is not always 1/2, but may be 0, 1/2, 1, 3/2, 2, etc., according to the number of free electrons involved and their orientations, parallel or anti-parallel, and the multiplicity r is no longer two in all cases, but is 1 when S = 0, 2, when S = 1/2, 3 when S = 1, 4 when S = 3/2, 5 when S = 2 and so on. Thus, the possible values of J for a given value of L may be one, two, three, four, etc., which means **each state may be a singlet** or **multiple**.

In the *two-electron system*, since S = 0 or 1, the state can be only a *singlet* or *triplet*. In the *three-electron system*, as S = 1/2 or 3/2, each state can be & doublet or a quartet. In the *four-electron system* since S = 0, l or 2, *singlet, triplet and quintent* are possible. In general, *odd electron systems* have even number of terms, while *even electron systems have odd number of terms*. Accordingly, the number of spectral terms of successive atoms in the periodic table of elements will alternate between even and odd.

But the ground terms is always a singlet in the system also. For, it corresponds to L = 0 and J = S. Now, since S can assume values 0, 1/2, 1, 3/2; etc., S is never less than L. For S = L = 0, the possible value of J is one ; when S > , the multiplicity of the state is given by (2L + 1) which leads again to the possible value of J as one.

Limitation of Term Multiplicity due to Pauli's Principle

Pauli's Exclusion Principle causes, however, the multiplicity of terms not to attain always the maximum (2S + 1) values which are to be expected from the number of free electrons involved. Taking, for instance, the case of an atom with 6 electrons in the L shell, at first sight, one expects the multiplicity r = (2S + 1) to attain the maximum value of 7 on the assumption that the spins of all the six electrons are similarly directed there by leading to a value of 3 for S. But such an arrangement of spins cannot take place without violation of Pauli's Principle. For, in this case, n = 1, l = 0 and 1; m_l can have only four values, *viz.*, 0 corresponding to l = 0, and 1, 0, –1, corresponding to l – 1. Now if we associate with these 4 values of m_l *six like* values of m_l *i.e.*, all either +1/2 or –1/2, the result would be that more than one electron will have identical quantum numbers as shown below :

L_I (n = 2, l = 0)	L_{II} (n = 2, l = 1)
m_l = 0	+1, 0, –1
m_s = +1/2, +1/2	+1/2, +1/2 +1/2 +1/2
The correct arrangement is	
m_l = 0	+1, 0, –1
m_s = +1/2, –1/2	+1/2, +1/2, +1/2, –1/2

Hence only four of the six electrons can have similarly directed spins (*i.e.*, same sign for m_s), while the other two must be directed oppositely. In consequence, the value of S reduces to 1, from which it follows that the actual multiplicity is only 3. Another inference is that, as a shell is gradually filled by the addition of successive electrons, the term multiplicity first rises and then falls, the variation being symmetrical with respect to the middle of the shell.

Notations of Spectral Terms

The states of the atom, in which the values of its L vector are 0, 1, 2, 3, 4, etc., are symbolically represented by the capital letters,

S, P, D, F, G, etc., respectively. The value of J is appended to this symbol as a subscript and the multiplicity of the state r written as a left superscript.

Thus, the spectral terms corresponding to L =1 and S = 1/2 are written as ${}^2P_{1/2}$ and ${}^2P_{3/2}$ For, since L = 1, the capital letter which represents the state is P, the multiplicity of the state r = (2 × 1/2 + 1) = 2 which is put as a superscript; since J = L ± S = 1 ± 1/2, *i.e.*, 1/2 or 3/2, these are put as subscripts.

To take another examples, the notation ${}^4D_{5/2}$ specifies a spectral term belonging to a state whose L = 2, J = 5/2 multiplicity r = 4 and hence S = 3/2 (from r = 2S + 1). Since the possible values of J are four, *viz.*, 7/2, 5/2, 3/2, 1/2, the above notation refers only to one of the four terms of the given state.

In the atoms belonging to the one-electron system, since every state is a doublet, except the ground state, the notations for the spectral terms in these atoms are

$${}^2S_{1/2},\ {}^2P_{1/2},\ {}^2P_{3/2},\ {}^2D_{3/2},\ {}^2D_{5/2},\ {}^2F_{5/2},\ {}^2F_{7/2},\ \text{etc.}$$

In the many electron system atoms, the notations will be more complex, but they can be readily written, provided the necessary data are available.

It may be noted that the *multiplicity symbol of the system is used on all terms even if all of them are not present.* Thus, the ground term of all alkali atoms is written as ${}^2S_{1/2}$, although the multiplicity is one and we should strictly speaking write ${}^1S_{1/2}$. But this method has the advantage of indicating to which system the ground term belongs; in the present case, for instance, to the one-electron system, giving rise to doublet terms in general.

It is likewise important to note that the vector model, where only the vectors L, S and J are involved, cannot determine the magnitude of the energies corresponding to the different spectral terms, since, the total quantum number n, necessary for such a determination, does not directly enter in the fixing up of the spectral terms. Yet the three vectors, L, S and J characterise the different energy states in such a definite manner that they may be used, without the intervention of n, to predict which transitions between the various energy states are permitted ;and hence which spectral lines may be expected.

1.34 FINE STRUCTURE

As spectral lines arise due to transitions of the atom from one spectral term to another, it is easy to understand, in a general way, how complex lines may appear when there are several spectral terms for a given atomic state defined by the vector L.

In the atoms belonging to the one-electron system all the states except the ground one are doublets. Transitions between these doublet states give rise to the fine structure. To account fully for the actually observed fine structure, as regards their number and intensity, the selection and intensity rules must be taken into consideration.

In the case of atoms pertaining to the many-electron system the fine structure may be expected to be much more complex on account of the greater term multiplicities involved. But here also, knowing the spectral terms and applying the selection and intensity rules, the observed fine structure can be accounted for. It is to be remarked that on account of the special property of closed shells the free electrons responsible for the spectra are limited to a small number. Hence the complexity of the fine structure does not go on increasing from one end to the other of the periodic table. The selection rules also, in their turn, put a limit on the complexity of fine structure. Likewise the limitations of term multiplicity arising from Pauli's exclusion principle may reduce the complexity still further.

These general principles may now be applied to the interpretation of the fine structure in the two simple cases already mentioned, *viz.*, the sodium D line and the hydrogen H_x line, to serve as illustrations.

Fine Structure of the Sodium D-Line

As we have already seen, ten out of the eleven electrons of the normal sodium atom are interlocked in closed shells, so that the eleventh free electron alone is responsible for the spectrum. The spectral lines arise from the various orbits which this single electron may occupy beyond the two closed shells and the state of the atom as a whole is determined by the state of that electron. For the same reason, all the states of the sodium atom except the ground one are doublets.

With the simple theory of Bohr it has been shown that the D-line belongs to the principal series which are due to transitions from the P state to the S state. Now for *the upper P state*, L = 1, J = 3/2 or 1/2 so that two terms are possible : $^2P_{3/2}$ and $^2P_{1/2}$. For the *lower S state*, L = 0, L = 1/2, so that only term is possible : $^2S_{1/2}$.

The transitions that can take place between the two terms of the P state and the single term of the S state are two:

$^2P_{1/2} \rightarrow {}^2S_{1/2}$ and $^2P_{3/2} \rightarrow {}^2S_{1/2}$

Now applying the selection rules $\Delta L = \pm 1$ and $\Delta J = 0$ or ± 1 (excluding O → 0), both the transitions are allowed, which explains the doublet line structure of the sodium D-line. The D_1 component (5896Å) is due to the transition $P_{1/2} \rightarrow S_{1/2}$ and the D_2 component (5890 to $P_{3/2} \rightarrow$ Å) $S_{1/2}$ (Fig. 1.27).

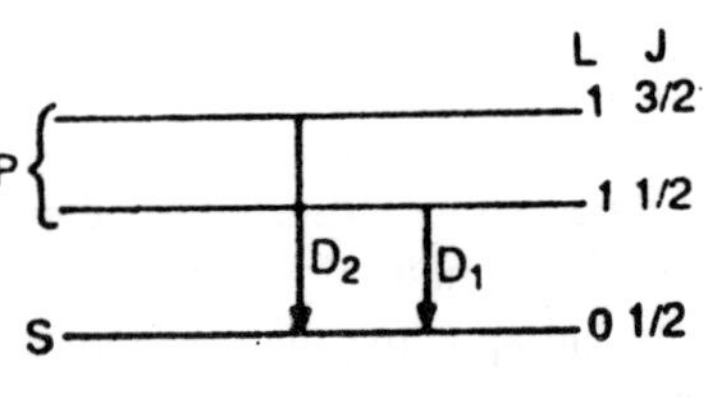

Fig. 1.27

According to the general intensity rules, D_2 will be more intense than D_1, since for D_2, $\Delta L = -1$, $\Delta J = -1$, while for D_1, $\Delta L = -1$, but $\Delta J = 0$.

Applying the total intensity (2J + 1) rule it can be shown that the ratio of the intensifies of D_2 to D_1 is 2 : 1, which has been fully confirmed .by observation.

Note: *Fine structure of lines in the other series of sodium. As* the *sharp* series arises due to transitions between the S and P states, the lines in it will be *doublets*, as in the principal series.

The *diffuse* series arises due to transitions between the D and P states and hence the lines in it should be *triplets*, since both the states have two sub-terms each: $D_{5/2}$ and $D_{3/2}$, $P_{3/2}$, and $P_{1/2}$. Four transitions are therefore possible: $D_{5/2} \rightarrow P_{3/2}$, $D_{5/2} \rightarrow P_{1/2}$, $D_{3/2} \rightarrow P_{3/2}$ and $D_{3/2} \rightarrow P_{1/2}$. Of these, $D_{5/2} \rightarrow P_{1/2}$ is forbidden, since $\Delta J = 2$. The lines $D_{5/2} \rightarrow P_{3/2}$ and $D_{3/2} \rightarrow P_{3/2}$ will be intense, since $\Delta L = -1$ and $\Delta J = -1$, while $D_{3/2} \rightarrow P_{3/2}$ will be weak,, since $\Delta L = -1$ and $\Delta J = 0$. Experimentally, with ordinary resolving power instruments the weak line is not seen. When it was first discovered it was called a "satellite" of the apparent doublet formed by the two brighter lines.

Thus, the chief spectrum of the alkalis should consist of doublets and triplets and theory accounts for them very well.

Fine Structure of the H_α-Line

As the hydrogen atom, in whose spectrum the H_α-line appears, belongs to the one-electron system, L = I, S = s, and J = j and all the quantum states are doublets, except the singlet ground state. According to the simple theory of Bohr, the H_x-line the first of the Balmer Series, arises, due to transition from the third quantum state (n = 3) to the second (n = 2).

For the *upper state* (n = 3), L can take any of the values 0, 1, 2. Using the relation J = L ±S, since S in this case has the value 1/2, the possible number of terms are five, *viz.*, $^2D_{5/2}$, $^2D_{3/2}$, $^2P_{3/2}$, $^2P_{1/2}$ and $^2S_{1/2}$

For the lower state, (n = 2), L can have two values 0 and 1, which gives the possible number of terms as three, *viz.*, $^2P_{3/2}$, $^2P_{1/2}$ and $^2S_{1/2}$

Fifteen transitions are theoretically possible between the five terms of the upper state and the three terms of the lower state. But the selection rules $\Delta L = \pm 1$, and $\Delta J = 0$ or ± 1 reduce the permitted transitions to the following, seven:

$D_{5/2} \rightarrow P_{3/2}$, $D_{3/2} \rightarrow P_{3/2}$, $D_{3/2} \rightarrow P_{3/2}$, $D_{3/2} \rightarrow P_{1/2}$, $P_{3/2} \rightarrow S_{1/2}$

$P_{1/2} \rightarrow S_{3/2}$, $S_{1/2} \rightarrow P_{3/2}$ and $S_{1/2} \rightarrow P_{1/2}$

Two pairs in separate cases of these seven allowed transitions are identical *viz.*

$D_{3/2} \rightarrow P_{1/2}$ and $P_{3/2} \rightarrow S_{1/2}$, $P_{1/2} \rightarrow S_{1/2}$ and $S_{3/2} \rightarrow P_{1/2}$

since they represent transitions between *coincident levels*, *i.e.*, levels whose **L** values differ by units, but **J** values are the same.

Taking into account these identical lines, *the fine structure of the H_α-line should have five components.*

Applying the intensity rules, the two strongest components are:

$$D_{5/2} \rightarrow P_{3/2} \text{ and } \begin{Bmatrix} D_{3/2} & \rightarrow & P_{1/2} \\ P_{3/2} & \rightarrow & S_{1/2} \end{Bmatrix}$$

since they are of the type $\Delta L = -1$, $\Delta J = -1$.

The two components $D_{3/2} \rightarrow P_{3/2}$ and $P_{1/2} \rightarrow S_{1/2}$ are less intense being of the type $\Delta L = -1$, $\Delta J = 0$. The component $S_{1/2} \rightarrow P_{3/2}$ is, still weaker, since it is of the type $\Delta L = +1$, $\Delta J = +1$. The weakest line would be $S_{1/2} \rightarrow P_{1/2}$ ($\Delta L = +1$, $\Delta J = 0$), but it is fused with $P_{1/2} \rightarrow$ for reasons given above.

Energy level diagram (Fig. 1.28). The *coincident levels* are represented by horizontal lines drawn close together. There are two such levels in the upper state and one in the lower. The other lines represent, the *neighbouring* levels. If we call those vertical lines which terminate at the upper level of the lower state as group I and those which terminate at the lower level as group II and if we number the initial levels of the upper state as a, b and c, in no way as they do not coincide, then the two identical pairs are II_b, and II_c. The five components are marked 1, 2, 3, 4 and 5, corresponding to I_a, I_b, I_c, II_b and II_c respectively.

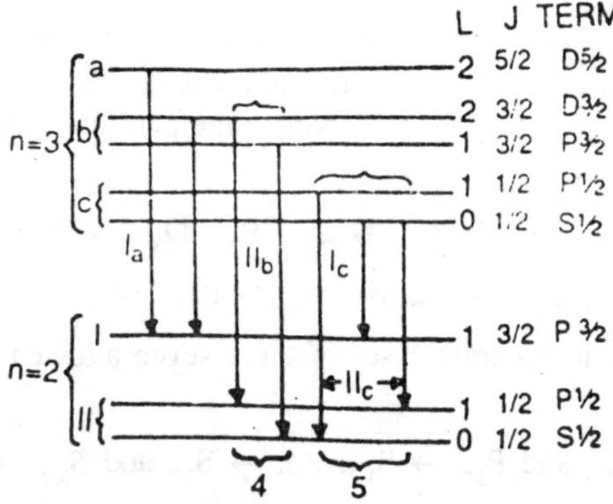

Fig. 1.28

According to the intensity rules, the strongest components are I_a and II_b, (1 and 4). The components I_b and II_c (2 and 5) are less strong. I_c (3) is still weaker.

In practice, not all the five components but only doublets can be observed. This is due to the fact that the five components are merged together by the Doppler broadening caused by the thermal motion of the molecules. Very careful experiments conducted by R.C. Williams, in 1938, where the Doppler effect was reduced to a great extent by the use of a liquid air-cooled discharge tube containing deuterium, revealed three components.

1.35 X-RAY SPECTRA

Spectral terms. We have seen that each X-ray energy level represents a state of an atom which has one electron missing from a closed shell. Pauli pointed out that in a configuration, in which an electron is missing from a completed shell, the spectral term is the same as if that one electron occupied the shell. This means that the term scheme for X-ray spectra corresponds to the one-electron system, such as hydrogen, alkali metals, etc. Moseley's researches on the characteristic X-rays also suggest the same Theory of Atomic Structure, *viz.*, that the terms involved in X-ray spectra are of the hydrogen type.

The notations used are, however, different. The X-ray energy states are designated by the capital letters K, L, M, etc., specifying the shells to which they refer. According to the vector model, not only each of these states is characterised by a total quantum number n (being 1 for the K state, 2 for the L state, 3 for the M state, etc.), but also all the states, except K, are further split into sub-groups, L into two, M into three, N into four, etc.

For, using the fact that an atom emitting X-rays is assimilated to the one-electron system, it follows that in the mechanism of X-ray emission, L = 1, S = s, J = j, so that $J = j = l \pm 1/2$, and that *every distinct state of an atom except the lowest is a doublet*. Hence the *K state* (n = 1, l = 0) corresponding to the lowest ground state is a *singlet*. The *L state* (n = 2, l = 0, 1) have two sub-groups. The first sub-group (n = 2. l = 0) gives one term while the second sub-group (n = 2, l = 1) two terms corresponding to the two different values of s, *viz.*, 3/2 and 1/2. The *L state* thus, having three possible terms is a *triplet*. The *M state* (n = 3, I = 0, 1, 2 with three sub-groups gives rise to *quintet*, one corresponding to l = 0, and j = 1/2 (first sub-group), two others to l = 1, and j = 5/2, 3/2 (third sub-group). The *N state* (n = 4, l = 0, 1, 2, 3) with four sub-groups is a *septet* composed of a singlet for T = 0 and three *doublets* for l = 1, 2, 3. The term multiplicity does not, however, go on increasing. Pauli's principle putting a limit, the next *O state is a quintet and P state a triplet*. There are in all 24 different terms which are quite sufficient to represent completely the X-ray spectra of all the elements. They are denoted by the Roman numerals attached to the capital letters representing the states, *e.g.*,

K, L_I L_{II} L_{III}, M_I M_{II} M_{III} M_{IV} M_V, etc.

They indicate all the possible combinations of l and j for given values of n.

Fine Structure

The X-ray lines, in general,, will arise due to transitions among the 24 different X-ray terms. But by the selection rule transitions can only occur subject to the conditions $\Delta L = \pm 1$ and $\Delta J = 0$ or ± 1, so that the number of possible transitions between two states is thereby considerably reduced and in consequence the number of fine structure components of the X-ray lines will be limited.

If the transition ends in the K state the corresponding radiation is called a K line; if it ends in the L state, a L line and so on. Fine structure of these general classes are distinguished by small Greek letters in accordance with the following recognised convention : If the transition takes place between the different terms of the L state and the single term of the K state, the lines are called $K_{\alpha 1}$, $K_{\alpha 2}$, etc., if from the different terms of the M state to the K state, $K_{\beta 1}$, , $K_{\beta 2}$ etc. Thus a general classification is made as K, L, M, lines etc. according to the final state of transition, while the particular classification of the individual fine structure components of each line is made according to the terms of the initial state of transition.

These points are illustrated in Fig. 1.29. The term scheme of the four states K, L, M, N alone is drawn. The values of n, l and j for each term are indicated. Some of the fine structure of the K and L lines are marked. The continuous lines represent the permitted transitions giving rise to the fine structure components, while the dotted lines indicate transitions forbidden by the selection rules. As one proceeds to elements of higher atomic number, the number of terms involved and consequently the number of the fine structure lines will increase. This clear and simple interpretation of the fine structure of X-ray lines in a single scheme was obtained in 1927, thanks chiefly to the work of four physicists, Smekal, Coster and Bohr. Thus, the fine structure of spectral lines, both in the optical and X-ray regions, finds an adequate explanation on the basis of spectral terms determined by the vector model and of certain selection rules. It should be emphasized that the fine structure of spectral lines can be satisfactorily accounted for only by the introduction of electron spin.

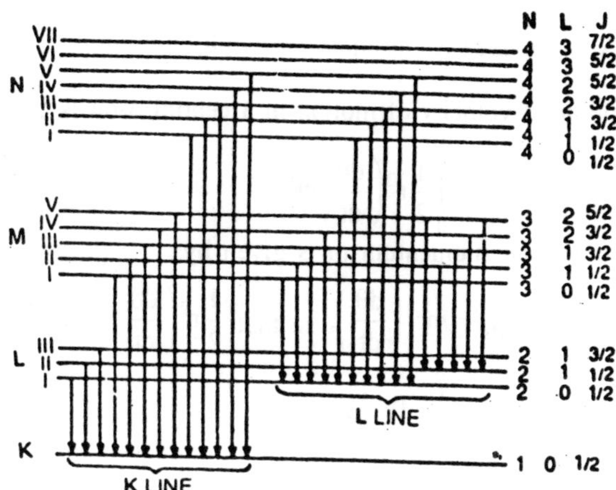

Fig. 1.29: Diagram of X-ray fine structure lines.

1.36 EXPERIMENTAL CONFIRMATION OF THE VECTOR MODEL STERN AND GERLACH EXPERIMENT

Experimental study of the behaviour of atoms in non-homogeneous magnetic fields, initiated by Stern and Gerlach in 1921 and since then elaborated in many directions by different workers, gave a very direct and most convincing confirmation of the essential features of the vector model, *viz.*, spatial quantisation, spinning electron and quantised atomic magnetic moment.

Principle

The magnetic moments due to the orbital and spin motions of the electrons in an atom may be estimated as follows:

Considering the simple case of an electron moving in a central orbit, such as an ellipse, the current i due to the motion of the electron round the orbit is given by $i = e/T$, where e is the charge and T the periodic time.

Applying Ampere's Theorem, this current gives rise to a magnetic moment μ_1 given by

$$\mu_1 = \left(\frac{e}{T}\right)A \qquad ...(1)$$

where A is the area enclosed by·the orbit.

Since the areal velocity of motion in a central orbit is $\frac{1}{2}r^2\left(\frac{d\theta}{dt}\right)$

$$A = \int_0^T \frac{1}{2}r^2\left(\frac{d\theta}{dt}\right)dt$$

Now the angular momentum pi is given by

$$p_l = mr^2\left(\frac{d\theta}{dt}\right), \text{ so that } \frac{1}{2}r^2\left(\frac{d\theta}{dt}\right) = p_l / 2m$$

Hence $$A = \int_0^T \left(\frac{p_l}{2m}\right)dt$$

Since p_l and m are constants

Substituting this value of A in equation (1)

$$\mu_1 = \frac{e}{T}.\frac{p_l T}{2m} = \frac{e}{2m}p_l \qquad ...(2)$$

As p_l, according to the quantum theory, is equal to $l\left(\frac{h}{2\pi}\right)$,

$$\mu_1 = \frac{eh}{4\pi m}l$$

Hence the orbital magnetic moment μ_1 is directly proportional to the orbital quantum number l.

if $l = 0$, $\mu_1 = 0$; if $l = 1$, $\mu_1 = \frac{eh}{4\pi m}$, etc.

The quantity $\frac{eh}{4\pi m}$ represents the *smallest unit*, in terms of which the magnitude of atomic and sub-automic magnetic moments are measured. It is known as the **Bohr Magneton**, whose value can be readily computed from the known value of h and e/m; it is found to be 9.253×10^{-24} **SI Units**.

For various reasons, both theoretical and experimental, it has been found necessary to attribute *one Bohr magneton to the magnetic moment due to electron spin.*

The atom, with its magnetic moment arising from the orbital and spin motions of the electrons in it, may be regarded as an elementary magnet, whose dimensions, though small, are yet finite. If this atomic magnet be placed in a magnetic field, it will be acted upon by the field. But the nature of this action will depend on the nature of the field.

If the field is homogeneous, the atomic magnet will experience merely a couple which will rotate its axis into the direction of the field, since the external field acts upon its poles with equal and opposite forces. *If the atomic magnet moves in such a field* in a direction normal to the field, although its magnetic axis will be rotated into the direction of the field, it *will trace a straight line path* without any deviation, *i.e.*, without any transitional displacement.

On the other hand, *if the field is non-homogeneous*, the forces on the two poles will not be equal, which would result in a translatory displacement of the atom as a whole over and above the rotation of its axis into the direction of the field. *If then, the atomic magnet flies across such a non-homogeneous field* normal to the field direction, *it will be deviated away from its rectilinear path.*

An expression for the amount of the deviation thus produced may be obtained as follows:

Let the magnetic field be non-homogeneous along the X-direction, so that the field gradient is dB/dX and is positive (Fig. 1.30).

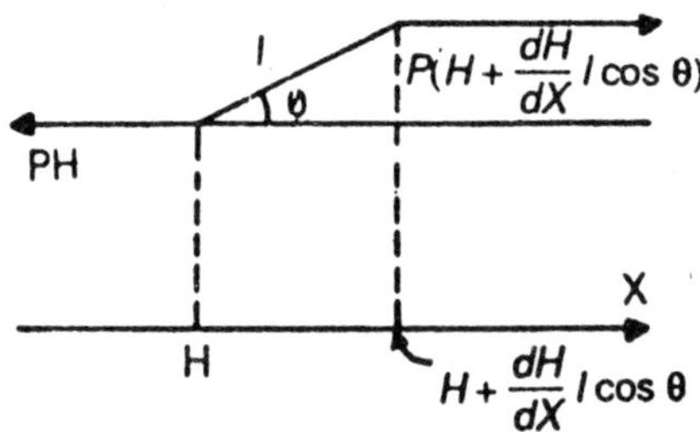

Fig. 1.30.

Let the atomic magnet, whose pole strength is p, length *l* and magnetic moment *M be placed* in the non-homogeneous field with its axis inclined at an angle 9 to the field direction.

If the field strength at one of the poles be B then at the other it will be

$$B+\left(\frac{dB}{dX}\right)l\cos\theta$$

Hence the force on the two poles of the atomic magnet due to the field are pB and p{B + (dB/dX). l cos θ} respectively. This means that, over and above the equal and opposite forces pH constituting the rotating couple, there is an extra force equal to p . (dB/dX). l cos θ acting on one of the poles alone; it is this force which displaces the atom as a whole; calling this translatory force F_x, we have

$$F_x = pl\left(\frac{dB}{dX}\right)\cos\theta = M\cos\theta\left(\frac{dB}{dX}\right) \qquad ...(1)$$

Supposing that the atomic magnet flies across the non-homogeneous field at right angles to the lines of force, it will be displaced from its straight path in the field direction. To find the amount of this displacement, let υ be the velocity of the atomic magnet of the mass m as it enters the field, L the length of its path in the field and t the time of flight through the field.

The acceleration α_x imparted to the atom along the field direction by the translatory force F_x is given by F_x/m.

The displacement D_x of the atom along the field direction at the end of time t is given by

$$D_x = \frac{1}{2}\alpha_x,\ t^2\,\frac{1}{2}\left(\frac{F_x}{m}\right).\left(\frac{L}{\upsilon}\right)^2$$

since t = (L/υ), the velocity of the atom perpendicular to the field being unaffected by the force in the field direction.

Substituting the value of F, from (1)

$$D_x = \frac{1}{2}.\frac{M\cos\theta}{m}.\frac{dB}{dX}.\left(\frac{L}{\upsilon}\right)^2$$

If μ be the resolved component of the magnetic moment in the field direction, then

$$D_x = \frac{1}{2}.\frac{\mu}{m}.\frac{dB}{dX}.\left(\frac{L}{\upsilon}\right)^2 \qquad ...(2)$$

From this relation we see that if D_x, m, dB/dX, L and υ are known calculated. If dB/dX = 0, *i.e.*, if the field is made homogeneous, $D_x = 0$, *i.e.*, there, is no deviation.

Since $D_x \propto dB/dX$, the amount of deviation is determined by the degree of inhomogeneity of the field, in fact, to produce even a small measurable displacement; the inhomogeneity must be so marked that the field changes even within the small length (of the order of 10^{-8}cm.) of the atomic magnet. Stern and Gerlach succeeded in producing a sufficient inhomogeneity by a suitable construction of the pole-pieces of a magnet, one piece being shaped as a knife-edge, while the other had a flat face provided with a groove, as shown in the Fig. 1.31, the magnetic lines of force, in consequence, crowd together at the knife-edge, so that the field strength is considerably greater there that the other pole-piece.

Apparatus

The experimental arrangement devised by Stem and Gerlach diagrammatically in Fig. 1.31. A sampler of the substance to the examined is heated in an electric oven O and, n consequence, sends out atomic rays in all directions with a velocity corresponding to the temperature of vaporisation. By the use of slits S_1 and S_2 a sharp linear beam of atoms is obtained, which then passes between the specially shaped pole-pieces of the electromagnet M M The magnetic field is made as intense and non-homogeneous as possible, its lines of force being perpendicular to the direction of the beam. Finally the beam is made to strike a suitably prepared plate P, placed normal to the initial direction of the team. The

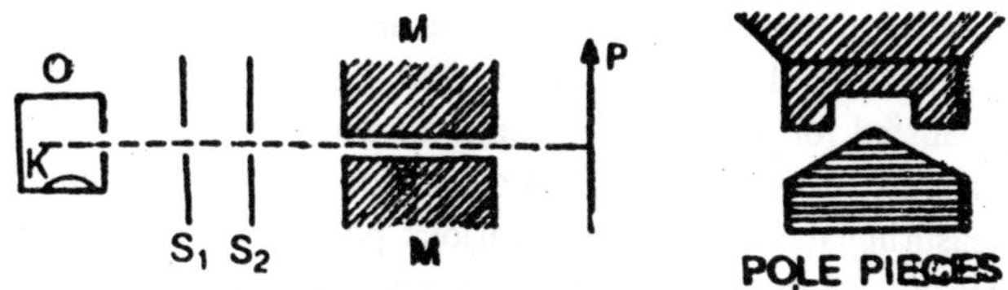

Fig 1.31 : Apparatus used in the Stern and Gerlach experiment.

entire apparatus is in an evacuated chamber. The arrangement, though simple in design, involves a very delicate technique in which a large number of details have to be carefully attended to. The different parts of the apparatus are quite small. In some cases, the pole-piece is about 5cm long and the recording plate 3 mm square. The distance between the knife-edge and the plane of the slotted pole-piece is about 1 mm.

Experimental Procedure

In performing the experiment, the following precautions have to be taken:

(a) As the deflections obtained are small, the different parts of the apparatus must be carefully and accurately aligned.

(b) The oven must be capable of withstanding a high temperature and this must be under delicate control as regards constancy over long periods.

(c) The velocity of the atoms depends on the oven temperature which must be measured thermoelectrically or optically. For atoms in thermal equilibrium, the kinetic theory of gases gives the relation.

$$\frac{1}{2}m\upsilon^2 = \frac{3}{3}kT \quad \text{or} \quad \upsilon = \sqrt{\frac{3kT}{m}}$$

where m and u are the mass and velocity of the atom, k the Boltzmann's constant and T the absolute temperature. Some preliminary experiments, however, showed that the mean velocity of the atoms issuing through slits was somewhat greater than this and hence the average value is to be taken as

$$\upsilon = \sqrt{\frac{3.5kT}{m}}$$

(d) In order to prevent collisions of the atoms in the beam with the atoms and molecules of any residual gas, the chamber must be exhausted for a long time, liquid air and charcoal being used.

(e) Accurate measurement of dB/dX is not easy. It is obtained by measuring the repulsion, at various points between the pole-pieces, of a thin bismuth wire mounted parallel to the edge of the knife-edge: pole-pieces. A correction is always found necessary for the variation in the value of dB/dX along the path of the deflected beam, as the beam is nearer the pole-piece at entrance into the field than at exist.

(f) Another problem is the development of the traces to render them Visible. Receiving plates of different kinds are used for the study of different atoms.

On account of these several technical difficulties to be overcome it is not surprising that of the many experiments begun, few were carried to a successful Theory of Atomic Structure.

Silver was the first to be investigated very completely, many series of experiments being made. With other elements, it was then possible to test whether the apparatus was functioning properly by interpolating experiments with silver.

Meissner and Scheffers improved the technique in the following manner which enabled the deviation D_x to be measured very accurately. Working with alkali metals, they caused the deflected beam to impinge on a heated tungsten filament instead of the receiving plate. The atoms of the alkali metals, on striking the hot filament, lost their electrons and were liberated as positive ions. The ionisation current thereby produced was measured in different positions and from the position of the maximum rent the magnetic deviation was accurately found.

Phipps and Taylor successfully carried out the difficult but important experimental determination of the magnetic moment of the hydrogen atom. The hydrogen atoms were produced in a long discharge tube under appropriate conditions or by dissociating the molecules by means of a hot filament. A stream containing a fairly high proportion of atomic hydrogen was passed through three special slits in a glass tube between which two high vacuum pumps were operating so that tine emergent beam contained effectively free, atoms. This atomic beam after passing through a powerful non-uniform magnetic field was received on a plate coated with molybdenum trioxide. The atoms reduce the oxide and so produce a visible trace. A sharply defined blue line against a white background was the result in the absence of the filed, while when the field was on, the beam was separated into two, and two traces were obtained. For determining the exact positions of the deviated lines microphotographs were taken. From the deviation thus, measured the magnetic moment of the hydrogen atom was calculated to be one Bohr magneton within the limits of experimental error.

Results

In the case of silver, with the field off, a fairly sharp line was traced by the atomic beam on the receiving plate. On establishing the non-homogeneous field a double trace was obtained, one on either side of the original trace. The traces obtained were somewhat as shown in Fig. 1.32. The irregularity in the right hand trace is due to the irregularity of the magnetic field near the knife-edge pole-piece. The convergence at the top and bottom arises from the fact that the field gradient decreases transversely. The traces actually obtained are much smaller than what is shown here, but they can be magnified for easy measurement. A certain diffusiveness is unavoidable owing to the velocity distribution (Maxwellian) of the atoms, but the magnetic splitting of the beam is relatively great enough, that the diffusiveness becomes of secondary importance. The mean deviation D_x of the beam caused by the field can be measured from the traces with fair accuracy. Thus, D_x being known, and the other quantities such as dB/dX, L and u being readily measured, μ can be calculated. In the case of silver, it was found to be equal to one Bohr Magneton. The results, obtained with other elements experimented upon, may be summarised as follows:

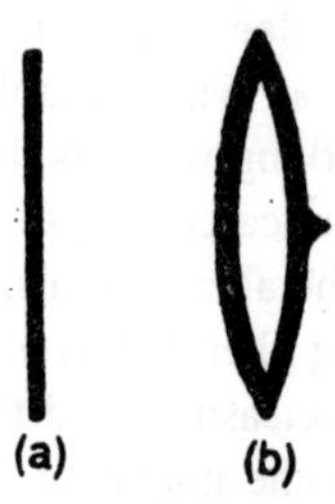

Fig. 1.32

Cu, Au, H, Li, Na, K: all gave double traces, the separation leading to the value of $\mu = \pm 1$ Bohr magneton, like silver.

Zn, Cd, Hg, Sn, Pb: no effect was produced by the field. Hence $m = 0$.

Tl (Thallium): the trace was divided into two, there being a definite absence of undeviated atoms. The deviation gave the value of

$$\mu = +1\frac{1}{3} \text{ Bohr magneton.}$$

Sb, Bi: Sb showed only an undeviated trace. The Bi trace obtained by Stern and Gerlach was remarkable in showing a one-sided continuous broadening indicating not only undeviated but also strongly attracted carriers. But later, a symmetrical splitting was obtained by Leu who showed that the distribution of density was consistent with the presence of atoms having values for μ in the ratio of 1 : 3.

Ni, Co, Fe: Nickel showed curious results in that three clearly defined traces appeared, one of them in the undeviated position itself, which meant that some of the atoms have no magnetic moment ($\mu = 0$). For the deviated traces μ was somewhat greater than one magneton. In addition to the three pronounced traces, there were indications of the presence of the atoms might having a value of μ greater than six. In the case of cobalt, a value $\mu = 6$ could just be established.

1.37 INTERPRETATION OF RESULTS

It is important to consider, in the first place, whether the beam consists of single atoms or molecules, since the experiments directly give the resolved moments of the carriers, whether atoms or molecules. At the temperature employed, there is little doubt that the vapours of Cu, Ag, Au, Sn, Pb, and Tl are monoatomic, so that the values of μ deduced for them do correspond to atomic moments. Ni, Co, Fe also probably monoatomic. But with Bi and Sb there is a large proportion of molecules in the beam.

Considering only the sure cases, it is possible to show how the results obtained offer a beautiful confirmation of the fundamental postulates of the vector model:

(a) Spatial Quantisation

The classical theory lays no restriction, on the orientation of the atomic magnet with respect to the field direction, all values from 0° to 180° for the angle 6 in the relation $\mu = M \cos \theta$ being permissible. Hence the displacement of the atoms in the non-homogeneous field should cover a continuous range, so that a *continuous diffuse band* should he obtained on the receiving plate. Further, on account of the Maxwellian distribution of velocities of the atoms in the beam, the densest part of the trace should be in the centre, since the number of atoms whose axes make angles between θ and $(\theta + d\theta)$ with the field is nearly proportional to $\sin \theta$ and hence a large fraction of the atoms would be oriented nearly transversely to the field and so would suffer very small deflections.

On the other hand, *according to the spatial quantisation theory*, not all settings are possible for the atomic magnet with respect to, the field direction, but only a certain discrete number. If the total angular momentum number of the atom is J the permitted orientations are (2J + 1). Hence on the plate, instead of obtaining a continuous band, (2J + 1), distinct traces should be obtained. Further, if the atom under test belongs to the *one electron, system* and is in the *ground state*, one can have higher values of J, since although L = 0, S = 1/2 = J, so that (2J + 1) = 2. In such a case, therefore, only two orientations are permitted and hence two traces should be obtained on the plate. In the case of many *electron system*, the atoms in the *ground state* can have higher values of J, since although L = 0, S can assume values 0, 1/2, 3/2, 2, etc., and

J = S. The number of permitted orientations may be greater than two, so that more complex setting of the beam could be expected: *e.g.*, for atoms of $^6S_{5/2}$term J being equal to 5/2, six traces should be obtained on the plate.

All these Theory of Atomic Structures of the spatial quantisation theory are exactly verified by the results of the Stern and Gerlach experiment. Thus,

(i) *H, Li, Na, K, Cu, Ag, Au, all belonging to the first column of the periodic table*, have In the ground state the spectral term $^2S_{1/2}$, which means that 2 J + 1 = 2. They all give only two traces as expected from theory. As soon as the atoms of these elements enter the magnetic field their magnetic axes orient themselves in the two permitted directions parallel and antiparallel to the field direction, no intermediate position being allowed. Further, in this case it can be shown that $\mu = \pm 1$ Bohr magneton, actually obtained from experimental data.

(ii) *Zn, Cd, Hg, belonging to the second column*, have two s electrons outside closed shells, and their normal state, as indicated by their spectra, is $^1S_0$. This means that J = 0 and hence the atom as whole has no resultant angular momentum and no magnetic moment. That is why no effect is produced on the atoms of these elements by the field.

(iii) Tl belonging to the third column, has one p electron in addition to completed groups. This gives a $^2P_{1/2}$ ground term *i.e.*, L = l, J = 1/2, 2J + 1 = 2. Hence it gives rise to two traces and calculations show that m = 0, which agrees with the experimental results.

(iv) Sn, Pb, belonging to the *fourth column* have two p electrons not forming a completed group, but the total magnetic moment vanishes in their normal $^3P_0$ state. This is why the atoms of these elements are undeviated by the field in the experiment.

(b) Electron Spin

Considering the well established case of silver, experiment clearly shows that the atomic beam of silver is split up by the magnetic field into two parts of approximately equal intensity. Further the value of the magnetic moment obtained is ±1 Bohr Magneton. From theory it is known that for the normal state of the silver atom L = O. Hence, if the spin of the electron does not exist, S would be zero, so that J also would be equal to zero and there would be no splitting of the atomic beam into two components. If, on the contrary, the existence of electron spin is admitted and the value of 1/2 be given to S in this case, there is perfect agreement between theory and experiment.

(c) Atomic and Quantum Nature of Magnetism

The idea of atomic nature of magnetism goes back to the days of Weber. The view that paramagnetic substances have a permanent molecular magnetic moment, while diamagnetic substances possess no such moment is a long established fact in the physical theory of magnetism. Weber was the first to develop this idea on the molecular current hypothesis of Ampere. It was next rendered certain by Langevin's treatment of paramagnetism based on the kinetic theory of gases and of diamagnetism on the basis of the classical electron theory. Weiss, in 1911, from experimental data then available, concluded that there was a fundamental unit of magnetic moment, of which all atomic and molecular moments were multiples. This unit was called after him the **Weiss Magneton**, whose value was estimated to be 1123.5 gauss cm. per gram-atom or per gram-molecule, which gives 1.85×10^{-21} gauss-cm. per atom or per molecule. With the advent of the quantum theory, another fundamental unit, known as the **Bohr Magneton**, came to be recognised. Its value was found to be 5,590 gauss-cm. per gram-atom or 9.21×10^{-21} gauss-cm. per atom, which is therefore, almost five times as large as the older unit.

The Stern and Gerlach experiment not only confirms the general Theory of Atomic Structures of the classical theory about para and diamagnetism, but goes much further and establishes the quantum nature and atomic origin of magnetism. For, in the first place, according to the results of the experiments, substances which are found to be diamagnetic have no atomic magnetic moment ($\mu = 0$), *e.g.*, Zn, Cd, Hg. Pb, Sn, etc. Substances which are paramagnetic are made up of atoms with one valency electron in the ground state, thereby giving rise to a magnetic moment of 1 Bohr Magneton due to the spin of the single free electron: *e.g.*, H, Li, the alkalis, Cu, Ag, Au. Ferromagnetic substances (Fe, Ni, Co) are made up of atoms with intermediate incomplete electronic shells and have a large value for the atomic moment.

Secondly, the results establish also that the true fundamental unit is the Bohr magneton and not the Weiss magneton. The latter based on the classical theory, according to which all possible orientations of the atomic axis with respect to the field are permissible, has only an apparent existence. The former is the one given by the experiment of Stern and Gerlach and conforms to the quantum theory; it alone, therefore, has a real existence.

There is no doubt that the vector atom model is a superb and many sided conception, which enables one to solve most of the important and intricate problems concerning atomic structure and spectral phenomena.

1.38 WAVE MECHANICAL ATOM MODEL

The great drawback of the vector atom model is that it does not present itself as a single unified system, containing the necessary theoretical justification of its principles and postulates. These are introduced somewhat piecemeal, partly on empirical data and partly by analogy. No adequate reasons are given for the use of electron spin with half-integer value; the assumption of discrete energy states and of emission of radiation when the atom passes from one of these states to another are arbitrary as in older models. The model therefore lacks a sound theoretical foundation. The most recent theory, known as the *wave mechanical atom model*, supplies this information and presents a picture of the peripheral electronic structure of the atom, the best that can be had in the present state of our knowledge.

A detailed study of this model is beyond the scope of this book. A masterly treatment by *Dirac* of the relativistic theory of the electron based on wave mechanics offers, almost without any special effort, theoretical justification of the postulates and Theory of Atomic Structures of the vector model. We shall limit ourselves to a brief survey of the essential features of this new conception, indicating the advances made over the older models.

According to wave mechanics, in order to have a complete picture of the electron, we should consider it under the double aspect of *particle* and *wave*. The change of finding a moving electron at a given point is therefore governed by a wave, which necessarily involves a certain degree of uncertainty about this exact position. The older models, which placed the electron at a precise point is a well-defined orbit, did not take into account this correct nature of the electron and to this fact all their short-comings are ultimately to be traced.

The-wave mechanical theory of the atom can be explained in a simple manner, *i.e.*, without entering into complicated mathematical treatment, analogy with other forms of waves. Free electrons emitted from some source and spreading outwards through vacuum are like the ripples on the surface of a large pond, which advance from the source of disturbance in ever-widening circles, becoming feebler as the distance from the starting point increases, until finally they are hardly perceptible. On the other hand, electrons in an atom, experiencing the electrostatic attractive force of the nucleus and hence prevented from leaving its neighbourhood, are like the stationary wave set up by the reflection of the ripples at the walls of a narrow vessel in which the ripples are produced, or better still like the stationary waves set up in a fixed string with a number of loops and nodes giving the various modes of vibration. On this analogy the electron waves may be said to be reflected back when they reach the boundary of the atom so that a system of stationary waves may be imagined within the atom.

With this picture of the atom a number of important Theory of Atomic Structures may be drawn:

1. Removal of Individualistic Particle Picture of the Electrons in the Atom

Giving to the electron a fixed position in a definite orbit is no longer correct. The electron probability is distributed over the whole range of the atom though corresponding to die loops and nodes in the stationary wave system the distribution is not uniform. For the same reason, when several electrons are contained in the atom, they lose their individualities to a certain extent, as it is not possible to say that any particular part of any of the "probability waves" representing the electrons belongs to any particular electron.

2. Nebulous Picture of the Atom

The atom loses much of its former vividness of well-defined electronic orbits, since the electron probability does not vanish sharply at any distance from the nucleus; it becomes zero only at infinity, so that theoretically there is a chance of finding the electron anywhere in space. But there is a region where the probability falls steeply to a small value and this corresponds to what is ordinarily called the boundary of the atom.

3. Quantum Stationary States of the Atom

The stationary electron waves can vibrate in different modes which imply different wavelengths and frequencies, and therefore, different amounts of energy in the system. In the mathematical treatment of the problem, this means that the wave equation representing the stationary .waves in the atom can be solved only when certain coefficients are given appropriate values, which must be whole numbers or occasionally simple fractions, like the different harmonic of a fundamental vibration. These coefficients: correspond to the *quantum numbers* of the older models, but they appear here quite naturally as characteristic solutions of the wave equation, whereas in the older theories they were introduced somewhat artificially, with the only reason that they were found necessary for a satisfactory interpretation of the experimentally observed atomic and spectral phenomena.

4. Non-Radiative Character of the Stationary States

In a stationary wave system the amplitude at any given point is constant and does not change with time. Now, considering the wave mechanical atom from the point of view of what is known as the *electronic cloud round* the nucleus or the density distribution of the electric charge in the atom, and interpreting square of the amplitude of the electron wave as measuring the electric charge density, it follows that the charge density at any point must be no matter how great the proper energy value, the charge distribution must be static, *i.e.*, must not vary with time. There being no movement of electric charge in the permitted energy states, the atom in such states cannot radiate electromagnetic energy, in consonance with classical electrodynamics. Thus, we arrive quite naturally at a result, which in the older theories had to be assumed as an arbitrary postulate contrary to the classical theory.

5. Emission of Radiation Governed by Bohr's Frequency Condition

When two vibrations of somewhat different frequencies are superimposed they give rise to the so-called "beat" phenomena, in which the amplitude waxes and wanes with a frequency equal to the difference between the two superposed frequencies. In a similar manner, when two different modes of the wave system, corresponding to two permitted energy states of the atom, can be excited simultaneously so that they are superimposed, radiation whose frequency is given by the number of beats per second between the frequencies of the two electron waves superposed, will be emitted. For, if we once again interpret the square of the amplitude of vibration in the wave mechanical model as a measure of the electric charge density, it is readily seen that when different energy states are simultaneously excited, the charge density at any given point is no longer constant but varies with time.

Hence, there is a variation of the electric charge, which must give rise to the emission of electromagnetic waves. The frequency of these emitted waves will be the best frequency. If v_L be the frequency of radiation emitted and v_1 and v_2 the frequencies of the two superposed vibrations of electron waves, then:

$$v_L = v_1 - v_2$$

According to the wave mechanical theory the energy W of a stationary state is given by W = hv, where v is the frequency of the electron waves corresponding to that state. Hence $v_1 = W_1/h$ and $v_2 = W_2/h$ where W_1, and W_2 are the energies of the two states involved.

$$v_L = \frac{W_1 - W_2}{h}$$

Thus, we arrive logically at Bohr's frequency condition which was introduced arbitrarily as a fundamental postulate of his theory, at variance with the classical standpoint.

Thus, the wave mechanical atom model is able to account adequately for all the postulates assumed by the older models. Further, it has the supreme merit, in a scientific theory of reducing the number of fundamental postulates and presenting itself as a single unified system.

SOLVED EXAMPLES

Example 1:

Find the radius and speed of the electron in the first bohr orbit of the hydrogen atom. How will the radius and speed of electron change with the increase in atomic number of the atom.

Solution:

(i) The radius of the Bohr orbit is given by

$$r = \frac{\epsilon_0 n^2 h^2}{\pi m Z e^2}$$

For hydrogen,

$$Z = 1,\ n = 1$$

$$r_H = \frac{\epsilon_0 h^2}{\pi m e^2}$$

Here $\epsilon_0 = 8.85 \times 10^{-12}\ C^2/N - m^2$

$$m = 9.1 \times 10^{-31}\ kg$$

$$e = 1.6 \times 10^{-19}\ C$$

$$h = 6.624 \times 10^{-34}\ J\text{-}s$$

$$\therefore \quad r_H = \frac{8.85 \times 10^{-12} \times 6.624 \times 10^{-34})^2}{\pi \times 9.1 \times 10^{-31} \times (1.6 \times 10^{-19^2})^2}$$

$$\mathbf{r_H = 5.30 \times 10^{-11}\ m.}$$

Also $r = \frac{r_H}{Z}$

Therefore, radius of the orbit will decrease with increase in atomic number Z.

(ii) velocity of the electron

$$\upsilon = \frac{Ze^2}{2\,\epsilon_0\, nh}$$

In hydrogen $Z = 1,\ n = 1$

$$\upsilon_H = \frac{Ze^2}{2\,\epsilon_0 h}$$

$$= \frac{(1.6 \times 10^{-19})^2}{2 \times 8.85 \times 10^{-12} \times 6.624 \times 10^{-34}}$$

$$\mathbf{= 2.2 \times 10^6\ m/s}$$

Also $\upsilon = Z\upsilon_H$

Therefore velocity of the electron in the orbit ill increase ith increase in atomic number Z.

Example 2:

The ionization potential of atomic hydrogen is 13.6 V. Calculate the wavelength of light emitted in a transition starting at the first excited state of hydrogen atom? **(IAS, 1985)**

Solution:

Ionization potential = 13.6 V

Energy of electron in the first orbit

$$U_1 = -13.6 \text{ eV}$$

Energy of electron in the second orbit

$$U_2 = \frac{-13.6}{n^2} = \frac{-13.6}{4} = -3.4\,\text{eV}$$

$$U_2 - U_1 = -3.4 + 13.6$$

$$h\nu = 10.2 \text{ eV}$$

$$\frac{hc}{\lambda} = 10.2 \times 1.6 \times 10^{-19} \text{ J}$$

$$= 1.632 \times 10^{-18} \text{ J}$$

$$\lambda = \frac{hc}{1.632 \times 10^{-18}}$$

$$= \frac{6.624 \times 10^{-34} \times 3 \times 10^{8}}{1.632 \times 10^{-18}}$$

$$= 1.217 \times 10^{-7} \text{ m}$$

$$= \mathbf{1217\ \text{Å}.}$$

Example 3:

The wavelength of the first number of Balmer series of hydrogen is

6563×10^{-10} m

Calculate the wavelength of its second number **[Delhi, 1982]**

Solution:

For the first number

$$\bar{v}_1 = R\left[\frac{1}{2^2} - \frac{1}{3^2}\right]$$

$$\bar{v}_1 = \frac{5}{36}R$$

or $$\frac{1}{\lambda_1} = \frac{5}{36}R \qquad ...(i)$$

For the second number,

$$\bar{v}_2 = R\left[\frac{1}{2^2} - \frac{1}{4^2}\right]$$

or $$\frac{1}{\lambda_2} = \frac{3}{16}R \qquad ...(ii)$$

Dividing (i) by (ii)

$$\frac{\lambda_2}{\lambda_1} = \frac{20}{27}$$

$$\lambda_2 = \frac{20}{27}\lambda_1$$

But $$\lambda_1 = 6563 \times 10^{-10} \text{ m}$$

$$\therefore \quad \lambda_2 = \frac{20 \times 6563 \times 10^{-10}}{27}$$

$$= \mathbf{4861 \times 10^{-10} \ m.}$$

Example 4:

A beam of electrons is used to bombard gaseous hydrogen. What is the minimum energy in **electron-volts** *the electrons must have if the first number of the Balmer of the Balmer series corresponding to a transition from $n_2 = 3$ state to $n_1 = 2$ state is to be emitted?*

$$h = 6.6 \times 10^{-34} \textit{ joule second.}$$

(Bombay, 1981, Calicut, 1992)

Solution:

$$\text{Energy required} = \frac{me^4}{8\epsilon_0^2 h^2}\left[\frac{1}{n_1^2} - \frac{1}{n_2^2}\right]$$

Here $n_1 = 2, n_2 = 3$

$\therefore$ Energy required $= \left(\frac{me^4}{8 \epsilon_0^2 h^2}\right) \times \left(\frac{5}{36}\right)$ joule

But $1 \text{ eV} = 1.6 \times 10^{-19}$ joule

$$\therefore \text{Energy required} = \frac{5\, me^4}{8 \times 36 \times \epsilon_0^2 h^2 \times (1.6 \times 10^{-19})} \text{ electron-volts}$$

$$= \frac{5 \times 9 \times 10^{-31} \times (1.6 \times 10^{-19})^4}{8 \times 36\,(8.85 \times 10^{-12}) \times (6.6 \times 10^{-34})^2 \times (1.6 \times 10^{-19})}$$

$= \mathbf{1.88\ eV.}$

Example 5:

Calculate the ionization potential in electron volts for hydrogen atom given that

$e = 1.6 \times 10^{-19}$ coulomb

$m = 9 \times 10^{-31}$ kg

$h = 6.6 \times 10^{-34}$ joule-second

$\epsilon_0 = 8.85 \times 10^{-19}$ coulomb^2/newton – m^2.

Solution:

Work function,

$$\phi = \frac{me^4}{8 \epsilon_0^2 h^2} \text{ joules}$$

$$\phi = \frac{9 \times 10^{-31} \times (1.6 \times 10^{-19})^4}{8 \times (8.85 \times 10^{-12})^2 \times (6.6 \times 10^{-34})^2} \text{ joules}$$

But $1 \text{ eV} = 1.6 \times 10^{-10}$ joule

$$\therefore \quad \phi = \frac{9 \times 10^{-31} \times (1.6 \times 10^{-19})^4}{8(8.85 \times 10^{-19})^2 \times (6.6 \times 10^{-34}) \times 1.6 \times 10^{-19}}$$

or $\phi = 13.51$ eV

Example 6:

The wavelength of the second line of the Balmer series in the hydrogen spectrum is 4861 Å. Calculate the wavelength of the first line.

(Rajasthan, 1975)

Solution:

For the second line,

$$\bar{v}_2 = R\left[\frac{1}{2^2} - \frac{1}{4^2}\right]$$

$$\bar{v}_2 = \frac{3}{16}R$$

$$\frac{1}{\lambda_2} = \frac{3}{16}R \qquad \text{...(i)}$$

For the first line,

$$\bar{v}_1 = R\left[\frac{1}{2^2} - \frac{1}{3^2}\right]$$

$$\bar{v}_1 = \frac{5}{36}R$$

or $$\frac{1}{\lambda_1} = \frac{5}{36}R \qquad \text{...(ii)}$$

Dividing (i) by (ii)

$$\frac{\lambda_1}{\lambda_2} = \frac{27}{20}$$

$\therefore$ $$\lambda_1 = \frac{27\,\lambda_2}{20}$$

But $$\lambda_2 = 4861\text{Å}$$

$$\lambda_1 = \frac{27 \times 4861}{10}$$

or $$\lambda_1 = \mathbf{6563\ Å.}$$

Example 7:

The wavelength of sodium D_1 line is 590 nm. Calculate the difference in energy levels involved in the emission or absorption of this line.

Solution:

Here $E_{(n_2-n_1)} = h\nu = \frac{hc}{\lambda}$

Here $h = 6.624 \times 10^{-34}$ J-s

$c = 3 \times 10^{8}$ m/s

$h = 590 \text{ nm} = 590 \times 10^{-9}$ m

$$\therefore E_{(n_2-n_1)} = \frac{(6.624 \times 10^{-34}) \times 3 \times 10^{8}}{590 \times 10^{-9}}$$

$$= \mathbf{3.37 \times 10^{-19}\ J.}$$

Example 8:

Wavelength of Balmer H_α line is 6563 Å. Calculate the wavelength of H_β line. **(Delhi (Hons.), 1982)**

Solution:

For H_α line of Balmer series *i.e.,* the second number,

$$\bar{\nu}_1 = R\left[\frac{1}{2^2} - \frac{1}{3^2}\right]$$

$$\frac{1}{\lambda_1} = \frac{5}{36}R \quad \text{...(i)}$$

For H_β line of Balmer Series *i.e.,* the second number,

$$\bar{\nu}_2 = R\left[\frac{1}{2^2} - \frac{1}{4^2}\right]$$

$$\frac{1}{\lambda_1} = \frac{3}{16}R \quad \text{...(ii)}$$

Dividing (i) by (ii)

$$\frac{\lambda_2}{\lambda_1} = \frac{20}{27}$$

$$\lambda_2 = \left(\frac{20}{27}\right)\lambda_1$$

But $\lambda_1 = 6563$ Å

$$\lambda_2 = \left(\frac{20}{27}\right)6563$$

$$\lambda_2 = \mathbf{4861\ Å.}$$

Example 9:

Calculate the energy required to excite the hydrogen atom from the ground state (n = 1) to the first excited state (n = 2).

(Delhi (Hons.), 1990)

Solution:

Energy required $U = \frac{me^4}{8\epsilon_0^2 h^2}\left[\frac{1}{n_1^2} - \frac{1}{n_2^2}\right]$

$$U = \frac{me^4}{8\epsilon_0^2 h^2}\left[\frac{1}{1^2} - \frac{1}{2^2}\right]$$

$$U = \frac{3me^4}{32\epsilon_0^2 h^2}$$

Here $m = 9.1 \times 10^{-31}$ Kg, $e = 1.6 \times 10^{-19}$ C

$\epsilon_0 = 8.85 \times 10^{-12}\ C^2/N - m^2$

$h = 6.624 \times 10^{-34}$ J-s

$$U = \frac{3 \times 9.1 \times 10^{-31} \times (1.6 \times 10^{-19})^4}{32 \times (8.85 \times 10^{-12})^2 \times (6.624 \times 10^{-34})^2} \text{ joules}$$

$$U = \frac{3 \times 9.1 \times 10^{-31} \times (1.6 \times 10^{-19})^4}{32 \times (8.85 \times 10^{-12})^2 \times (6.624 \times 10^{-34})^2 \times 1.6 \times 10^{-19}} \text{ eV}$$

$$\mathbf{U = 10.13\ eV.}$$

Example 10:

Calculate the radius of the first Bohr orbit for (i) H and (ii) He atoms and also the velocity of the electron in these orbits as compared to the velocity of light.

Solution:

For hydrogen atom, the radius of the orbit is

$$r_H = \frac{\epsilon_0 n^2h^2}{\pi mZe^2}$$

Here $n = 1, Z = 1$

$\epsilon_0 = 8.85 \times 10^{-12}$ $C^2/N - m^2$

$m = 9.1 \times 10^{-31}$ kg

$e = 1.6 \times 10^{-19}C$

$h = 6.624 \times 10^{-34}$ J-s

$$\therefore \quad r_H = \frac{8.85 \times 10^{-12} \times 1 \times (6.624 \times 10^{-34})^2}{3.142 \times 9.1 \times 10^{-31} \times (1.6 \times 10^{-19})^2}$$

$r_H = 5.30 \times 10^{-11}$ m.

For helium, atom, $n = 1, Z = 2$

$$r_H = \frac{r_H}{Z} = \frac{5.30 \times 10^{-11}}{2}$$

$= \mathbf{2.65 \times 10^{-11} m.}$

(ii) velocity of the electron,

$$\upsilon = \frac{Ze^2}{2 \epsilon_0 nh}$$

In hydrogen $Z = 1$, and $n = 1$

$$\upsilon_H = \frac{e^2}{2\epsilon_0 h}$$

$$= \frac{(1.6 \times 10^{-19})^2}{2 \times 8.85 \times 10^{-12} \times 6.624 \times 10^{-34}}$$

$= \mathbf{2.2 \times 10^6 \ m/s}$

Also, $\frac{\upsilon_H}{c} = \frac{2.2 \times 10^6}{3 \times 10^8} = 7.33 \times 10^{-3}$

For helium

$$\upsilon_{He} = \frac{Ze^2}{2 \epsilon_0 nh}$$

Here $n = 1, Z = 2$

$$\upsilon_{He} = \frac{2e^2}{2 \epsilon_0 h} = 2\upsilon_H$$

$$\upsilon_{He} = 2 \times 2.2 \times 10^6$$

$$= \mathbf{4.4 \times 10^6 \ m/s}$$

$$\frac{\upsilon_{He}}{c} = \frac{4.4 \times 10^6}{3 \times 10^8}$$

$$= \mathbf{1.466 \times 10^{-2}}.$$

Example 11:

In hydrogen atom the electron is replaced by a muon whose mass is 200 times that of an electron and charge is same as that of electron, calculate the ionization potential on the basis of Bohr's theory.

[IAS, 1989]

Solution:

In the case of hydrogen atom, having an electron, the ionization potential

$$\phi = \frac{me^4}{8 \epsilon_0^2 h^2} \qquad \text{...(i)}$$

when electron is replaced by muon.

$$m_1 = 200 \ m$$

$$\phi_1 = \frac{m_1 e^4}{8 \epsilon_0^2 h^2} = \frac{(200 m) e^4}{8 \epsilon_0^2 h^2} \qquad \text{...(ii)}$$

Dividing (ii) by (i)

$$\frac{\phi_1}{\phi} = 200$$

But $\phi = 13.6$ eV

$\therefore$ $\phi_1 = 200 \times 13.6$

$$= \mathbf{2.72 \times 10^3 \ eV}.$$

Example 12:

Calculate the difference in wavelength in the spectra of hydrogen and heavy hydrogen corresponding to the first line on the long wave side of Balmer series.

$$\begin{bmatrix} R_H = 1.097 \times 10^7/\text{m} \\ m = 0.000549\ M_H \end{bmatrix}$$

Solution:

$$R_H = \frac{R}{1 + \frac{m}{M_H}} \qquad \text{...(i)}$$

$$R_D = \frac{R}{1 + \frac{m}{M_D}}$$

$$\frac{m}{M_H} = 0.000549$$

$$\frac{m}{M_D} = \frac{m}{2M_H} = 0.000274$$

$$\frac{R_D}{R_H} = \frac{1 + \frac{M}{M_H}}{1 + \frac{m}{M_D}} = \frac{1 + 0.000549}{1 + 0.000274}$$

$$\frac{R_D}{R_H} = 1.00027$$

$$R_D = (R_H)\ 1.00027$$

$$R_D = 1.097 \times 10^7/\text{m}$$

In the case of Balmer series, for the first member

$$n_1 = 2,\ n_2 = 3$$

$$\frac{1}{\lambda_H} = R_H\left[\frac{1}{n_1^2} - \frac{1}{n_2^2}\right] = \left(\frac{5}{36}\right) R_H$$

$$\lambda_H = \frac{36}{5R_H} = \frac{36}{5 \times 1.097 \times 10^7}$$

$$\lambda_H = 6.5633 \times 10^{-7} \text{ m}$$

Similarly,

$$\frac{1}{\lambda_D} R_D \left[\frac{1}{n_1^2} - \frac{1}{n_2^2}\right] = \left(\frac{5}{36}\right) R_D$$

$$\lambda_D = \frac{36}{5R_D}$$

$$\lambda_D \frac{36}{5 \times 1.0973 \times 10^7}$$

$$\lambda_D = 6.5615 \times 10^{-7}\text{m}$$

$$\therefore \quad \Delta\lambda = \lambda_H - \lambda_D$$

$$= 6.5633 \times 10^{-7} - 6.5615 \times 10^{-7}$$

$$= 0.0018 \times 10^{-7}\text{m}$$

$$= \mathbf{1.8\ \text{Å}.}$$

Example 13:

What is the energy, momentum and wavelength of the photon emitted by a hydrogen atom when an electron makes a transition from $n = 2$ to $n = 1$ state. Given ionization potential = 13.6 eV.

Solution:

Energy of electron in the first orbit of hydrogen atom,

$$E_1 = 13.6 \text{ eV}$$

Energy of electron in second orbit

$$E_2 = \frac{E_1}{n^2} = \frac{E_1}{4} = \frac{-13.6}{4}$$

$$E_2 = -3.4 \text{ eV.}$$

(1) Energy of photon emitted = $E_2 - E_1$

$$E = -3.4 + (13.6)$$

$$= \mathbf{10.2\ eV}$$

$$= 10.2 \times 1.6 \times 10^{-19} \text{ J}$$

$$= 16.32 \times 10^{-19} \text{ J}$$

(2) Momentum, $p = \frac{E}{c} = \frac{16.32 \times 10^{-19}}{3 \times 10^{8}}$

$$p = 5.44 \times 10^{-27} \text{ kg} - \text{m/s}.$$

(3) Wavelength

$$\lambda = \frac{h}{p}$$

$$= \frac{6.624 \times 10^{-34}}{5.44 \times 10^{-27}}$$

$$= 1.218 \times 10^{-7} \text{m}$$

$$= \mathbf{1218 \text{ Å}.}$$

Example 14:

The wavelength of the first line of Balmer series is 563 Å. Calculate the wavelength of the second line of Balmer series. **(Delhi, 1990)**

Solution:

For the first member,

$$\bar{v}_1 = R\left[\frac{1}{2^2} - \frac{1}{3^2}\right]$$

$$\bar{v}_1 = \frac{5}{36} R$$

or $$\frac{1}{\lambda_1} = \frac{5}{36} R \qquad ...(i)$$

For second member

$$\bar{v}_2 = R\left[\frac{1}{2^2} - \frac{1}{4^2}\right]$$

$$\bar{v}_2 = \frac{3}{16} R$$

or $$\frac{1}{\lambda_2} = \frac{3}{16} R \qquad ...(ii)$$

Dividing (i) by (ii)

$$\frac{\lambda_2}{\lambda_1} = \frac{20}{27}$$

or $$\lambda_2 = \left(\frac{20}{27}\right)\lambda_1$$

But $$\lambda_1 = 6563 \times 10^{-10}\text{m}$$

$\therefore$ $$\lambda_2 = \left(\frac{20}{27}\right) 6563 \times 10^{-10}\,\text{m}$$

$$\mathbf{\lambda_2 = 4861 \times 10^{-10}m}$$

$$\mathbf{\lambda_2 = 4861\ \text{Å}.}$$

Example 15:

The radiations given off by the Hg atoms returning to their normal states were studied by Frank and Hertz and a line was observed at 2537Å. Calculate excitation potential for Hg. **(Bombay, 1991)**

Solution:

Here $\lambda = 2537\ \text{Å} = 2537 \times 10^{-10}\text{m}$

Let excitation potential be V

$\therefore$ $$eV = h\nu = \frac{hc}{\lambda}$$

$\therefore$ $$V = \frac{hc}{e\lambda}$$

Here $h = 6.624 \times 10^{-34}$ J-s

$c = 3 \times 10^{8}$ m/s

$e = 1.6 \times 10^{-19}$ C

$\lambda = 2537 \times 10^{-10}\text{m}$

$\therefore$ $$V = \frac{6.624 \times 10^{-34} \times 3 \times 10^{8}}{1.6 \times 10^{-19} \times 2537 \times 10^{-10}}$$

V = 4.9 volts.

Example 16:

What are the energy, momentum and wavelength of the photon emitted by hydrogen atom in transition 2s → Is? **(Osmania, 1992)**

Solution:

In hydrogen,

Energy of 1s state, $E_1 = -13.6$ eV

Energy of 2s state, $E_2 = \frac{-13.6}{2^2} = -3.4$ eV

(1) Energy of photon,

$$E = E_2 - E_1 = -3.4 - (-13.6)$$

$$\mathbf{E = 10.2\ eV}$$

$$E = 10.2 \times 1.6 \times 10^{-19}\ J$$

$$\mathbf{E = 1.632 \times 10^{-18}\ J.}$$

(2) Momentum $= \frac{E}{c} = \frac{1.632 \times 10^{-18}}{3 \times 10^8}$

$$\mathbf{= 5.44 \times 10^{-27}\ kg\text{–}m/s.}$$

(3) $E = h\nu = \frac{hc}{\lambda}$

$$\lambda = \frac{hc}{E}$$

$$\lambda = \frac{6.624 \times 10^{-34} \times 3 \times 10^8}{1.632 \times 10^{-18}}$$

$$\mathbf{\lambda = 12.18 \times 10^{-8}\ m = 1218 \times 10^{-10} m}$$

$$\mathbf{\lambda = 1218\ Å.}$$

Example 17:

Calculate the energy required to create a vacancy in

(i) K shell of copper atom

(ii) L shell of copper atom

Given, Ionization potential of hydrogen atom = 13.6 eV

For copper Z = 29.

Solution:

(i) The energy required to create a vacancy in the first orbit (n = 1) *i.e.,* K shell of copper atom.

$$E_1 = \frac{(Z^2) \times 13.6}{n^2}$$

$$= \frac{(29)^2 \times 13.6}{1}$$

$$= 11.44 \times 10^3 \text{ eV.}$$

(ii) Energy required to create a vacancy in the second orbit (n = 2) *i.e.,* L shell of copper atom

$$E_2 = \frac{(29)^2 \times 13.6}{2^2}$$

$$\mathbf{E_2 = 2.86 \times 10^3 \text{ eV.}}$$

Example 18:

In a μ - mesonic atom, a muon with a charge-e and mass 200 m_e moves in a circular orbit around the nucleus of charge + 3e. Assuming that Bohr's model of atom is applicable to this system.

(i) Derive the expression for the radius of n th Bohr orbit

(ii) Find the value of n for which the radius of the orbits is approximately the same as that of Bohr's first orbit of hydrogen atom. Bohr radius = 0.53×10^{-10}m. **(Calcutta, 1991)**

Solution:

(i) Here Z = 3

$$m = 200m_e$$

Here $$\frac{m\upsilon^2}{r} = \frac{Ze^2}{4\pi \epsilon_0 r^2}$$

$$m\upsilon^2 = \frac{Ze^2}{4\pi \epsilon_0 r} \qquad ...(i)$$

Also $$mvr = \left(\frac{nh}{2\pi}\right) \qquad \text{...(ii)}$$

$$\therefore \quad v = \left(\frac{nh}{2\pi mr}\right)$$

$$\therefore \quad m = \left(\frac{nh}{2\pi mr}\right)^2 = \frac{Ze^2}{4\pi \epsilon_0 r}$$

$$r = \frac{\epsilon_0 n^2h^2}{\pi mZe^2} \qquad \text{...(iii)}$$

(ii) $$r = \frac{\epsilon_0 n^2h^2}{\pi mZe^2}$$

Here $$m = 200 \ m_e$$

$$Z = 3$$

$$r = \frac{\epsilon_0 n^2h^2}{\pi \times 200 m_e \times 3e^2}$$

$$r = \frac{n^2}{600}\left[\frac{\epsilon_0 h^2}{\pi m_e e^2}\right]$$

For hydrogen atom, for the first Bohr orbit

$$r_H = \left(\frac{\epsilon_0 h^2}{\pi m_e e^4}\right) = 0.53 \times 10^{-10} m$$

$$r = \left(\frac{n^2}{600}\right) r_H$$

But $$r = r_H \text{ (given)}$$

$$\therefore \quad \frac{n^2}{600} = 1, \ n^2 = 600$$

$$\mathbf{n = 24.}$$

(iii) $$hv = \frac{me^4Z^2}{8\epsilon_0^2 h^2}\left(\frac{1}{n_1^2} - \frac{1}{n_2^2}\right)$$

Here $$n_1 = 1, \ n_2 = 2$$

$$U = \frac{200 m_e e^4 (3^2)}{8 \epsilon_0^2 h^2} \left(\frac{1}{1} - \frac{1}{4} \right)$$

$$U = \left(\frac{1800 \times 3}{4} \right) \left(\frac{m_e e^4}{8 \epsilon_0^2 h^2} \right)$$

$$U = \left(\frac{1800 \times 3}{4} \right) \left[\frac{9.1 \times 10^{-31} \times (1.6 \times 10^{-19})^4}{8 \times (8.85 \times 10^{-12})^2 \times (6.624 \times 10^{-34})^2} \right] J$$

$$U = \left(\frac{1800 \times 3}{4} \right) \left[\frac{9.1 \times 10^{-31} \times (1.6 \times 10^{-19})^4}{8 \times (8.85 \times 10^{-12})^2 \times (6.624 \times 10^{-34})^2 \times 1.6 \times 10^{-19}} \right] eV$$

$$U = 1.8234 \times 10^4 \text{ eV}$$

Excitation potential

$$\mathbf{V = 1.8234 \times 10^4 \text{ Volts.}}$$

Example 19:

An X-ray line of wave length 0.53832 Å is found to be emitted from an X-ray tube with Zinc (Z = 30) target in addition to the characteristic K_α line of Zinc of wavelength 1.43603 Å. If the unknown line is due to an impurity in the target, obtain the atomic number of impurity.

(Calcutta, 1992)

Solution:

Here $\lambda_1 = 1.43603$ Å

$Z_1 = 30$

$\lambda_2 = 0.53832$ Å

$Z_2 = ?$

$$\frac{1}{\lambda_1} \propto Z_1^2$$

and $$\frac{1}{\lambda_2} \propto Z_2^2$$

$$\therefore \quad \left(\frac{Z_2}{Z_1} \right)^2 = \left(\frac{\lambda_1}{\lambda_2} \right)$$

$$Z_2 = Z_1\left[\frac{\lambda_1}{\lambda_2}\right]^{1/2}$$

$$Z_2 = 30\left[\frac{1.43603}{0.53832}\right]^{1/2}$$

$$\mathbf{Z_2 = 49.}$$

Example 20:

Calculate the screening fact or for K shell for copper (Z = 29) if the K shell binding energies in hydrogen and copper are respectively 13.5 eV and 8940 eV. **(IAS, 1997)**

Solution:

For hydrogen

$$U_H = \frac{me^4Z^2}{8\epsilon_0^2 h^2}$$

$$Z = 1 \text{ and } n = 1$$

$$U_H = \frac{me^4}{8\epsilon_0^2 h^2} = 13.5 \text{ eV}$$

$$= 13.5 \times 1.6 \times 10^{-19} \text{ J} \qquad \text{...(i)}$$

For copper

$$U_{cu} = \frac{me^4Z_{eff}^2}{8\epsilon_0^2 h^2} = 8940 \text{ eV}$$

$$= (8940 \times 1.6 \times 10^{-19})\text{J} \qquad \text{...(ii)}$$

$$\therefore \quad (13.5 \times 1.6 \times 10^{-19}) = Z_{eff}^2\ (8940 \times 1.6 \times 10^{-19})$$

$$Z_{eff}^2 = 662.2$$

$$Z_{eff} = 25.73 \approx 26$$

$$Z = 29$$

$$\therefore \quad Z_{eff} = (Z - x)$$

$$26 = (29 - x)$$

$$\mathbf{x = 3.}$$

Example 21:

Calculate the energy and wavelength of the photon emitted from a tungsten target Z = 74 when electron jumps from n = 3 state to n = 1 state due to a vacancy in n = 1 state. Given Energy of electron in n = 1 state of hydrogen atom = –13.6 eV

Solution:

Here $Z = 74$

For $n = 1, \quad Z_{Eff} = 74 - 1 = 73$

$$E_1 = \left(\frac{-13.6 Z_{eff}^2}{n^2}\right) \text{eV}$$

$$= \left[\frac{-13.6 \times (73)^2}{1^2}\right] \text{eV}$$

$$= -7.2474 \times 10^4 \text{ eV}.$$

For n = 3 state, the electron is subject to an effective nuclear charge that depends on the number of electrons in orbits, n = 1 and n = 2. There are eight electrons in n = 2 state and one electron in n = 1 state as there is one vacancy in n = 1 state.

Therefore $Z_{eff} = (Z - 9)$

$$E_3 = \left[\frac{-13.6 \times (Z-9)^2}{3^2}\right] \text{eV}$$

$$E_3 = \left[\frac{(74-9)^2 \times 13.6}{9}\right] \text{eV}$$

$$= -\left[\frac{(65)^2 \times 13.6}{9}\right] \text{eV}$$

$$= \mathbf{-6384 \text{ eV}.}$$

(1) Energy of emitted photon,

$$E = E_3 = E_1$$
$$= -6384 - (-72474)$$
$$= \mathbf{6.609 \times 10^4\ eV} = 6.609 \times 10^4 \times 6 \times 10^{-19}\ J$$
$$= 1.057 \times 10^{-14}\ J.$$

(2) $$E = \frac{hc}{\lambda}$$
$$\lambda = \frac{hc}{E}$$
$$= \frac{6.624 \times 10^{-34} \times 3 \times 10^8}{1.057 \times 10^{-14}}\ m$$
$$= \mathbf{1.88 \times 10^{-11}\ m}$$
$$= \mathbf{0.188\ Å.}$$

Example 22:

Calculate the wavelength of K_α line in case of copper atom. Take Moseley law into consideration.

Solution:

Take $R = 1.096776 \times 10^7$ per meter.

Here $Z = 29$

Taking Mosely's Law into consideration

$$Z_{eff} = (Z - 1) = 29 - 1 = 28$$
$$n_1 = 1 \text{ and } n_2 = 2$$
$$\frac{1}{\lambda} = R(Z_{eff})^2\left[\frac{1}{n_1^2} - \frac{1}{n_2^2}\right] = \frac{3R[Z-1]^2}{4}$$
$$\lambda = \frac{4}{3R(Z-1)^2}$$
$$\lambda = \frac{4}{3 \times 1.096776 \times 10^7 \times (28)^2}$$
$$= \mathbf{1.55 \times 10^{-10}m}$$
$$= \mathbf{1.55\ Å.}$$

Example 23:

The measured value of the total kinetic energy of the fission fragments from the thermal neutron fission of U^{92}_{235} *is 196 MeV. Assuming that the respective values of Z and A of the fission fragments as (35, 72) and (57, 162). Calcvulate the distance r between the fragments at the instant of separation. Compare this value with the sum of the radio of the two fragments. Nuclear radi* $R = r_0 A^{1\ 3}$ *where* $r_0 = 1.2 \times 10^{-15}$*m.*

(Calcutta, 1992)

Solution:

Here $Z_1 = 35, Z_2 = 57$

$A_1 = 72, A_2 = 162$

$$U = \frac{(Z_1\ e)\ (Z_2\ e)}{4\pi \in_0 r}$$

$$U = 196 \text{ MeV} = 196 \times 10^6 \times 1.6^6 \times 1.6 \times 10^{-19} \text{ J}$$

$$U = 313.6 \times 10^{-13} \text{ J}$$

$$r = \frac{Z_1 Z_2\ e^2}{4\pi \in_0 U}$$

$$r = \frac{35 \times 72 \times (1.6 \times 10^{-19})^2 \times 9 \times 10^9}{313.6 \times 10^{-13}}$$

$$\mathbf{r = 1.466 \times 10^{-14} m} \qquad ...(i)$$

(ii) $R_1 = r_0\ A_1^{1/3}$

$$R_2 = r_0\ A_2^{1/3}$$

$$R = R_1 + R_2 = r_0 \left[A_2^{1/3} + A_2^{1/3}\right]$$

$$R = 1.2 \times 10^{-15}\ [(72)^{1/3} + (162)^{1/3})]$$

$$= 1.2 \times 10^{-15}\ [4.16 + 5.45]$$

$$= \mathbf{1.27 \times 10^{-14} m}$$

Therefore r is greater than R.

Example 24:

Find the energy (in eV) and wavelength of K_α - X-ray of Co-27.

(GNDU, 1992)

Solution:

$$R = 1.09678 \times 10^7/\text{m}$$

For Co, $Z = 27$

$$Z_{eff} = (Z - 1) = 26$$

$$\bar{v} = (Z - 1)^2 R \left[\frac{1}{n_1^2} - \frac{1}{n_2^2}\right]$$

$$\bar{v} = (26)^2 \times 1.09678 \times 10^7 \left[\frac{1}{1} - \frac{1}{4}\right]$$

$$\bar{v} = 5.56 \times 10^9 \text{ per metre}$$

$$\frac{1}{\lambda} = 5.56 \times 10^9 \text{ per metre}$$

(i) $\lambda = 1.799 \times 10^{-10}$ m

$$\lambda = \mathbf{1.799\ \text{Å}}$$

Energy $= \dfrac{hc}{\lambda}$

$$= \frac{6.624 \times 10^{-34} \times 3 \times 10^8}{1.799 \times 10^{-10}}$$

$$= \frac{6.624 \times 10^{-34} \times 3 \times 10^8}{1.799 \times 10^{-10} \times 1.6 \times 10^{-19}} \text{ eV}$$

$$= \mathbf{6.9 \times 10^3\ eV.}$$

Example 25:

Calculate the wavelength of the K_α line from a copper target using Moseley's law. Rydberg constant for hydrogen = 1.09687×10^7 per metre.

(Gauhati, 1997)

Solution:

For Copper $Z = 29$

$$Z_{eff} = (Z - 1) = 28$$

$$\bar{v} = (Z-1)^2 \; R\left[\frac{1}{n_1^2} - \frac{1}{n_2^2}\right]$$

$$\bar{v} = (28)^2 \times 1.09678 \times 10^7 \left[\frac{1}{1} - \frac{1}{4}\right]$$

$$\bar{v} = 6.449 \times 10^9 \text{ per metre}$$

$$\frac{1}{\lambda} = 6.449 \times 10^9 \text{ per metre}$$

$$\boldsymbol{\lambda = 1.55 \times 10^{-10} m}$$

$$\boldsymbol{\lambda = 1.55 \text{ Å}.}$$

Example 26:

The experimental value of Bohr magneton is 9.21 × 10^{-24} SI units and Planck's constant h = 6.6 × 10^{-34} joule-second.

Calcuate the value of e/m of an electron.

Solution:

Here $\beta = 9.21 \times 10^{-24}$ units

$h = 6.6 \times 10^{-34}$ joule-second

$$\beta = \frac{eh}{4\pi m}$$

or
$$\frac{e}{m} = \frac{4\pi\beta}{h} = \frac{4 \times 22 \times 9.21 \times 10^{-24}}{7 \times 6.6 \times 10^{-34}}$$

$$\frac{e}{m} = \mathbf{1.755 \times 10^{11} \; C/kg.}$$

Example 27:

What is Bohr Magneton? Calculate its value.

(Delhi (Hons.), 1996)

Solution:

Bohr magneton

$$\beta = \frac{eh}{4\pi m}$$

Here $e = 1.6 \times 10^{-19}$ C

$h = 6.624 \times 10^{-34}$ J-s

$m = 9.1 \times 10^{-31}$ kg

$$\beta = \frac{1.6 \times 10^{-19} \times 6.624 \times 10^{-34}}{4 \times 3.142 \times 9.1 \times 10-31}$$

$\beta = 9.27 \times 10–24$ SI Units.

Example 28:

Calculate the wavelength separation between the two component lines which are observed in the normal Zeeman effect. The magnetic field used is 0.4 weber/m^2, the specific charge = 1.76×10^{11} coulombs/kg and λ = 6000 Å. **(Delhi (Hons.), 1999)**

Solution:

Here $\lambda = 6000$ Å $= 6000 \times 10^{-10}$m

or $\lambda = 6 \times 10^{-7}$m

$$\frac{e}{m} = 1.76 \times 10^{11} \text{ coulombs/kg}$$

$B = 0.4$ weber/m^2

$c = 3 \times 10^8$ m/s

The wavelength separation,

$$\lambda_1 - \lambda_2 = 2\,(d\lambda) = \frac{2Be\lambda^2}{4\pi mc}$$

$$\lambda_1 - \lambda_2 = \frac{Be\lambda^2}{4\pi mc}$$

$$\lambda_1 - \lambda_2 = \frac{0.4 \times 1.76 \times 10^{11} \times (6 \times 10^{-7})^2}{2 \times 3.14 \times 3 \times 10^8}$$

$\lambda_1 - \lambda_2 = 1.335 \times 10^{-11}$m = 0.1335 Å.

Example 29:

A 1 MeV proton is sent aganist a gold leaf (Z = 79). Calculate the distance of the closest approach for aq head on collison.

(Rajastan, 1994)

Solution:

$$d = \frac{Ze^2}{4\pi\epsilon_0\left(\frac{1}{2}\right)m\upsilon^2} = \frac{Ze^2}{(4\pi\epsilon_0)U}$$

$$U = 1\ \text{MeV} = 10^6 \times 1.6 \times 10^{-19}\ \text{J} = 1.6 \times 10^{-13}\ \text{J}$$

$$e = 1.6 \times 10^{-19}\ \text{C}$$

$$Z = 79$$

$$d = \frac{79 \times (1.6 \times 10^{-19}) \times 9 \times 10^9}{1.6 \times 10^{-13}}$$

$$\mathbf{d = 1.137 \times 10^{-13} m.}$$

Example 30:

An α particle of energy 2×10^{-13} J is scattered of an aluminium atom through an angle of 90°. Calculate the distance of closest approach.

Atomic number of Al is 13, $e = 1.6 \times 10^{-19}$ coulomb and $4\pi\epsilon_0 = 1 \times 10^{-10}$.

(Delhi (Hons.) 1995)

Solution:

$$d = \left(\frac{2Ze^2}{4\pi\varepsilon_0 m\upsilon^2}\right)\left(1 + \operatorname{cosec}\frac{\theta}{2}\right)$$

Here

$$Z = 13$$

$$e = 1.6 \times 10^{-19}\ \text{C}$$

$$\frac{1}{2} m\upsilon^2 = 2 \times 10^{-13}\ \text{J}$$

$$m\upsilon^2 = 4 \times 10^{-31}\ \text{J}$$

$$\theta = 90°$$

$$\operatorname{cosec}\frac{90°}{2} = \sqrt{2} = 1.414$$

$$4\pi\varepsilon_0 = 1 \times 10^{-10} \text{ (given)}$$

$$d = \left[\frac{2 \times 13 \times (1.6 \times 10^{-19})^2}{10^{-10} \times 4 \times 10^{-13}}\right](1 + 1.414)$$

$$\mathbf{d = 4.169 \times 10^{-14} \ m.}$$

Example 31:

The Zeeman components of a 5461 Å spectral line are 0.417 Å apart when the magnetic field is 1.5 T. From these data, calculate the value of e/m of an electron. **(Delhi (Hons.), 1997)**

Solution:

$$\lambda_1 - \lambda_2 = \frac{Be\lambda^2}{2\pi mc}$$

Here $\lambda_1 - \lambda_2 = 0.417 \text{ Å} = 0.417 \times 10^{-10} \text{ m}$

$$\lambda = 5461 \text{ Å} = 5461 \times 10^{-10} \text{ m}$$

$$B = 1.5T$$

$$C = 3 \times 10^8 \text{ m/s}$$

$$\therefore \quad \frac{e}{m} = \frac{(\lambda_1 - \lambda_2)\, 2\pi c}{B\lambda^2}$$

$$\frac{e}{m} = \frac{(0.417 \times 10^{-10}) \times 2 \times 3.14 \times 3 \times 10^8}{1.5 \times (5461 \times 10^{-10})^2}$$

$$\frac{e}{m} = \mathbf{1.756 \times 10^{11} \ C/kg.}$$

Example 32:

In observing the Raman spectrum of a sample, using 2537 Å as the exciting line, one gets stoke line at 2683 Å, Deuce the Raman shift in cm ^{1} units. compute the wavelength in Å for corresponding stokes and antistokes lines if the exciting line is 5461 Å. **(IAS, 1999)**

Solution:

(i) Here $\upsilon_s = \upsilon_0 = \upsilon_m$

$$\upsilon_m = v_0 = v_s$$

$$= \frac{c}{\lambda_0} - \frac{c}{\lambda_s}$$

$$\frac{c}{\lambda_m} = c\left[\frac{1}{\lambda_0} - \frac{1}{\lambda_s}\right]$$

$$\frac{1}{\lambda_m} = \frac{\lambda_s - \lambda_0}{\lambda_0 \lambda_s}$$

Here $\lambda_0 = 2537 \text{ Å} = 2537 \times 10^{-10}$ m

$\lambda_s = 2683 \text{ Å} = 2683 \times 10^{-10}$ m

$$\frac{1}{\lambda_m} = \frac{2683 \times 10^{-10} - 2537 \times 10^{-10}}{2537 \times 10^{-10} \times 2683 \times 10^{-10}}$$

$$\frac{1}{\lambda_m} = 2.145 \times 10^5 \text{ per metre}$$

Raman Shift,

$$\upsilon_m = \frac{1}{\lambda_m} = 2.145 \times 10^5 \text{ per metre}$$

$$\mathbf{= 2.145 \times 10^3 \text{ per cm}}$$

(i) Here $\lambda_0 = 5661 \text{ Å} = 5461 \times 10^{-10}$ metre

$$\frac{1}{\lambda_m} = 2.145 \times 10^5 \text{ per metre}$$

$\lambda_s = ?$

$\upsilon_s = \upsilon_0 - \upsilon_m$

$$\frac{c}{\lambda_s} = \frac{c}{\lambda_0} - \frac{c}{\lambda_m}$$

$$\frac{1}{\lambda_s} = \frac{1}{\lambda_0} - \frac{1}{\lambda_m}$$

$$\frac{1}{\lambda_s} = \left(\frac{1}{5461 \times 10^{-10}}\right) - 2.145 \times 10^5$$

$$\frac{1}{\lambda_s} = 18.31 \times 10^5 - 2.145 \times 10^5$$

$$= 16.165 \times 10^5$$

$$\lambda_s = \frac{1}{16.165 \times 10^5}$$

$$= \mathbf{6186 \times 10^{-10} m}$$

$$= \mathbf{6186\ Å.}$$

(ii) Similarly, for antistokes line

$$\upsilon_{as} = \upsilon_0 + \upsilon_m$$

$$\frac{c}{\lambda_{as}} = \frac{c}{\lambda_0} + \frac{c}{\lambda_m}$$

$$\frac{1}{\lambda_{as}} = \frac{1}{\lambda_0} + \frac{1}{\lambda_m}$$

$$\frac{1}{\lambda_{as}} = \left(\frac{1}{5461 \times 10^{-10}}\right) + 2.145 \times 10^5$$

$$\frac{1}{\lambda_{as}} = 18.31 \times 10^5 + 2.145 \times 10^5$$

$$\frac{1}{\lambda_{as}} = 20.455 \times 10^5$$

$$\lambda_{as} = \frac{1}{20.455 \times 10^5}$$

$$= \mathbf{4889 \times 10^{-10} m}$$

$$= \mathbf{4889\ Å.}$$

Example 33:

In an experiment with benzene the wavelengths of a pair of Stokes and anti-Stokes lines were measured to be 4554 Å and 4178 Å respectively. Calculate the wavelength of the corresponding infra-red absorption line.

(Calcutta, 1992)

Solution:

Here $\upsilon = \upsilon_0 \pm \upsilon_m$

$\therefore$ $\upsilon_1 = \upsilon_0 + \upsilon_m$

and $\upsilon_2 = \upsilon_0 - \upsilon_m$

$\therefore$ $\upsilon_1 - \upsilon_2 = 2\upsilon_m$

$$\frac{c}{\lambda_1} - \frac{c}{\lambda_2} = \frac{2c}{\lambda_m}$$

$$\lambda_m = \left[\frac{2\lambda_1\lambda_2}{\lambda_2 - \lambda_1}\right]$$

Here $\lambda_2 = 4554$ Å

$\lambda_1 = 4178$ Å

$$\lambda_m = \frac{2 \times 4554 \times 4178}{4554 - 4178}$$

$$\lambda_m = \mathbf{1.012 \times 10^5} \textbf{ Å.}$$

Example 34:

Calculate the frequency of Oscillation of a hydrogen molecule if its force constant is 4.8×10^2 N/m and mass of hydrogen atom $= 1.67 \times 10^{-27}$ kg. **(GNDU, 1991)**

Solution:

$$\upsilon = \frac{1}{2\pi}\sqrt{\frac{K}{\mu}}$$

Here μ is the reduced mass of the system

$$\frac{1}{\mu} = \frac{1}{m} + \frac{1}{m} = \frac{2}{m}$$

$$\mu = \left(\frac{m}{2}\right)$$

Here $K = 4.8 \times 10^2$ N/m

$$\mu = \frac{1.67 \times 10^{-27}}{2} = 0.835 \times 10^{-27} \text{ kg}$$

$$\therefore \quad \upsilon = \frac{1}{2\pi}\sqrt{\frac{4.8 \times 10^{2}}{0.835 \times 10^{27}}}$$

$$\upsilon = \mathbf{1.2 \times 10^{-14}\ Hz.}$$

EXERCISES

1. What is the effect of magnetic field on orbital electron motion. Explain the normal Zeeman Effect **[Delhi (Hons.), 1982]**
2. Calculate the value of Bohr's Magneton. **[Delhi (Hons.), 1982]**
3. Derive Rutherford's scattering formula and estimate the "size of the nucleus. **[Rajasthan, 1980]**
4. Give the theory of the Rutherford scattering of a particles and show that

$$N\alpha = \frac{1}{\sin^4\left(\frac{\varphi}{2}\right)}$$

5. Explain the term "quantization of angular momentum". State Wilson Sommerfeld Quantization Rule.
6. Give briefly Wilson Sommerfeld quantization theory of elliptical orbits. Explain how the theory helps in explaining the fine structure of spectial lines.
7. Discuss the effect of the variable mass of the electron on the energy of the electron in the orbit.
8. Discuss the correction for the finite mass of the nucleus.
9. Explain Raman Effect. **[Delhi, 1983]**
10. Give briefly the experimental arrangement to study Raman effect.
11. On the basis of Bohr's Theory, deduce expressions for the radius, frequency and energy of an orbit of hydrogen atom. **[Delhi (Hons.) (1980)]**
12. What is meant by ionization potential. Calculate its value in the case of hydrogen atom. **[Delhi, 1980]**

13. State and explain the fundamental postulates of Bohr's Theory of the hydrogen atom. Derive an expression for the energy of an electron in the nth orbit. Explain the various series in the hydrogen spectrum on the basis of Bohr's Theory.
[Delhi, 1981, 83]

14. State Bohr's postulates for the theory of the hydrogen atom. Derive an expression for the energy of this atom when the electron is in the nth orbit. What is binding energy?
[Delhi, 1982]

15. Explain hydrogen spectrum on the basis of Bohr's Theory.
[Delhi, 1982]

16. Outline Bohr's theory of the hydrogen atom. What is the justification for assuming quantized stationary orbit? Explain the formation of Balmer lines. **[Delhi (Hons.), 1982]**

17. Explain correspondence principle on the basis of Bohr's Theory.
[Delhi (Hons.), 1982]

18. What do you understand by energy level diagram. Compare the energy levels of hydrogen with those of singly ionized Helium.
[Delhi, 1980, 1983]

19. Calculate the velocity of the electron in the first orbit of the hydrogen atom. **[Delhi, 1980]**

20. Discuss normal Zeeman Effect. **[Delhi, 1982]**

21. State and explain Bohr's postulates regarding the structure of the hydrogen atom. Derive an expression for the energy of the electron in the first orbit is (1/137)c, where c is the velocity of light. **[Delhi, 1992]**

22. (a) Using Bohr's Theory, obtain an expression for the Rydberg constant for a singly ionised helium (He^+) atom, taking into account motion of the nucleus.
    (b) Calculate the radius of the hydrogen atom. Show that the velocity of the electron in the first orbit is (l/137)c, where c is the velocity of light. **[Delhi (Hons.), 1992]**

23. (a) Discuss the concept of spin and space quantisation as introduced in the vector atom model.
    (b) What is a Bohr's magneton? Find an expression for it.
[Delhi, (Hons.), 1992}

24. How is Raman Effect theoretically explained?
25. Describe briefly o particle scattering experiment. **[Delhi, 1983]**
26. Assuming p = b/2 cot (φ/2) obtain (for Rutherford theory of a α-particle scattering) an expression for the number of particles scattered on the a unit area of a screen at a given distance from the foil at a given angle f from the original direction.

    [Delhi, 1983]
27. What are the main assumptions of Rutherfords's α-particle scattering formula? Assuming p = b/2 cot (φ/2) obtain an expression for the number of particles scattering on to a unit area of a screen at a given angle from the foil at a given angle from the original direction.

    [Delhi, 1985; Delhi (Hons.), 1985]
28. Give an account of Bohr-Sommerfeld model of elliptical electron orbits of an atom and show that the total energy of an electron moving in Sommerfeld orbit of the same quantum number is the same and identical with that of the corresponding Bohr orbit.

    [Delhi (Hons.), 1984; Delhi, 1985]
29. Describe and explain Stern-Gerlach experiment and indicate its importance in atomic physics. **[Delhi (Hons.), 1984]**
30. Give an account of various quantum numbers used to specify completely the state of an electron in an atom.

    (Delhi (Hons.), 1984]
31. State and explain Pauli's Exclusion Principle. Apply it to acco ·.ıt for the periodic classification of elements.

    [Delhi (Hons.), 1984]
32. Show that the quantum and classical concepts of the hydrogen atom make identical predictions in the limit of large quantum numbers. **[Delhi (Hons.), 1985]**
33. Give an account of normal Zeeman Effect.

    [Delhi (Hons.), 1985]
34. The wavelength of the second member of Balmer series of the hydrogen is 4861 Å. Calculate the wavelength of its first number.
35. Calculate the value of Planck's Constant from the following data.

$e = 1.6 \times 10^{-19} C$

$m = 9 \times 10^{-31} kg$

$\epsilon_0 = 8.85 \times 10^{-12}\ C^2/N - m^2$

Work function for hydrogen, $\varphi = 13.5 eV$.

36. A beam of electrons is used to bombard gaseous hydrogen. What is the minimum energy (in electron volts) the electrons must have if the first member of the Balmer Series corresponding to a transition from n = 2 state n = 1 state.

37. The experimental value of Bohr magneton is 9.21×10^{-24} SI units. Calculate the specific charge of the electron.

38. (a) Comment the need of vector atom model. Describe the vector model of atom and explain different quantum numbers associated with it. **[Kanpur, 1990]**

    (b) Stating principle describe the Stern and Gerlach experiment and explain the importance of the results obtained. **[Kanpur, 1990]**

39. (a) What are shortcomings of Bohr-Sommerfeld Theory? How are these overcome in Vector Atom-Model? Explain the fine structure of D-lines in the Sodium spectrum on the basis of Vector Atom Model. **[Kanpur, 1991]**

    (b) What is Zeeman Effect? Describe the experimental arrangement for studying the effect. How Vector Atom Model helps to the explanation of Zeeman Effect? **[Kanpur, 1991]**

40. Derive an expression for the Magnetic Moment of Hydrogen Atom. What is Bohr Magneton and calculate its value. **[Kanpur, 1991]**

41. Describe the vector atom model. Explain the various quantum numbers, coupling schemes, and selection rules associated with it. **[Calicut, 1992]**

42. (a) What is space quantisation? Explain how the Stern-Gerlach experiment established this.

    (b) Explain the Lande Vector Model of the atom. Derive an expression for the magnetic moment of an atom in terms of the Lande g factor. **[Gauhati, 1991]**

43. What are Stokes and anti-Stokes lines in Raman scattering? How is Compton scattering different from Raman scattering? Explain from elementary quantum theory why the anti-Stokes line become more pronounced as the temperature of the scattering substance in increased. **[Calcutta, 1992]**

44. (a) (i) What is LS Coupling?

    (ii) If one of the two valency electrons is in the 4s state and the other in the 4p state what new spectral terms may result? Indicate by notations.

    (iii) Explain why the lines of the singlet series exhibit the normal Zeeman Effect.

    (b) Describe the Stern Gerlach experiment. Explain the importance of the results of the experiment.

    [Calcutta, 1992]

45. Derive Rutherford's formula for the scattering of alpha particles by a heavy nucleus of charge Ze. Why is it necessary that the scattering foil be thin? Describe briefly how the formula was tested in the laboratory. **[Calcutta, 1992]**

46. Explain the vibrational rotational spectra of diatomic molecules. State the importance of Mosley's law. **[Osmania, 1992]**

47. What is Normal Zeeman effect? Give the quantum mechanical explanation of that. **[Osmania, 1992]**

48. What is the difference between electronic and mu-mesonic orbits in atom? **[GNDU, 1992]**

49. Give for the acceptance of the neutron-proton hypothesis of constitution of the nucleus. Why don't the electrons exist inside the nuclei? **[GNDU, 1992]**

50. Give the main assumptions of the liquid drop model of the nucleus. Write down various terms in the semi-empirical mass formula giving their significance. **[GNDU, 1992]**

51. What are magic number nuclei? Give reasons in support of shell structure of nuclei. State the coupling rules. **[GNDU, 1992]**

52. What is the qualitative difference between single particle and independent particle models? **[GNDU, 1992]**

53. (a) What are symmetric and antisymmetric wave function?

Obtain symmetric and antisymmetric wave functions of Helium atom.

(b) What is meant by LS coupling? Under what conditions does it hold.

(c) Find S, L and J values that correspond to each of following states: $^1S_6$, $^3P_2$, $^2D_{3/2}$, and $^2S_{1/2}$ **[GNDU, 1992]**

54. What is normal and anomalous Zeeman Effect? **[GNDU, 1992]**

55. (a) What is meant by fine structure of spectral lines? How is it explained?

(b) What is meant by space quantization of electron spin? Describe, briefly the experiment which established the space quantization of spin.

(c) State and explain Pauli's Exclusion Principle. **[GNDU, 1992]**

56. What are stokes and anti-stokes lines in Raman Spectra ? **[GNDU, 1992,**

57. (b) Determine the longest wavelength found in the Paschen series of hydrogen. Neglect the effect of nuclear motion. **[GNDU, 1992]**

58. Explain LS and j-j coupling on the basis of vector atom model. **[GNDU, 1992]**

59. What is Moseley diagram for X-ray spectra? Interpret Moseley diagram on the basis of Bohar's Theory. **[GNDU, 1991]**

60. What is Raman Effect? Give quantum mechanical explanation of Raman Effect. Define Stokes and Anti-Stokes lines. **[GNDU, 1991]**

61. Give the experimental evidences for the magic numbers and discuss the successes of nuclear shell model. **[GNDU, 1991]**

62. What is meant by nuclear motion and reduced mass? Explain the importance of reduced mass. **[Bombay, 1991]**

63. Determine the longest wavelength found in the Balmer Series of the hydrogen atom. Neglect the effect of nuclear motion. What is meant by hydrogenic atoms ? **[Bombay, 1991]**

64. Calculate the magnetic moment of electron in the presence of spin orbit interaction. Comment on the result.

[Bombay, 1991]

65. Explain the vector model for LS coupling of two valence electrons. Give an example for finding spectral terms in the case of LS coupling. Find the fine structure term in case of LS coupling by calculating the relevant factor. **[Bombay, 1991]**

66. Explain normal Zeeman Effect. Obtain an expression for the frequency shift in Zeeman pattern for two electron atom.

[Bombay, 1991]

67. What is Moseley diagram for X-ray spectra? Interprete Moseley diagram on the basis of Bohr's Theory. **[Bombay, 1991]**

68. What is Raman Effect? Describe an experimental arrangement to study Raman Spectra. Give quantum mechanical explanation of Raman Effect. State how Raman different from fluorescence and infra-red spectra. **[Bombay, 1991]**

2

Nuclear Detectors

2.1 INTRODUCTION

The detection of radiation or particles in nuclear physics is bases on the fact that charged particle. While passing through matter leaves along its path and counted. Most of the detectors measure the ionization produced by the passage of a charged particle through a suitable material. When an electric field is maintained across the material, the ion will be set in motion resulting an ionization current.

The various instruments which are used in the detection of nuclear radiations are as classified:

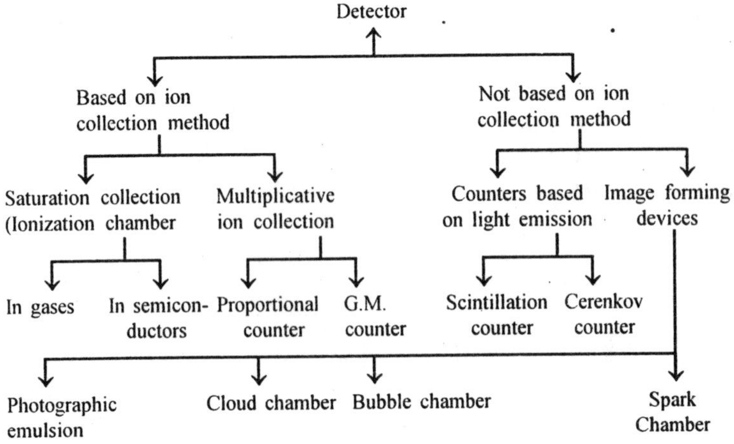

Detector are also as *electric* (Ionisation chamber, Proportional counter, G.M. counter, semiconductor counter), and *optical* (scintillation counter, Cerenkov counter, Photographic emulsion, Cloud chamber, Bubble chamber, Spark chamber). The same detector can be used in studies of various radiations. Interactions of the radiations in the detector may depend on different phenomena.

2.2 IONIZATION CHAMBER

An ionization chamber, in its simplest form consist of vessel usually field with gas and into which it inserted an insulated electrodes. The gas becomes ionized when charged particles pass through it. If no voltage is applied to the container there will be no directing force acting the ions and no current will flow through the gas. If we applied a small potential difference to the electrodes, the –ve ions (or electrons) will be attracted to the anode and the +ve ions will go to the cathode. At the surface of the anode there will be a high concentration of negative ions. Recombination in this region will be a small because the rate of recombination is proportional to the product n^+n^- and here n^+ is very small. All positive ions produced in this region move out promptly towards the cathode. Therefore, the recombination will be maximum at the center of the space. A small current will flow through a resistor connecting the electrodes externally As the voltage between the electrodes is increased, the ions will move faster, the number of recombination will be minimum and the pulse size will thus be increased. The pulse size increases with the voltage, until it reaches to a value corresponding to the no recombination of ions. This value of current is called the *saturation current* and the chamber is said to reach in *saturation.*

The ionisation chamber works in the region of constant pulse size. In this region the applied voltage is high enough to prevent recombination of ions and low enough to prevent multiplication.

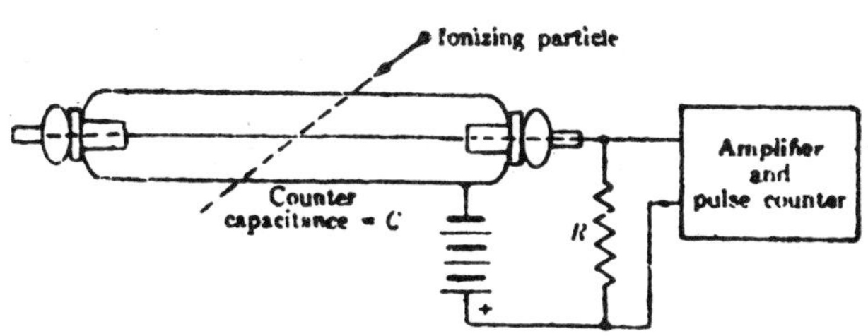

Fig. 2.1 : Ionization chamber.

In the initial stages, the gold-leaf electroscopes of great sensitivities or the quadrant electrometers were used. The next step in the development was a modification of the electroscope, designed by Lauritsen, where the gold leaf was replaced by a tiny metal coated quartz fibre. It could not give quick response to particles that follow one another in rapid succession. This difficulty was overcome to a great extent by a device known as linear or dc amplifier, which increases, without distorting, the feeble ionisation current from a single particle to a value which can easily register on an oscillograph or an electrical counter, dc amplifiers are more susceptible to disturbance and drift and more difficult to arrange with several successive stages of amplification than those designed for amplifying alternating currents.

Ionization chamber instruments may be grouped into two categories, depending upon the value of the time constant RC of the system relative to the frequency of arrival of the ionizing events. If RC is long compared to time between ionizing events a steady state is reached and a *direct current* may be measured. On the other hand, if RC is small each pulse produced by an ionizing event may be detected separately. These two types are also known as *integrating and non-integrating* respectively and the instrument as current and pulse ionization chambers.

Let a track of n ion pairs is formed parallel to and at a distance x_0 from the axial wire The electrodes are parallel and separated at a distance d. A voltage V_0 is applied to the electrodes, so that electrons and positive ions drift in opposite directions in a uniform electric field V_0/d. Let us first assume that the time constant RC (where C = capacitance of the chamber + input capacitance of the amplifier) is much greater than the collection time of ions. So that the current through R may begin as soon as the electrons and +ve ions move apart in the chamber.

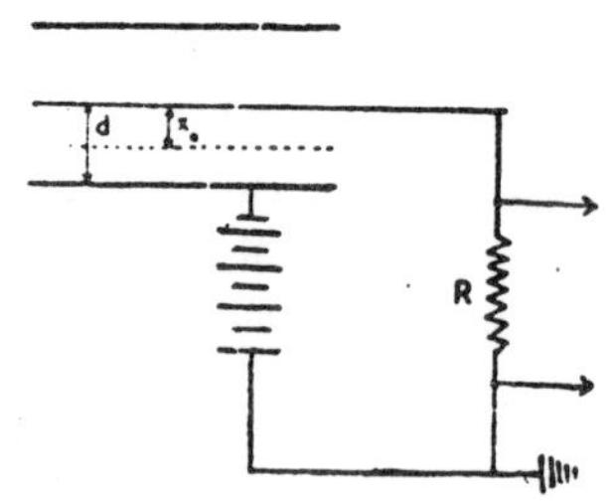

Fig. 2.2 : Ionization in an ionization chamber.

The change of potential at the positive electrode at the time t, due to a track of n ion pairs formed at t = 0 is given as

$$V(t) = n\,[q_e(t) + q_p(t)]/C, \qquad ...(1)$$

where $q_e(t)$, the charge induced on the positive electrode at time t by a single electron, and is given as

$$q_e(t) = -e.\frac{\text{P.D. between the electron at time t and the -ve electrode}}{\text{P.D. between the electrodes}}$$

$$= -e.\frac{d - x_e(t)}{d} = -e\frac{[d - (x_0 - v_e t)]}{d} \qquad ...(2)$$

and $q_p(t)$, the charge induced on the positive electrode at time t by a single positive ion, is given as

$$q_p(t) = +e.\frac{\text{P.D. between the +ve ion at time t and the -ve electode}}{\text{P.D. between the electrodes}}$$

$$= e.\frac{d - x_p(t)}{d} = e\frac{[d - (x_0 - v_p t)]}{d} \qquad ...(3)$$

The relations (2) and (3) are only true for parallel plate geometry in which potential gradient is uniform.

From eq. (2), (3) and (1) we get.

$$V(t) = -ne(v_e t) + v_p t)/Cd. \qquad ...(4)$$

As the electron drift velocities v_e are about a thousand times greater than the heavy ion velocities v_p, hence, for the time between 0 and t_e is the electron collection time $(x_0 = v_e t_e)$, the above relation transform to

$$V(t) = -nev_e t/Cd. \qquad ....(5)$$

For $t > t_e$, there is an abrupt decrease in the rate of rise of potential, the new rate being v_p/v_e times the original rate. The change of potential

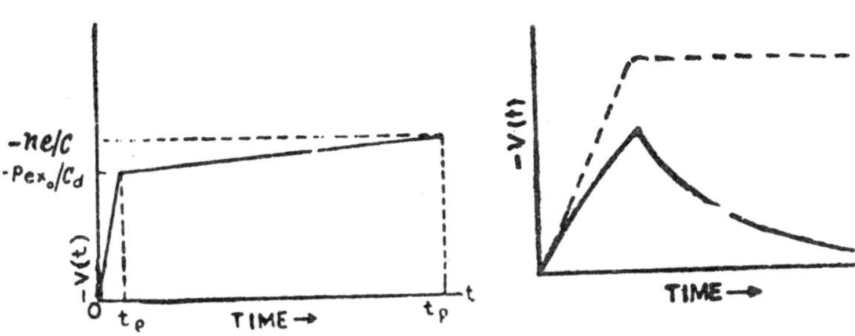

Fig. 2.3 : (a) Shape of potential pulse. (b) Pulse cliping.

becomes—ne/C when $t = t_p$, the time in which positive ions reach the negative electrode. This limit is proportional to n and is independent of where the ions were formed in the chamber. The shape of the collector potential curve does depend upon where the ions are formed.

With sufficient amplification, a large pulse will appear at the amplifier output terminal. To have the height of the output pulse proportional to the amount of ionization produced by the particle in the chamber, two kinds of proportional or linear amplifiers can be used. One, known as slow *amplifiers*, in which the shortest time constant is chosen long compared with the drift time of the +ve ions, have several disadvantages. The others, known as fast amplifiers, in which time constant RC is made short enough so that the induced potentials by the +ve ions are not of interest and clip the pulses Thus the slow linear rise, after the collection of electrons, is removed and the height of the recorded pulses depends on $V(t_e)$ and not on $V(t_p)$. If the complete range of an *ionizing particle is confined to the region between the negative electrode and grid*, the grid screens the +ve electrode from the electrons and +ve ions. Electrons, while drifting through the grid, induce a change of potential at the +ve electrode. As positive ions are screened from the +ve electrode by the grid, hence have no effect. Thus, the pulse height increases due to the induction effect of the moving electrons. When all the electrons are collected it attains a maximum value given by

$$V_{max} = -\frac{ne}{C} \qquad \text{...(6)}$$

It is also independent of the position and orientation of the ionization track.

The height of the amplified pulses from the ionization chamber

$$V = -\frac{AEe}{WC} \qquad \text{...(7)}$$

where A is the voltage gain of the amplifier, E the energy of the ionizing particle, e the electronic charge and W the mean energy required to produce an ion pair in the gas filled in the chamber.

2.3 SOLID-STATE DETECTORS

Crystal counters are essentially the ionization chambers using solid dielectrics between parallel plate electrodes. The first solid state nuclear detector was given by P.J. Van Heerden in 1945. A crystal of silver chloride (*intrinsic semiconductor*) was placed between two electrodes to which a voltage was applied. When the crystal was exposed to the radiations, the some of the valence band electrons got energy and entered the conduction band, leaving the (*positive*) holes. Under an electric field the electrons moved toward the negative electrode and holes moved toward the positive electrode and a pulse of current equivalent to the number of electrons transferred to the conduction band was produced. The first junction type detector was demonstrated by K.G. Mckay in the United States in 1949. A collimated beam of α-particles was allowed to fall on a n-p-germanium junction. He showed that nearly whole of energy of *α-particles absorbed in it was utilized in producing electron hole pairs and the average energy required to produce one electron hole pair was about 3eV which was very small in comparison to 34eV, the energy required to form an ion pair in a gas.*

J.D. Mayer and B. Gossick showed that *the size of the output pulse was proportional to the alpha-particle energy. In this respect, a solid state detector behaves like an ideal pulse type ionization chamber.* Most semi-conductor junction type detectors fall into three categories, according to the formation of junction.

1. Diffused Junction Detector

In this detector, a thin layer of the n-type is formed on the surface of a p-type silicon (or *germanium*) crystal, by allowing a donor impurity (*phosphorous*) to diffuse into one face of the crystal at a fairly high temperature. There is a tendency for electrons to diffuse from the n-region (*where they are in excess*) into p-region (where the electrons are in deficit). An equilibrium is reached when the fermi-level is constant throughout the material. Due to this an intermediate volume around the interface will be cleared from carriers of both signs and this region is called the *depletion layer*. This layer is of low conductivity and high field, hence offers ideal conditions for the detection of ionizing radiation.

When the negatively charged electrons diffuse away from the n-type region, they leave the material positively charged. Similarly, the filling of the hole in the p-region generates a corresponding negative charge. The small potential difference of about a volt, established in this manner, is enhanced by an additional potential of several hundred volts, referred to as a reverse-bias voltage. In the presence of this electric field the positive holes in the p-type side are pulled towards the –ve electrode, the electron towards the p-n junction and thus, a depletion layer is created in the p-type material. The result is an extremely low leakage current. The thickness of the depletion layer increases with the reverse-bias voltage and the limit is set by electrical break-down in the semi-conductor.

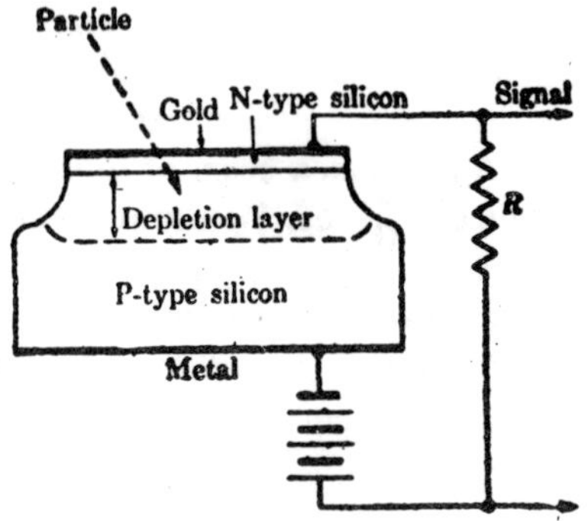

Fig. 2.4 : A p-n junction detector.

Thickness of Depleted Region

The thickness of the depleted region and the capacitance can be calculated by Poisson's equation. Assuming the variation of V in the direction ⊥ to the surface separating the n- and p-regions, we have

$$\frac{d^2V}{dx^2} + \frac{\rho}{\epsilon} = 0 \qquad ...(8)$$

ρ the volume charge density given by

$$\rho = q(p + N_{\&} - n - N_a) \quad ...(9)$$

where p, n, $N_{\&}$ and N_a denote, respectively, the concentrations of holes, electrons, donors and acceptors. In the completely depleted region n and p are negligible, hence we have

$$\rho = q\,N_{\&} \quad 0 < x < x_1$$

$$\rho = -q\,N_a \quad -x_2 < x < 0$$

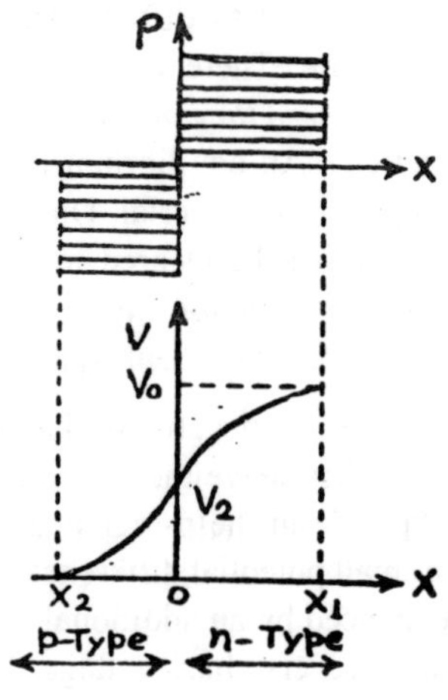

Fig. 2.5 : Charge and potential disturbance.

After successive integrations of eqn. (8), we get

$$V = -\frac{qN_d}{\in}\frac{x^2}{2} + \frac{qN_d}{\in} xx_1 + V_2$$

$\therefore \quad V(x_1) = V_0 = qN_d x_1^2/2\in + V_2$

or $\quad V_0 - V_2 = qN_d x_1^2/2\in$ ...(10)

Similarly $V(-x_2) = 0$ and $V_2 = qN_a \dfrac{x_2^2}{2\in}$ ...(11)

Hence the potential barrier height becomes

$$V_0 = q(N_d x_1^2 + N_a x_2^2)/2\in \qquad ...(12)$$

As charge is conserved, hence $N_d x_1 = N_a x_2$ and

$$V_e = \frac{q}{2\in}\left(1 + \frac{N_a}{N_d}\right) N_a x_2^2 = \frac{q}{2\in}\left(1 + \frac{N_d}{N_a}\right) N_d x_1^2$$

The thickness of the depletion layer

$$X_0 = x_1 + x_2 = \left[\frac{2\in V_0}{q}\frac{N_a + N_d}{N_a N_d}\right]^{1/2} \qquad ...(13)$$

If p-type region is more heavily doped, than the other region, we have $N_a > N_d$ and $x_2 < x_1$. Hence, we have

$$X_0 \simeq x_1 = \left[\frac{2\in V_0}{qN_d}\right]^{1/2} = [2\in \rho_n \mu_n V_0]^{1/2} \qquad ...(14)$$

where ρ_n is the resistivity of n-type region and μ_n the electron mobility.

The same treatment will give the value of X_0 in the presence of reverse bias, as

$$X = \left[\frac{2\in(V_0 + V)}{qN_d}\right]^{1/2} = [2\in\rho_n \mu_n (V_0 + V)]^{1/2}$$

If the applied voltage $V > V_0$, we get

$$X \simeq [2\in \rho_n \mu_n V]^{1/2}, \qquad ...(15)$$

where a is a constant of a medium.

The depletion layer having area A has a capacitance C given as

$$C = \frac{\epsilon A}{X} = A\left[\frac{q\,\epsilon N_a N_d}{2(N_a + N_d)\,(V_a + V)}\right]^{1/2}$$

For $N_a > N_d$ and $V > V_0$, we have

$$C \simeq A\,(\epsilon/2\;\rho_n\mu_n\,V)^{1/2}. \qquad ...(16)$$

If ρ is in ohm cm, V in volts, X in microns (μ), and C in pf/cm^2 then we have following values:

G_e	n-type	$X = (\rho V)^{1/2}$	$C = 1.37 \times 10^4\,(\rho V)^{-1/2}$
	p-type	$= 0.65\,(\rho V)^{1/2}$	$= 2.12 \times 10^4\,(\rho V)^{-1/2}$
Si	n-type	$X = 0.5\,(\rho V)^{1/2}$	$C = 2.2 \times 10^4\,(\rho V)^{-1/2}$
	p-type	$= 0.3\,(\rho V)^{1/2}$	$= 3.7 \times 10^4\,(\rho V)^{-1/2}$

These detectors should always be kept in complete darkness, otherwise the photosensitivity will bring out a *photoelectric current.*

2. Surface Barrier Detector

It has been very useful for charged particles detection. It is made by exposing a chemically etched surface of an n-type silicon crystal to air. The etched surface is thus oxidised and this oxidation laye-acts like a very thin p-type layer.

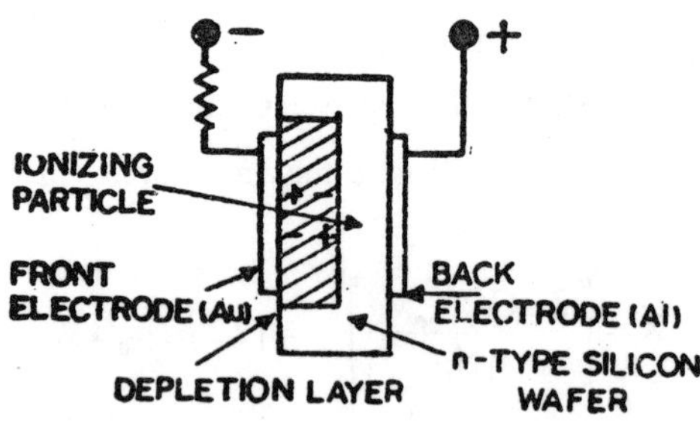

Fig. 2.6 : Surface barrier detector.

In analogy to an N-P junction, a depletion layer is formed, but it is clear that this layer only exists with in the semiconductor. The thickness of the depletion layer X_0 (or X with the reverse bias V) and the capacitance can be written as given in the case of N-P junction. Using 4000Ω silicon with the reverse bias of 1000 volts, this thickness reaches to 3 mm. In contrast to N-P junction, the temperature attained during the various stages of the construction of the surface barrier detectors does not exceed the boiling point of water, and the life time remains as that of the original material.

Detection Characteristics of Surface Barrier Detectors

(a) *Resolution:* The high energy resolution means a narrow distribution of pulse size around the mean, *i.e.,* FWHM (full width at half maxima). With monoergic radiations,, Siffert, *et al.*, found that the value of FWHM was smaller for protons and deuterons than for α-particles. The broadening of peak for α-particles was due to nuclear collisions. The FWHM is found to be ~10 keV for α-particles of several MeV. Mayer and others suggested FWHM of 2.1keV at 135°K. The cooling increases the resolution. It is thus usually necessary to cool the detector to –30°C.

(b) *Linearity:* Many experiments verify the linearity of the response to various particles with different energies. In addition to the defect related to a thin space charge region, other deviations arise for heavy particles and fission fragments.

(c) *High Conversion Efficiency:* The energy required to produce ion pair in silicon is 3.5eV. Thus, the number of pairs for a particular incident energy is about 10 times the number of pairs in air.

(d) *Differential Sensitivity:* The surface barrier detectors are relatively insensitive to neutrons or photons. They can detect charge particles in the presence of other radiations.

(e) *Small Size:* These detectors are of very small size and are important. In some cases the small sizes are a drawback.

(f) *High Speed of Response:* The carrier mobilities are high (~1500cm^{2}volt/sec) for electrons and 500 for holes. The ion pairs move only a short distance before the collection and the pulse widths are in the order of nanoseconds.

3. Lithium Ion Drifted Junction Detector

The above limitation was reduced by E.M. Pell in the United States in 1960 after introducing the lithium-ion drifted junction detector. The technique is to allow the lithium, which acts as an electron donor, to diffuse into the p-type silicon or germanium at an elevated temperature (120 to 150°C) and under reverse bias. Under these conditions the ions drift for a considerable distance into the crystal where the lithium donor exactly compensates the existing acceptor impurities in the p-region and an effective intrinsic semi-conductor layer is produced between the n- and p-regions.

Principle and Characteristics

The first step of Pells method is to form an N-P junction, by diffusing Lithium from a layer deposited on the surface of a p-type material. If a reverse bias V is applied to this diode, +vely charged Li-ions will move from n-region to p-region. The electric field E is assumed to be large enough to make drifting much more important than the diffusion. To obtain a sufficiently large mobility μ, the temperature of the material may be reached. In the case of a long drift time, the thickness

$$X \simeq (2\mu\ V^{t})^{1/2} \quad ...(17)$$

In the presence of a reverse bias, the electrons move towards N-region and holes towards P-region. A space charge is thus produced which varies linearly with distance, if recombination is neglected (Fig. 2.7b). This space charge causes a distortion of the electric field. The field becomes minimum at the centre of the intrinsic region (Fig. 2.7c), when the mobilities are equal. The distribution of Li-ions must be such that the space charge is compensated in the intrinsic region. It varies with distance as shown in (Fig. 2.7d). When the diode is cooled after the drift the thermally generated carriers become very small but the lithium distribution remains unchanged.

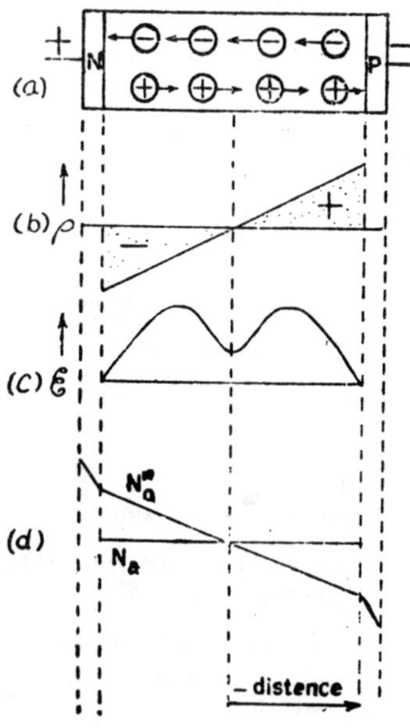

Fig. 2.7 : NIP-diode.

Lithium-drifted germanium detectors are more suitable than silicon detectors for the detection of γ-rays. It may be recalled that the photo-electric absorption cross section for γ-rays is proportional to Z^5 and therefore germanium is more efficient than silicon for γ-ray

Advantage

1. The elimination of windows and the requirement of low applied voltage to operate the counter are advantages over the gas counters.
2. These are more suitable than the grid ionization chambers coincidence experiments.
3. Very good energy resolution is possible.
4. Counting rates high as 5×10^4 counts/sec. are possible.
5. If the particle range is less than the junction thickness, the pulse height is proportional to the K.E. of the incident protons, deuterons, α-particles or any other ionizing particles.
6. These detectors have low sensitivity to gamma background radiation.

2.4 REGIONS OF MULTIPLICATIVE OPERATION

Ion current or pulse height in an ionization chamber increases above the saturation value at sufficiently high applied voltages because electrons moving in these high fields acquire enough energy to cause secondary ionization. The size of the current pulse depends on the number of initial ion pairs produced by the ionizing particles and on the applied voltage. For a given ion pairs, the pulse size is shown in Fig. 2.8 on a logarithmic scale as a function of the applied voltage.

By these results shows that the curve can be divided into six more distinct regions.

In region I, the voltage applied is so low that recombination of ion pairs takes place. As the voltage is increased there is a competition between the loss of ion pairs by recombination and the removal of ions by collection on the electrodes. The region is known as ion recombination region,.

In region II, the voltage is sufficiently high so that only a negligible amount of recombination takes place and the ion pairs move to the electrodes so rapidly that virtually every ion pair reaches the electrodes.

Thus, the pulse size remains independent of the potential as long as the voltage is increased upto the point at which secondary electrons arc produced by collision in the gas. This region is known as *ionization chamber region or saturation region.*

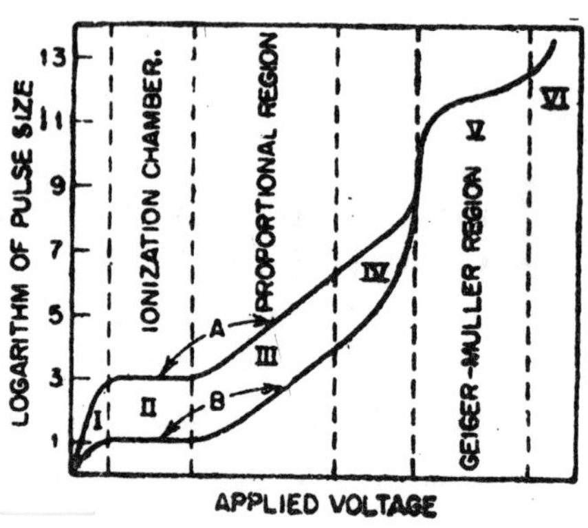

Fig. 2.8 : Variation of pulse size (log) with applied voltage.

In region III, voltage is sufficiently high so that electron approaching the central wire attains sufficient energy to produce new ions, due to collision with gas molecules. The original electron and the new one, thus, produced, are again accelerated by the field and more ions are still produced by collision. This process is cumulative and an avalanche of electrons is thus produced. The phenomenon fit frequently called the *Townsend avalanche*, in honor of J.S. Townsend who did pioneer work in this subject. The number of the ion pairs formed by collision as each electron travels toward the central wire is known as *multiplication factor M.* The voltage region in which M remains constant is called the *proportional region.* The apparatus operates in this region as a proportional counter. In this pulse height remains proportional to the amount of energy lost in the chamber by the primary ionizing particle.

If the voltage is increased further (*region IV*) pulse heights continue to increase. The heights are still related but no longer proportional to initial ionizing intensity. The voltage region in which M depends some what on the pulse size is called the region of *limited proportionality* and is not much used for measuring. The limitation may be because:

(a) Ultraviolet photons may be formed.

(b) New electrons may be formed when positive ions reach the cathode,

(c) The space charge may distort the electric field.

For still higher voltages the *region V* is reached, in which pulse height is very large and also substantially independent of the initial number of ions. This is called *Geiger region.* The device operated in this region is called a Geiger-Müller counter. In the Geiger region the avalanche associated with the drift of a single electron to one point on the anode initiates avalanches over the entire anode length.

At the end of the Geiger region the counter goes into continuous discharge region,

2.5 PROPORTIONAL COUNTER

In the parallel plate ionization chamber if the electrode voltage is increased by beyond the saturation interval up the primary ions produced secondary by collisions with the gas: consequently, the primary ionization is multiplied by a factor that depends on the Geometry of the apparatus and on the applied voltage v the chamber under these condition is ion own as proportional counter.

For a cylindrical geometry, the electric field required to produce secondary ionization is given by

$$E = \frac{V}{r \log_e b/a}$$

Where,

a = radius of central wire

b = inner radius of cathode cylinder

V = electrode or central wire voltage

r = distance of electric field E from the central wire.

The electrons ejected from gas molecules by ionizing particle drift towards the positive electrode. If they move in a region of high electric field E, they will produce additional ionization by collisions with molecules of the gas. In drifting a short distance ax, n electrons will increase the number of elections to n + dn, where

$$dn = \alpha \, n \, dx \qquad ...(18)$$

Here a is the first *Townsend coefficient.* If n_0 is the number of electrons set free by ionizing particle and x is the distance through which these electrons drift to reach the positive electrode, the number of electrons reaching this electrode is given by

$$n = n_0 \exp.\left[\int_0^x \alpha\, dx\right] \qquad ...(19)$$

The ratio n/n_0 is called the *gas amplification or gas multiplication factor M.* The radius of the cylinder and a the radius of an axial wire, the electric field strength at a radius x is

$$E_x^{\cdot} = \frac{V_0}{[x \log_e b/a]} \qquad ...(20)$$

Let us assume that every electron produced in the primary ionization gives rise to n-secondary electrons by collisions. The production of secondary electrons during the collision is associated with the photons which in turn produce photoelectrons. If p is the probability of production of photoelectrons, there will be np photoelectrons in all. If each photoelectron produces n-electrons by further collisions, we shall have second-generation avalanche of n^2p electrons, which in turn will give rise to more electrons.

$\therefore$ Total number of electrons $M = n + n^2p + n^3p^2 + ...$ ...(21)

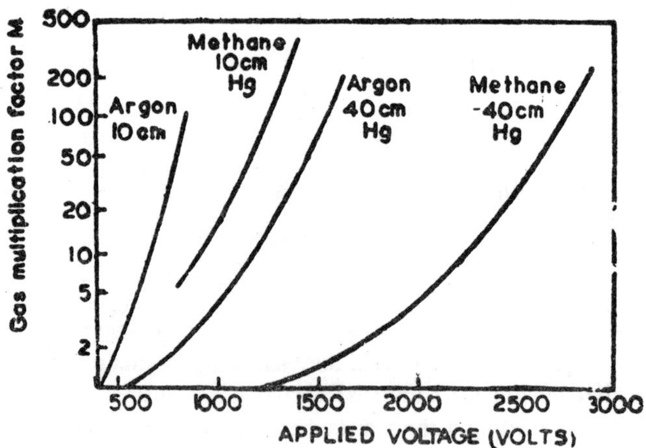

Fig. 2.9 : Variation of M with applied voltages.

For practical proportional counter $np < 1$, hence $M = n/(1 - np)$. It represents that in a proportional counter the total number of secondary electrons is proportional to the number of initial electrons. The variation of M with voltage in argon and methane at two different pressures is shown in the given Fig. 2.9.

The voltage-time dependence of the pulse is shown in Fig. 2.10. The upper curve shows the behaviour for large values of the time constant. The initial rise of the pulse is due to the high velocity of the positive ions as they recede from the anode. Positive ions then move in a weak field and the pulse approaches its final height extremely slowly. For the case of a finite time constant the pulse has the shape shown by lower curves. In practice, suitable clipping time constants (RC ~ 0.1μ sec.) are used in the output circuit so that better time resolutions are available.

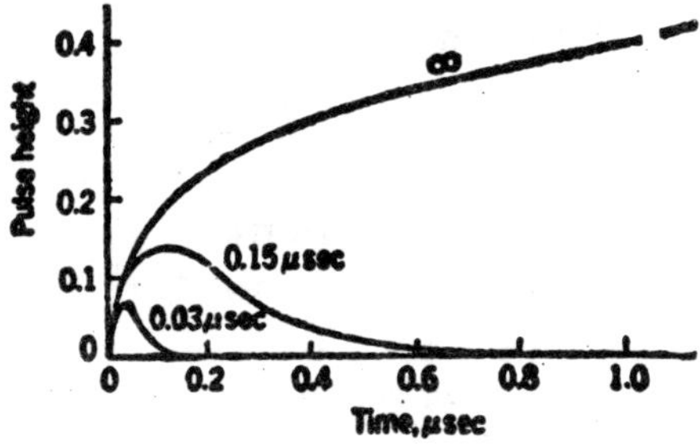

Fig. 2.10 : The pulse space.

If the primary ionizing particles arrive in rapid succession, mechanical counter can not record all the pulses. The output from the amplifier is therefore first passed through a scaling circuit before being fed into a mechanical counter. A sealer is an electronic circuit arrangement that feeds a predetermined fraction of the actual particle counts to the mechanical counter.

2.6 GEIGER-MÜLLER COUNTER

Geiger and Müller in Germany developed the counter in 1928. This counter is thus named as Geiger-Müller counter or G.M. counter. It works in Geiger region. The main difference between proportional and Geiger counter is that in the *former the avalanche is formed only at one point whereas in the latter the avalanche spreads along the whole length of the central wire. The amplification does not therefore depend on the initial ionization produced by the ionizing particle.*

Pulse Formation and Decay

Let us Consider the situation when the space charge has moved outside to a radius r_0. Let there be q charges per unit length on the wire and Q charges per unit length in the space charge. The field strengths inside and outside the sheath are

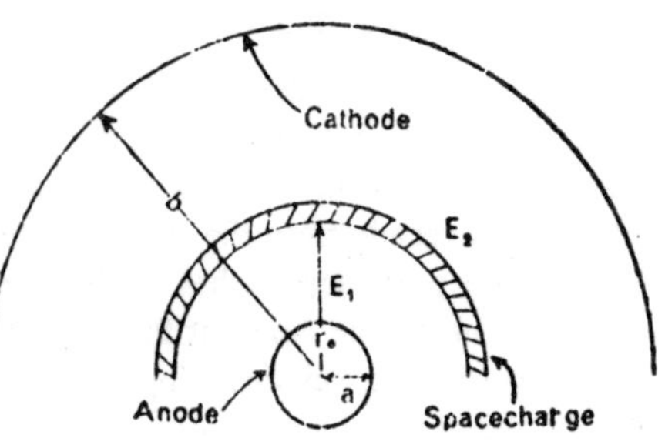

Fig. 2.11 : Positive ion sheath.

$$E_1 = q/2\pi\epsilon r \text{ and } E_2 = (q + Q)/2\pi\epsilon r.$$

Voltage applied to the counter

$$V = \int_a^{r_0} \frac{q}{2\pi \epsilon r} dr + \int_{r_0}^{b} \frac{q+Q}{2\pi \epsilon r} dr$$

Integrating we have

$$= \frac{q}{2\pi \epsilon} \log_e \frac{b}{a} + \frac{Q}{2\pi \epsilon} \log_e \frac{b}{r_0}$$

$\therefore$ Charge
$$q = \frac{2\pi \epsilon}{\log_e b/a} \left[V - \frac{Q}{2\pi \epsilon} \log_e \frac{b}{r_0} \right] \qquad ...(22)$$

The negative term in the bracket shows that q and hence E_1 is reduced during a pulse. As the positive ion sheath moves toward the cathode, r_0 increases, negative term decreases and the field strength attains its maximum initial value. The positive ion sheath will reach the cathode in about 10^{-4}sec. As the positive ions reach the wall and are about 10^{-7}cm from it they cause electron to be liberated from the metal of the cathode by *field emission*. These electrons then neutralize the positive ions, leaving the excited molecules. From where they promptly go to the ground state radiating characteristic spectral lines. Some of these lines are in the ultraviolet region of the spectrum and hence liberate photoelectrons from the cathode. These secondary electrons are accelerated toward the central wire. When the electrons reach the wire, they start a new avalanche which produces a new sheath and the discharge will continue indefinitely. This continuous discharge is not really continuous in time but consists of a series of pulses at close and some what overlapping intervals. The growth of the potential pulse at the anode is determined by the motion of the +ve ion sheath as it is repelled away from the high field region. When the ion sheath has travelled a distance equal to about half the radius of cathode cylinder, the field close to the anode wire is below the Geiger threshold. Hence the second pulse can not be recorded. The time, during which the counter is incapable of responding to a second ionizing event, is known as *dead time* (~200μ sec.). As the space charge moves further towards the cathode, the counter sensitivity gradually returns to its original value and the second pule can now be recorded. This time, during which pulses of reduced size are produced, is known as *recovery time*. The dead time plus recovery time may corresponds to the time which +ve ion sheath takes to reach the cathode, and is known as the *resolving time* of the counter. The high counting rate is generally reduced by reducing the high potential supply to the counter for a definite time Interval called *paralysis time*. It is achieved by an external electrical circuits generally a relaxation oscillator triggered by the step rise of Geiger pulse;

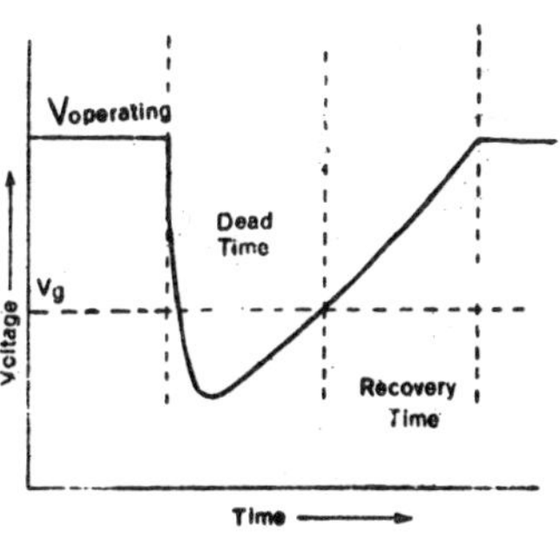

Fig. 2.12 : Dead time and recovery time.

Quenching of the Discharge

The process used to prevent a continuous series of pulses from taking place in the G.M. counter are called quenching. There are three types of quenching (*suppressing*) involved in the operation of a counter:

(a) quenching of photons in the initial avalanche,

(b) electrostatic quenching of the avalanche by the +ve ion space charge, and

(c) quenching of the secondary emission when the +ve ions reach the cathode. The quenching of the discharge involves both the electrostatic quenching and the quenching of secondary emission. There are two procedures for quenching the discharge in order to improve the resolving power of G M. counters.

In the *nonself-quenched* (externally quenched) type of G. M. counter, the quenching of the discharge is achieved by means of an external resistance or by use of an auxiliary electronic circuit. In the first method counter can be quenched by using the load resistor ($10^{10}\Omega$) in series with the tube. High voltage is applied across the tube which reduces as the current pulse flows through the resistance, the gas amplification factor is thus decreased preventing a second discharge. With a chamber of 10pf capacitance this method results in a counter dead time of the order of 0.1 second. In the second method the load resistance is about $10^7\Omega$ and at the same time the voltage across the chamber is reduced, by the application of a negative voltage pulse to the anode from a monostable multivibrator, every time a count takes place. The dead time of the counting circuit can be adjusted electronically to be a chosen value.

In the *self-quenched* (internally quenched) type of G.M. counter, quenching is achieved by the inclusion of a suitable gas in the counter, such as:

(a) organic molecules, and

(b) the halogen gases. In the first case the filling is a mixture of argon and 10% of a polyatomic organic gas or vapour such as methane, ethane or ethyl alcohol as a quenching gas. As the quenching gag ionizes more readily than argon due to its lower (11.3 eV instead of 15.7 eV) ionization potential, hence the ion cloud reaching the cathode consists almost entirely of alcohol ions. These ions are neutralized by pulling electrons from the metal surface of cathode and again form largely excited molecules. *These excited molecules usually prefer decomposition rather than photon emission and hence no photons are available to start a second avalanche.* The photons if emitted by excited molecules are not likely to travel more than a millimeter through the gas before being absorbed by the photoelectric process. The process of charge exchange during collisions prevents any argon ion reaching the cathode, thereby preventing the release of any electrons from the cathode which would trigger off a second discharge.

The each transfer of ionization from argon to alcohol is accompanied by the emission of photons of energy 15.7 – 11.3 = 4.4eV. This is too low to eject an electron from a cathode.

A satisfactory quenching gas must nave three main properties:

1. When in an excited state it must prefer to dissociate rather than to de-excite by the emission of photons.
2. Its ionization potential should be lower than that of the main counting gas in the tube.
3. It must have broad and intense ultraviolet absorption bands.

The halogen quenching action seems to be similar to that of the organic agents in that some of the molecules dissociate at each discharge. However, the dissociated halogen molecules combine at the end of the discharge, so the tube has unlimited life. Halogen molecules are chemically so active that the materials of which the tube and its electrodes are made must be chosen carefully. The halogen quenched counters usually operate at 300 to 700 volts. They can be used over a temperature range of from –50°C to +60°C, whereas the organic quenched counters often cease to work at 0°C because the organic gas condenses. They are not damaged by a continuous discharge whereas poly-atomic organic gases filled counters damaged. Due to their lower voltages, long life and robustness, the halogen quenched counters are recommended for use in colleges.

In the self quenching type counters the discharge is quenched due to the internal mechanism and hence no electronic quenching circuit or high resistance is necessary. The disadvantage of high resistance is that it lengthens the recovery time of the counter. The equivalent electronic circuit permits a short time constant to be used but adds the inconvenience of an extra stage in the circuits. The advantage gained in the elimination of the quenching resistor or electronic circuit is secured at the price of stability.

G.M. Tube Construction

G. M. *tubes* have been made in a great variety of sizes and shapes, from 1 cm. to 100 cm. in length and from 0.3 cm. to 10 cm. in diameter. The walls can be of metal (copper), a metallic cylinder may be supported inside a glass tube, or the interior surface of the glass tube may be coated with a thin layer of an electrical conductor) *e.g.*, silver). The central wire is usually of tungsten with a thickness of 2×10^{-3} to 5×10^{-3}cm. Various observers have tried a large number of different compounds in counters and found that any polyatomic gas molecule would produce the quenching action. The molecule need not be organic. There does not appear to be any great difference in merit between various possible polyatomic molecules, each having certain advantages and disadvantages as compared

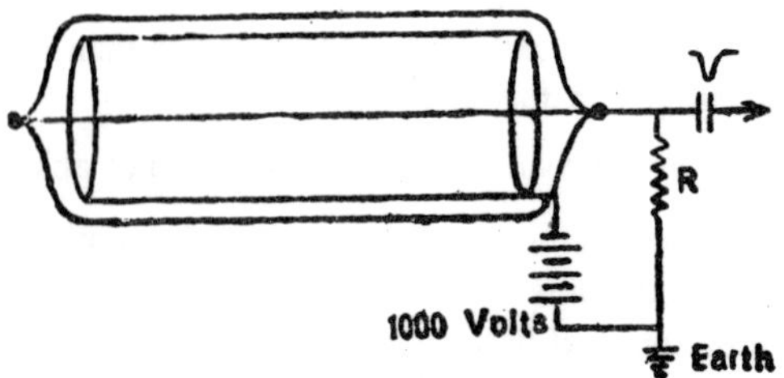

Fig. 2.13 : Basic circuit of G.M. counter.

to other. For example, amyl acetate shows a some what longer useful life but a slower resolving time than does methane. To extend the life time of the tube, bromine vapour can be introduced instead of the alcohol. To detect radiation of very small penetrating power counters with very thin end windows are used. Thin end windows of mica, cellophane or glass, only 1 or 2 mg/cm^2. thick will allow a- and P-particles to pass into the counter. For detecting P-radiation thin walled glass counters (the cathode is often a thin layer of colloidal carbon deposited on the inner wall of the glass envelope), which allow the entry of β-rays from all directions, can be used. Detection of γ-rays is mainly through the ejection of photoelectron from the cathode cylinder. The probability of this process varies as Z^4. hence, counters using high Z (bismuth or lead) cathode are used for the detection of γ-rays. For detecting cosmic rays more robust counters sometimes two or three feet long are used. For our purposes:

(i) *G.M. tube should be capable of responding to the radiation to be detected,*

(ii) *It should operate at fairly low voltages,*

(iii) *It should have long life, and*

(iv) *Its cost should be minimum.*

In India G.M. tube of different shape and size are designed in Hyderabad. These are marked as I 1000, I 1002 (Helogen filled); I 1010 (Thin walled); I 1030, I 1031, I 1032 (end windows); I 1200 (liquid sample); I 1300 (gas flow detector windowless and I 1350 (gas flow detector end window).

Characteristic Curves

There are two important characteristic curves exhibited by Geiger counters. The first is the familiar *plateau curve.* This is a plot of the recorded counting rate when the voltage applied to the counter is varied. The G. M. counter, similar to proportional counter, is connected with a device capable of indicating only relatively large pulses but no small ones. Until the voltage reaches the value indicated as the *starting potential* the pulses are too small to be detected. But with rising potential, the gas amplification increases and number of pulses rise rapidly to a flat portion of the curve called *the plateau.* This is the Geiger tube region for which the *count rate is nearly independent of the potential difference across the tube.* Beyond the plateau the applied electric field is so high that a continuous discharge takes place in the tube and the count rate increases very rapidly. It should be emphasized that the shape of the knee of the plateau curve is partly a property of the associated recording circuit.

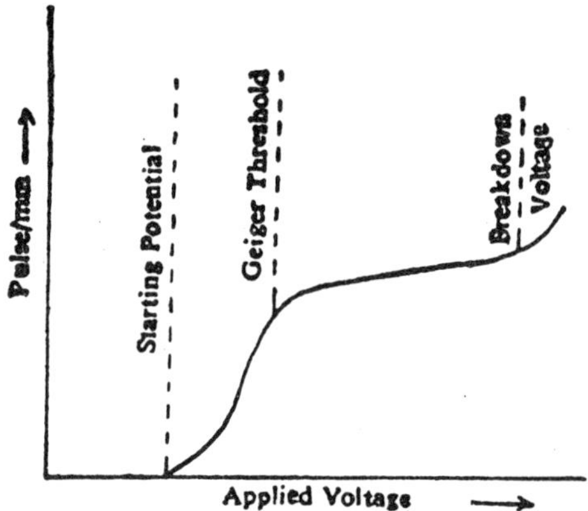

Fig. 2.14 : Plateau of G.M. tube.

$$\text{Slope of plateau } \frac{n_2 - n_1}{n_{av}} \times \frac{100}{V_2 - V_1} \times 100\% \text{ per 100 volts.}$$

The plateau has a width (or length) of a few 100 volts. A good tubes often have a plateau length of 100 to 200 volts or more with a slope of about 5% counts per 100 volts applied. *The slope for halogen filled tubes is however greater than that for organic gase filled tubes.* When the plateau becomes notably shorter and steeper, it is a sign that a counter tube is nearing the end of its useful life. The other type of curve is that of the starting potential as a function of the pressure of the gas in the counter. It can be shown from these curves that the *starting potential increases, practically linearly with the pressure of the filled gas.*

The *efficiency of the counter* is defined as the ratio of the observed counts/sec to the number of ionizing particles entering the counter per sec. Counting efficiency is defined as the ability of its counting if at least one ion pair is produced in it.

$$\therefore \qquad \text{Counting efficiency} = (1 - e^{-s/p}).$$

where s is the specific ionization at one atmosphere, p the pressure in atmospheres and l the path length of the ionization particle in the counter.

Precautions in Using G.M. Tube

1. The tube must be disconnected when measurements arc over.
2. Introduction of light must be prevented in order to avoid photo electric effects.
3. The background counting rate must be subtracted.
4. The tube should never operate at a voltage higher than its normal voltage.
5. In case a continuous discharge is produced, the voltage should be decreased.
6. The high voltage must be relatively stabilized.

Main Features of a G.M. Counter

(i) Constant output pulse size, independent of initial ionization.

(ii) Sensitivity to the production of even a single ion pair,

(iii) A long insensitive time to allow entry of each particle.

(iv) Infinite life (specially with halogen quenched),

(v) Can detect α, β, γ and cosmic rays.

Pulse Counting Circuits

The pulses obtained from the G.M. tube are too small to count direct. It requires amplification, which is done frequently in two steps. The counter is immediately followed by a *preamplifier*, which has little gain but produces an output signal across a relatively low impedance. This signal is then fed into the main amplifier, which has a gain strictly independent of the size of the input pulse. The output signal is then fed into the scaling circuit.

Resolving Time and Actual Counts

Let us assume that a counting system with a resolving time τ responds at a rate n counts per unit time when exposed to N initiating events per unit time. In unit time the total insensitive time will be $n\tau$ and the number of counts missed will be $Nn\tau$.

No of counts missed = error in counting or $Nn\tau = N - n$

$\therefore$ Actual count rate $N = n/(1 - n\tau)$

Thus, the actual count rate can be calculated by knowing τ. The most usual method for it involves the measurement of four counts using two radioactive sources. The background count (counts without any source) rate B is first found. With one source, count rate is determined. Let the observed count rate be $n_1 + B$ and expected $N_1 + B$. With the first, source the second source is introduced and adjusted until the count rate is approximately doubled. We now have $N_1 + N_2 + B$ expected and $n_{12} + B$ observed. The first source is removed and the second is left in position. If the counts rate is $N_2 + B$ expected and $n_2 + B$ observed. Solving equations

$$N_1 = \frac{n_1}{(1+n_1\tau)}, \; N_1 + N_2 = \frac{n_{12}}{(1-n_{12}\tau)} \text{ and } N_2 = \frac{n_2}{(1-n_2\tau)},$$

We get
$$\tau = \frac{n_1 + n_2 - n_{12}}{2n_1n_2} \quad ...(23)$$

Standard Deviation

Because of the random nature of radioactive decay it is found that if a long lived source is counted for a number of equal intervals of time the result shows considerable variations. This deviation is the square root of the average of the squares of the deviations from the mean. If N is the total number of counts in time t, then the

$$\text{Standard deviation } \sigma = \sqrt{\frac{(N)}{t}} \quad ...(24)$$

The coefficient of Variation $V = \dfrac{\text{Standard deviation } \sigma}{\text{Rate of counting R}}$

$$= \frac{\sqrt{(N)/t}}{N/t} = \frac{1}{\sqrt{(N)}} \qquad ...(25)$$

Account must be taken of the count rate due to background, unless the count rate of the sample is very large. Hence actual counting rate

$$RT = R_s + R_B,$$

where the subscripts refer to the total, sample and background respectively.

The standard deviation $\sigma s = \sqrt{(\sigma T^2 + \sigma n^2)}$

or $$\sigma s = \sqrt{\left(\frac{N_T}{tT^2} + \frac{N_B}{t_B{}^2}\right)} = \sqrt{\left(\frac{R_T}{tT} + \frac{R_B}{t_B}\right)} \qquad ...(26)$$

and $$V_S = \frac{\sigma s}{Rs} = \frac{\sigma s}{R_T - R_B} \qquad ...(27)$$

2.7. SCINTILLATION COUNTERS

One of the earliest methods of detecting nuclear radiations was by the luminescence, they produced in certain substances. In 1903, Crookes in England and Elster and Geitel in Germany independently reported that alpha particles impinging on zinc sulphide screens produced individual flashes which could be observed through a microscope. The first attempt to count α-particles, by observing flashes of light (scintillations) generated in diamond by them, was made by E. Regener in Germany in 1908. In the same year *Rutherford and Geiger compared the number of scintillations seen with the number of pulses recorded by an ionization chamber. The numbers were same in both cases. Hence, if each particle caused a single pulse in the counter then it also gave rise to one scintillation.* In this way this method was available for counting individual α-particle. The observations of scintillations through low power microscope were very tedious and limited to relatively low counting rates. For this reason early forms of scintillation method soon became obsolete and more attention was paid to the development of gas counters.

Since 1945, this old technique, is now playing a most important role. This has been due to the development of the photo-multiplier tube and the intensive study of the luminescent properties of many inorganic solids as well as organic compounds either in the solid state or in solution. A purely schematic diagram of a scintillation counter unit is shown in the following figure. A scintillation counter consists of the luminescent material (known is *scintillator*), reflecting layer such as aluminium foil enclosing the luminescent substance to facilitate the collection of light, light pipes, the photomultiplier tube, amplifier, voltage discriminator and an electrical circuit to record the output pulses. Before considering the performance of the scintillation counter let us first examine the behaviour of some of its components.

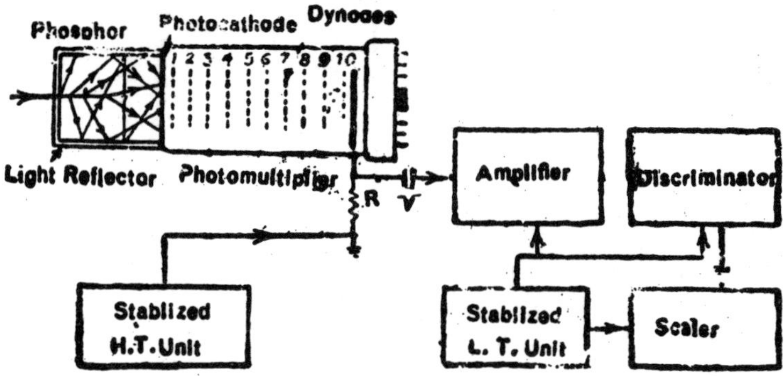

Fig. 2.15 : Schematic diagram of Scintillation Counter.

Photomultiplier Tube

In modern scintillation counters human eye is replaced by an electronic device called a *photomultiplier tube*. It is a highly sensitive photocell, converting light energy into electrical energy. In one of the commonest types of photomultipliers a semi transparent Cs – Sb cathode is deposited on the inside of the end of a high vacuum envelope. Cathode's of this type have a maximum response to light at the blue end of the spectrum. The electrons ejected from this cathode are accelerated towards the electrode D_1 and are collected by the electrode where they liberate more secondary electrons. These are focused on an electrode D_2. The process is repeated again and again These collecting electrodes D_1, D_2, etc., are called *dynodes*. A photomultiplier tube may have as many as thirteen or sixteen dynodes, each one being maintained at a positive potential of 100 volts with respect to the previous one. The total voltage drop across the whole system is supplied from the rectifier power pack. The dynode has a *venetian blind structure*, and is covered with a layer of material with a high secondary emission coefficient, such as Cs – Sb or Ag – Mg. The number of electrons reaching the anode A is thus a million or more times the number of electrons reaching the first dynode. The anode A is connected to the positive voltage supply through a series resistance R. Because of the voltage drop across the load resistor, a negative output pulse is produced by a flash of light falling on the photo cathode. At room temperature there will be an appreciable emission of thermionic electrons from the cathode to produce undesirable current, known as *dark current*.

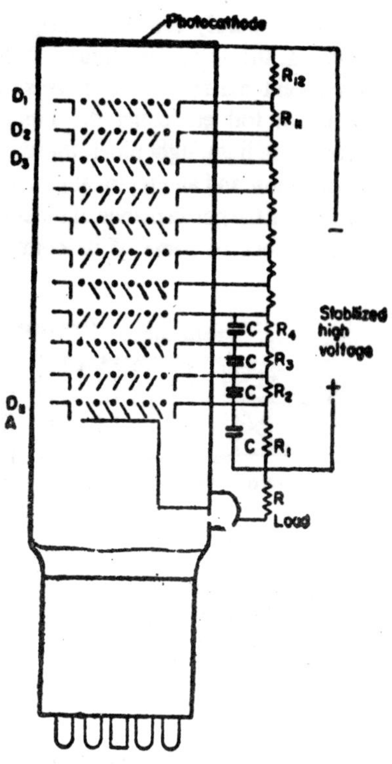

Fig. 2.16 : Photomultiplier tube.

Scintillators

The principal methods of excitation are *incandescence and luminescence.* The luminescence caused by photons are named as photo luminescence and are further divided into phosphorescence and fluorescence. When the visible or ultraviolet light is emitted within 10^{-8}sec or less after the radiation absorption the emission is called fluorescence. Phosphorescence refers to delayed light emission, which may follow the radiation absorption by minutes, day or even years. For the moment we shall consider fluorescence response. The substances scintillate when bombarded by radiations are named as scintillators and are mainly classified as:

(a) Organic Liquids

Some solutions of organic compounds are also capable of acting as phosphors, although, they are not as efficient as the best pure crystals. The liquid scintillator has two main components: the solvent, (*e.g.*, toluene or xylene) and the primary solute (*e.g.*, a few percent of diphenyloxazole or terphenyl). Most of the nuclear radiation energy is absorbed by the solvent which is not a scintillator and is then transferred to the solute which emits the light actually. When the scintillations occur in the extreme ultraviolet region one uses a light shifter to increase the wavelength of the emitted light.

(b) Organic Crystal

Kallmann employed a large clear crystal of the organic substance naphthalene as phosphor. Following this, it has been found that other related organic compounds, consisting of several linked benzene rings are better scintillators. Of the solid organic scintillators anthracene and stilbene appear to be the best. The organic crystalline scintillators have a faster decay time ($\sim 10^{-8}$ sec), high transparency and poorer conversion efficiency especially for heavy particles. For β-particles response is linear. Because of low Z and ρ, these crystals have low cross-sections for photoelectric and pair production processes, hence can not be used for the detection γ-rays and X-rays.

(c) Inorganic Crystals

Unlike the organic compounds, inorganic substances do not scintillate when they are pure. Best one, especially for gamma rays, is crystalline *sodium iodide activated with thallium* (about 0.1 per cent). A less sensitive alternative is the nonhygroscopic cesium iodide, also activated with thallium. Calcium iodide with a trace of europium is more efficient than NaI, although, it is also hygroscopic. Silver activated zinc sulphide is an excellent scintillator but can be used only in thin layers, because it soon becomes opaque to the light emitted by it.

(d) Plastics

Organic scintillator system, which lies between a solid and liquid has a plastic as the base. A solution is made of the primary solute in a solvent like vinyl toluene or styrene which can be readily converted into a solid plastic. The resulting transparent material can be made in large pieces and cut to any desired shape. Plastic scintillators are widely used in high energy physics.

In a crystal such as NaI (T*l*) an ion Tl^+ replaces an ion Na^+ and forms an emission center. The T*l* ion thus occupies a position of equilibrium in the lattice such that a displacement from that position produces an increase in potential energy for the entire lattice system near the Tl^+ ion. The absorption of radiation changes the Tl^+ ion to an excited state for which the potential energy curve has a minimum at a position different from that in the ground state. The absorption (transition A) leaves the system in a state of high vibrational energy. The excited particle makes many collisions and with in 10^{-5} to 10^{-8} sec dissipates its surplus vibrational energy to the lattice. The most intense emission thus corresponds to transition A'. If the temperature is high enough, so that the higher vibrational levels of the ground state have appreciable population. Now, it is possible that the absorption takes place along B which is followed by emission of radiation.

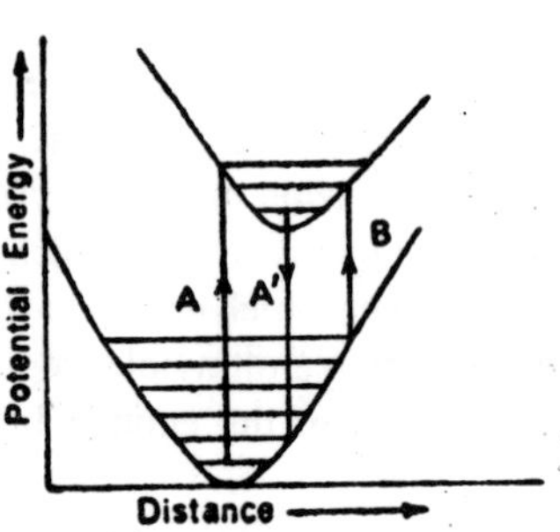

Fig. 2.17 : Luminescence proccess in NaI (Tl).

ZnS (activated with Cu or Ag)

1. The output is proportional to the energy of the incident particle.
2. C_{IP} is independent of the incident energy.
3. It has highest light conversion efficiency C_{IP} of any known phosphor.
4. The efficiency and emission spectra depend on the preparation and activator used.

For ZnS (activated with Cu)—conversion efficiency C_{IP} = 28% and the emission spectra from 4000-6000Å has a peak at 4500Å.

For ZnS (activated with Cu)— C_{IP} = 25% and λ_{peak} = 5200Å.

It has following limitations

1. Decay time is ~10^{-6} sec or longer.
2. A large crystal can not be prepared.
3. It has low transparency. Beyond thickness 25 mg/cm^2, the detector efficiency decreases.

4. It can not be used for the detection or γ-rays and electrons. Only the detection of slightly heavier particles such as a, p, d is possible.

NaI (activated with Tl)

1. Its absorption spectra lies in ultraviolet region, with the peaks at 2340 and 2930Å. Due to this reason, the crystals are highly transparent.
2. Its emission spectra has a average value of wavelength ~4100Å, and FWHM = 850 ± 100Å.
3. Its can be grown into large single crystal.
4. It has apart from ZnS, the highest.
5. Because of high values of p and Z, it has large cross sections for pair-production, photoelectric effect and compton effect. Due to this reason it is very suitable for the detection of gamma and X-rays.
6. The decay time at room temperature is lowest (~2.5×10^{-7}sec.).

C_{IP} decreases with increasing specific ionization causing a nonlinear response to α-particles. The response is linear to p, d and e. The crystal must be protected from moisture with surface film of liquid paraffin or be placed in dry atmosphere.

(e) Gases

Inert gases of the atmosphere, *e.g.*, argon, krypton and xenon when exposed to nuclear radiations produce scintillations in the ultraviolet region of the spectrum. To increase photomultiplier response the flashes are converted into visible light by means of a suitable wavelength shifter. An important advantage of the inert element scintillators iff that the light output is proportional very closely to the energy of the nuclear radiations.

Choice of Phosphors

NaI (Tl) is a highly effective scintillator as far as light output is concerned, but the light pulse has a relatively high decay time, compared with the organic phosphors. Hence *NaI (Tl) is less satisfactory when a very rapid response is required.* Iodine present in it gives a much greater stopping power for γ-rays. The size of the output pulse is proportional to the energy.

Organic scintillators give comparatively smaller light pulses, but have a shorter decay time. Hence, a higher counting rates are possible. The decay time in the liquid and plastic scintillators are even shorter than in the crystals Because of the small stopping power, the organic scintillators cannot be used for γ-rays unless the counter is large and are used for β-detection.

Screens with activated ZnS, in very thin layers, are useful for the detection of α-particles. Neutrons can be detected through the emission of a suitable charged particle resulting from a nuclear reaction initiated by the incident neutron.

Different Processes in the Scintillation Detector

(a) Absorption Process

If the dimensions of the scintillator are large compared to the range of the incident particle, the incident charged particle can dissipate all its energy to the scintillator. If the γ-ray is incident on a scintillator it may interact with the scintillator material in three ways: *photo-electric absorption, Comption scattering andpair production* each cases, electrons are produced. The all or part of the energy of incident γ-ray is transformed into the kinetic energy of these electrons.

(b) Scintillation Process

The electrons emitted will give up their kinetic energy in ionization or excitation in the scintillator material. This absorbed energy appears either as heat energy or as luminescence photons. In the latter cases the scintillator material de-excites by light emission within 10^{-8} seconds. This emitted light is known as scintillation. If τ is the mean life of the scintillator, the number of light photons in time t after the ionization radiation has arrived is given by

$$n = \text{const}\ (1 - e^{-t/\tau}) \qquad ...(28)$$

(c) Formation of Electrical Pulses

The light produced in scintillator falls on the photocathode of the photomultiplier tube, producing photoelectrons. These electrons fall on the first dynode, producing a bunch of secondary electrons. This process is repeated and a gain upto $10^{-7} - 10^{8}$ is achieved. The electrons finally fall on the electrode, known as anode, and thus, produce an electrical pulse across the load resistance R. The pulse height V at the output

Table 2.1 : Properties of Some Scintillators

Scintillator	Form	Density gm/cm^2	Refractive Index	Relative response	Decay constant (x10^{-9} + sec)	Wavelength of Maximum emission A°	General remarks
1. Inorganic Scintillators							
(a) ZnS (Ag)	powder	4.09	2.4	1-4	10^4	4500	Generally used in thin layers to detect α-particles.
(b) NaI (Tl)	crystal	3.67	1.77	2.10	250	4100	Mainly used to detect γ-rays. Hygroscopi
(c) CsI (Tl)	crystal	4.51	1.75	0.5	1100	white	Large absorption co-efficient for γ-rays. Non hygroscopic.
2. Organic Crystals							
(a) Stilbenc	crystal	1.16	1.62	0.60	6.4	4100	Commonly used for β-dctection.
(b) Anthracence	crystal	1.25	1.59	1.00	30	4470	Commnly used for β-detection
3. Organic Solutions							
(a) p-terphenyl in polystyrene	Plastic	1.06	1.59	0.39	4.0	4450	Useful for high energy β-detection
(b) p-terphenyl in Toluene (5gm/litre concentration of terphenyl)	solution	0.86	1.50	0.35	2.2	4500	Useful for shoft β-particle detection

related to the amount of charge Q collected at the output through the relation

$$V = \frac{Q}{C} \qquad ...(29)$$

where C is the capacity of the output point.

If E_i be the energy of the incident charged particle or γ-rays and A the probability of amount of energy absorbed by the phosphor.

$\therefore$ Energy dissipated in the phosphor = E_iA.

This energy is converted within efficiency C_{IP} into photons of average energy E_p.

$$nr = E_iA\ C_{IP}/E_p \qquad ...(30)$$

If T_p is the optical transparency and G is the fraction of the photons falling on the photocathode. Hence, the number of photons incident on the photocathode.

$$n_p/ = T_pGn_p = E_iAC_{IP}T_pG/E_p.$$

These photons are converted into photoelectrons with an efficiency C_{PE} f(v), where C_{PE} is the photoelectric conversion efficiency of the cathode and f(v) is the relative response at f.equency v.

$\therefore$ No. of photo electrons $n = C_{PE}/(v)\ np/$

If f_d be the fraction of photoelectrons reaching the first dynode and M the electron multiplication factor, then the total number of electrons reaching the anode

$$N = f_dMn = C_{PE}f(v)\ E_iAC_{IP}T_pGf_dM/E_p.$$

Hence total charge Q = eN and the pulse height

$$V = Q/C = eE_i\ AC_{IP}T_PGC_{PE}f(v)f_dM/CE_p \qquad ...(31)$$

Here, we see that the pulse height is proportional to the incident particle if the particle is absorbed in the scintillator completely.

Resolving Power

The height of the output pulse exhibits variations about the mean value for a given particle energy E. This variation is due to the statistical fluctuations in the various factors of eqn. (31). The resolution R of a scintillation counter can be defined as follows

$$\frac{1}{R} = \frac{\overline{Q}^2 - (\overline{Q}^2)}{(\overline{Q}^2)} \qquad \text{...(32)}$$

where $\overline{Q}$ is the mean charge collected at the anode and $\overline{Q}^2$ is the mean squared charge collected at the anode. The smaller the half width of the photo peak, the higher will be energy resolution R.

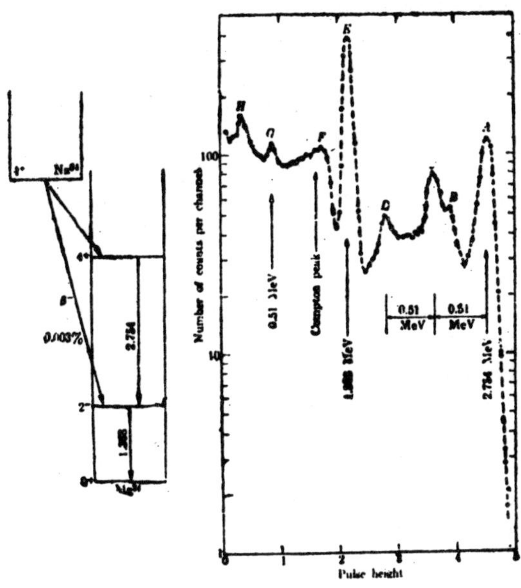

Fig. 2.18 : Decay scheme for Na^{24} (Left). Pulse height spectrum for gamma rays (Right).

The scintillation counter can be used as a spectrometer in conjunction with an electronic pulse height analyzer, which has found many applications in low energy, as well as in high-energy nuclear physics. Suitable circuits have been designed to sort out particles with in a given range of energy so that distribution in energy of the incident radiation may be measured. Unfortunately, even mono-energetic γ-rays produce a complex pulse height spectrum. The pulse height spectrum produced by γ-rays from the lower excited states of Mg^{24} is shown in Fig. 2.18 (right). This isotope is produced in low lying states by β-emission of

Na^{24} according to the decay scheme shown in Fig. 2.18 (left). Two γ-rays of energy, one at 2.754 MeV and other at 1.368 MeV are emitted. The peak A in this spectrum corresponds to the full energy of 2.754 γ-ray and is due to pair production process only. The next peak B corresponds to the electrons produced by compton process. The peak C is caused by pair-production processes in which one of the two annihiliation photons has escaped and peak D corresponds to the processes in which both the annihiliation photons have escaped. The central sharp peak E corresponds to the energy 1.368 MeV and is due to pair production or photo electric process. The compton peak corresponding to this peak is F. As 1.368 MeV pair production is relatively unimportant, hence first and second escape peaks corresponding to the energies 1.368 – 0.51 = .858 and 1.368 – 1.02 = .348 are not observed. Peak G corresponds to 0.51 MeV annihiliation photons produced outside the scintillator but absorbed by it. The peak H arises from back scattered photons from compton events.

The scintillation counter has a much higher counting efficiency for gamma rays due to the greater amount of energy dissipation by the gamma rays. The scintillation counter offers greater stability, greater accuracy, shorter resolving time and higher efficiency than G.M. counter. Provided all the particle energy is dissipated in the scintillator, the voltage pulse height produced can be used to measure the particle energy, once the counter is calibrated by means of a standard source. Because of their extremely rapid response scintillation counters are having many applications in atomic energy field. These are used for the accurate timing of nuclear particles moving with very high speeds.

2.8 CERENKOV COUNTERS

When a charged particle moves through a transparent dielectric medium with a velocity greater than the velocity of light in that medium, weak electro-magnetic radiation is emitted. These radiations are named Cerenkov radiations. The radiation is emitted mainly in the direction of the charged particle and has a continuous spectrum from red to ultra-violet at least.

A charged particle traversing a dielectric medium causes polarization of the molecules adjacent to its path, as the charged particle attracts the oppositely charged component of a molecule and repels the similarly charged component of the molecule (Fig. 2.19). For slowly moving particles, polarized regions are formed in step with the motion in the vicinity of the particle. The resultant dipole field is zero as the polarization field around the electron is symmetrical in the axial and azimuthal directions, hence no radiation is emitted. For fast moving particles the polarization is *axially asymmetric* but azimuthally symmetric and the resultant dipole field acts even at large distances from the electron's track. The radiation pulses are thus emitted in succession by the molecules along the path of the charged particle. These radiations are like the radiations emitted from a *damped oscillating dipole moment*.

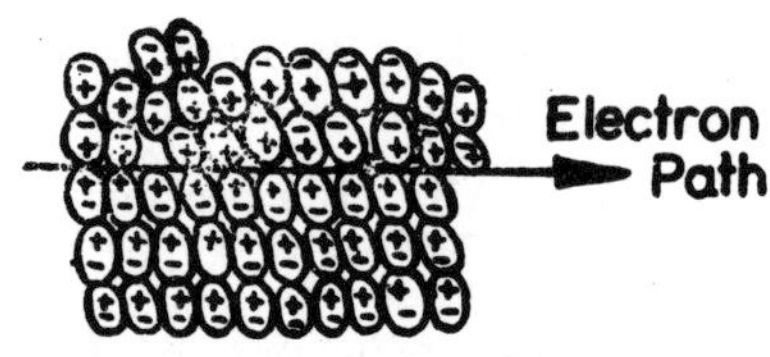

Fig. 2.19 : Polarization of molecules.

In a dielectric medium of refractive index μ, photons move with a velocity c/μ. Fig. 2.20 (left) shows the three interactions between a charged particle moving with velocity v (< c/μ) and the electrons in the medium. The wavefronts of the electromagnetic waves resulting from interaction do not interrelate and these interactions are entirely independent. Fig. 2.20 (right) shows the interactions when v > c/μ. The wavefronts from 1 and 2 lie inside 3 at the time of the third interaction.

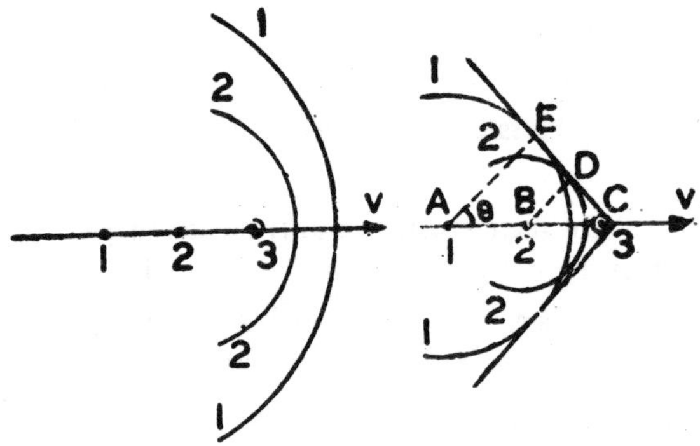

Fig. 2.20 : Wavefronts resulting from radiation losses: with v < c/μ (left) and with v > c/μ (right).

The Huygens construction of elementary wave optics shows that a conical wavefront can be drawn tangent to the spherical surfaces and the Cerenkov radiation can be observed at a particular angle with respect to the direction of motion of the particle. The particle moves along ABC with velocity v It radiates energy in all directions from the points along its path ABC. The time in which particle travels a distance AC, the radiations will move a distance AE, CE is thus the wavefront. The angle which the Cerenkov Radiation makes with the direction of charged particle is given by

$$\cos\theta = \frac{AB}{AC} = \frac{BD}{BC} = \frac{ct/\mu}{vt} = \frac{c}{\mu v} = \frac{1}{\beta\mu} \qquad ...(33)$$

The Cerenkov radiation is not confined to one plane but is propagated along the surface of a cone whose axis coincides with the direction of motion of the charged particle and whose semivertical angle is θ, given by eqn. (33), which shows that:

1. θ depends on the frequency (*i.e.*, wavelength) of the Cerenkov radiation. The radiation arc mainly of the high frequency region of the visible spectrum.
2. Un!ike the bremsstrahlung, θ is independent of the mass of moving charged particle.
3. The emission angle increases with increasing the particle velocity β.
4. As P cannot exceed unity the maximum angle of emission.

$$\theta_{max} = \cos^{-1}(1/\mu)$$

5. For a given constant refractive index μ, there is a threshold velocity $\beta_{th} = l/\mu$ below which no Cerenkov radiation is emitted and at which the radiations are along the particles direction of motion.

Radiation (33) for the constructive interference should satisfy the following conditions:

(a) The particle track length must be large compared with the wavelength λ of the Cerenkov radiation.

(b) The particle must move with the constant velocity in the medium.

Frank and Tamm have shown from classical theory that the total energy radiated in a short element of the particle's path ds is given as

$$\frac{dT}{ds} = \frac{\pi z^2 e^2}{\epsilon_0 c^2} \int \left(1 - \frac{1}{\beta^2 \mu^2}\right) v\, dv$$

where ze is the charge on the moving charged particles, v is the frequency of the emitted radiation, and the integral extends overall frequencies for which $\beta\mu > 1$. This loss is quite small (*e.g.*, about 10^3eV/cm for singly charged particles moving through glass or lucite) and is negligible compared to the ionization losses.

Consider the Cerenkov radiation emitted between two frequencies v_1 and v_2 as composed of quanta whose average energy is hv = 1/2h $(v_1 + v_2)$, then from eqn. (44) the average number of quanta emitted per unit length is

$$\frac{1}{h_v} = \frac{dT}{ds} = \frac{\pi z^2 e^2}{\epsilon_0 hc^2}(v_2 - v_1)\left(1 - \frac{1}{\beta^2 \mu^2}\right) \qquad ...(35)$$

where μ if the average refractive index of the medium over the frequency interval v_1 to v_2.

Relation (35) shows that for extreme relativistic electrons passing through glass ($Z = -1$, $\beta \simeq 1$, $\mu \simeq 1.5$) the (Cerenkov angle is 48° and the Cerenkov photons emitted per cm within the visible spectrum ($\Delta v = v_2 - v_1 \simeq 3 \times 10^{14}$ sec^{-1}) are 200. As the number of quanta per unit path length varies as $1/\lambda^2$, the short wavelength are thus predominantly present in Cerenkov radiation.

The properties discussed above have been extensively utilized in Cerenkov detectors which can be used not only to detect swiftly moving charged particles but also to determine their velocity. These detectors possess the following advantages:

1. Dependence of intensity on particle velocity.
2. Directional emission of light, with an angle dependent on velocity.
3. High efficiency and high counting rate.
4. The duration of the light pulse at any point with in the Cerenkov cone is vanishingly short ($\sim \leq 50^{-10}$ sec).

R.L. Mather has observed Cerenkov radiations radiated by a collimated beam of 340 MeV protons passing through an optically flat sheet of dense flint glass of 0.67 mm thickness with a 35 mm Leica camera. The glass sheet, whose rear surface was alumunized, was set

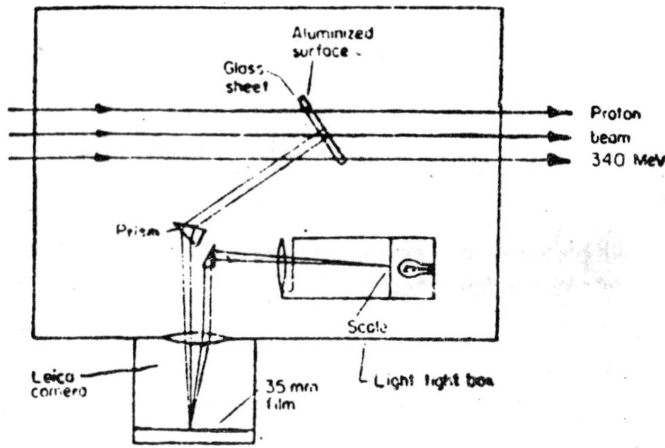

Fig. 2.21 : Measurement of velocity of high energy protons.

with respect to the proton beam so that part of the Cerenkov cone radiated in the glass would incident normally on the aluminium surface and reflect in such a way that there would be no refraction at the front surface of the glass. The angle of emission of radiations varies with wavelength λ because the refractive index depends on λ. This dispersion effect was decreased to nearly zero by Mather after introducing a prism. The recorded Cerenkov radiation on the camera film corresponded to a θ value of 38.5°. Using the relation (33) β comes out to be 0.680. The kinetic energy E_{kin} of the protons can be obtained as

$$E_{kin}\ m_pc^2\ (1-\beta^2)^{-1} - m_pc^2 + 341 \text{ MeV} \qquad ...(36)$$

This value was found to have 1% accuracy.

J. Marshall designed a good detector, in which the Cerenkov light was focused on to the cathode of two photomuitiplier tubes for recording high energy π-mesons (pions). A collimated beam of pions was allowed to pass along the axis of a large hemispherical perspex lens. The radiation radiated in this lens was reflected by a surrounding cylindrical mirror M_1 on to one of the plane mirrors M_2, M_3 and then on to the cathodes of the photomuitiplier tubes (lP 28), which were operated in coincidence. The position of the radiator lens was adjusted so that the Cerenkov light cone was correctly fọcused on to the photo tubes to give maximum coincidence counting rate. For 145 MeV pions the energy resolution is 13%.

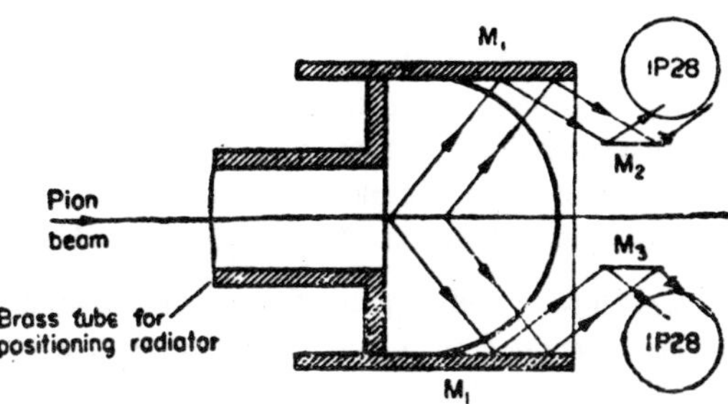

Fig. 2.22 : Marshall's focusing Cerenkov detector.

Cerenkov counters have been designed for charged particles in a narrow velocity range. Cerenkov detectors in which the light is radiated by a gas have been used for highly relativistic charged particles. Because the radiation in a given medium is produced only when the speed of the charged particles is greater than the velocity of light in that medium, this counter can be used as a *threshold detector, i.e.*, to detect only particles whose speed exceeds a certain value.

These counters are essentially *directional counters* became the Cerenkov radiations are emitted in a forward cone of an angle given by eqn. (33). The direction of the incoming particle can, therefore, be measured. The angle θ depends on the velocity v of the incident particle. This counter can thus be used to measure the velocity of incoming particle. Particles of *different masses* can be distinguished in two ways: first, for the same velocity, *i.e.,* the same Cerenkov angle, the brightness of the flash is proportional to the square of the atomic number and, second, for the same K.E. or range, the velocity of the particle is inversely related to the square root of the mass number. The particles of different charge can also be distinguished by these counters.

The blue glow seen around nuclear reactors of the swimming pool type reactor is almost due to *Cerenkov radiation*, produced by high speed electrons (Compton electrons produced in γ-ray scattering events). Since the core or fuel elements are in extended source and the emitted electrons show a distribution over a wide range of energies and angles. The radiation thus appears diffuse. The energy and momentum of charged particles (electrons) can thus be evaluated.

2.9 CLOUD CHAMBER

During the stay in the Scottish hills, in 1984, he started experimenting under laboratory conditions and intended to study the optical interesting phenomena appearing from illuminated fogs. After two years he discovered the fundamental principle of the cloud chamber and designed the first cloud chamber based on the principle in 1911.

Air mixed with saturated vapour of water or any other liquid, *e.g.*, alcohol or ether contained in a cylinder fitted with a piston can be expanded rapidly by a fast but relatively short motion of a piston. This adiabatic expansion will result in fall of temperature and super-saturation of vapour. The vapour will condense in cloud of water droplets. The production of such a cloud is however, impossible unless nuclei, on which the water vapour may condense are provided. If particles of dust are present in the air (as an ordinary cases) they will act as condensation nuclei. But in dust free air this phenomenon of condensation and formation of cloud will, therefore, be extremely difficult. C.T.R. Wilson discovered that the condensation cloud could be produced even in dust free air provided the air was ionized by an ionizing agent such as the X-rays,

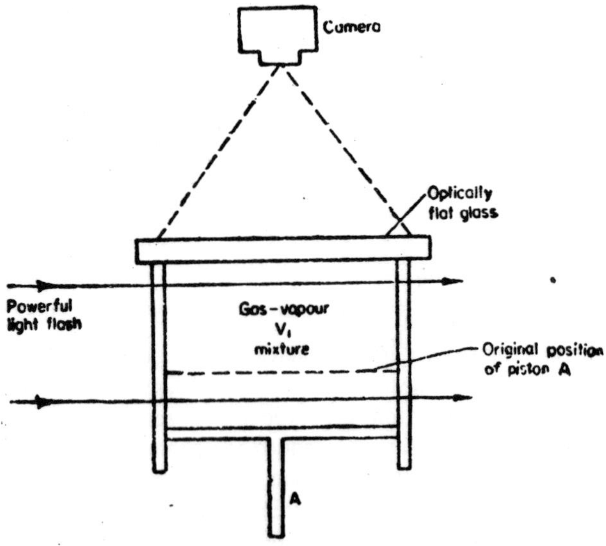

Fig. 2.23 : Schematic diagram of Wilson expansion cloud chamber.

cathode rays or any nuclear radiations. If ionizing particle entered the chamber either immediately before or immediately after the expansion, the ions left in its path would act as condensation nuclei. This experiment shows that as super saturation increases, the negative ions first serve as centres of condensation, then as the volume increases both positive and negative ions serve as the nuclei of droplets. A close array of fine droplets, *i.e.*, a kind of linear cloud, called a *cloud track*, will thus be formed. By using suitable strong illumination from the side, the track appears as a white line on a dark background. This can be photographed by means of a camera so that a record can be obtained.

Let V_1 be the volume, P_1 the saturated vapour pressure and T_1 the temperature of the vapour. Its mass m_0 (in moles) is determined by the relation

$$P_1V_1 = m_0RT_1 \qquad ...(37)$$

If the vapour is expended adiabatically from V_1 to V_2. The new temperature T_2 is given by

$$T_1V_1\gamma^{-1} = T_2V_2\gamma^{-1} \qquad ...(38)$$

If no vapour condenses, then the new pressure is given by

$$P_2V_2 = m_0RT_2 \qquad ...(39)$$

The vapour is in supersaturated condition. If ions are present in the space, the droplets will condense on them till the pressure falls to P_2. If m is the mass of the vapour remaining uncondensed, then

$$P_2V_2 = mRT_2 \qquad ...(40)$$

The degree of supersaturation, is the main determinant governing the growth of droplets about ions and may be defined as the ratio of the actual density of vapour to the saturation density, Thus, we have

$$S = \frac{m_0}{m} = \frac{P_2}{P_1} = \frac{P_1}{P_2}\left(\frac{V_1}{V_2}\right)^{\gamma} = \frac{P_1}{P_2} \times \frac{1}{(\text{Expansion ratio})^{\gamma}} \qquad ...(41)$$

The pressure of vapour in equilibrium with small drops of liquid is not the same as that for a plane liquid surface. For a drop of radius r and charge ze the saturated vapour pressure P_r is given by

$$\log \frac{P_r}{P_\infty} = \frac{M}{\rho TR}\left[\frac{2\theta}{r} - \frac{z^2e^2(k-1)}{32 \in_0 \pi^2 k^{\gamma 4}}\right] \qquad ...(42)$$

The formula for ze = 0 was derived by Lord Kelvin and the electrostatic term was later introduced by J. J. Thomson. In this relation P_∞ a and M are respectively the normal saturated vapour pressure for plane surface and the molecular weight of the vapour and ρ, ∞ and R are respectively the density, surface tension and dielectric constant of the liquid.

The Wilson cloud chamber makes it possible to study the behaviour of individual atoms, to photograph the actual path of ionizing radiations and to analyse at a spare time the complicated interactions which may take place between charged particles and individual atoms. Since the year 1911 the apparatus has been improved in many ways, although the fundamental principle remains unchanged.

P.M.S. Blackett and G.P.S. Occhialini invented the Geiger counter controlled expansion cloud chamber for the study of cosmic rays. Such chambers have been used extensively in T.I.F.R. Bombay and Bose Institute Calcutta, in India, for cosmic ray work. One such design is shown in Fig. 2.24.

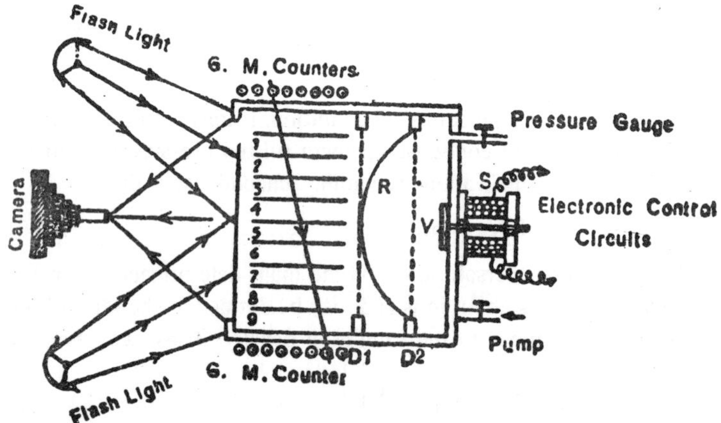

Fig. 2.24 : Schematic diagram of automatic.

In this arrangement two sets of G.M. counter trays are placed one at the top and other below the bottom of the active volume of the chamber. These trays are connected in coincidence so that any event due to the passage of a particle is triggered by these coincidence counters. The chamber is kept at a higher pressure with the help of compression pump which presses the membrane of rubber R. R is kept under tension and rests on the perforated metal disc D_1. R magnetic valve V, electrically controlled with the help of a solenoid S, is kept in position by energising the solenoid. When the counters (in coincidence) register an event, the electric pulse produces an impulse of current. The impulse is given to the solenoid in opposite direction. The valve V is now released and the gas escapes out bringing the membrane of rubber R to the position of the perforated disc D_2. After expansion, light is automatically turned to illuminate the tracks, the chamber volume is photographed. The ions are swept from the chamber by application of a clearing field and both chamber and camera are reset to await the next event.

In Wilson Cloud Chamber water is used to saturate the air but it becomes more common to employ ethyl or propyl alcohol or a mixture of alcohol and water. The use of alcohol gives better condensation on positive ions and the extent of expansion necessary for droplet formation is diminished from 1.25 to about 1.10 at ordinary pressure. Although, air is the usual gas, cloud chambers containing argon are sometimes employed. The higher pressures are desirable for the study of high energy particles. In cosmic-ray work dense metal plates have often been placed across the cloud chamber to ensure interactions of the high energy particle.

The heavy, slow particles like α-particles produce broad, densely packed, straight line tracks. Near the end of their path the particles may suffer sharp deflection as the result of impacts with the nuclei of oxygen or nitrogen present in the air. Slow electrons produce narrow, beaded, tortuous tracks, while fast particles both light and heavy produce narrow, beaded, straight tracks. By counting the drops in the cloud track the specific ionization can be determined and the nature of the particle identified. The sign of the electric charge and the momentum p of the particle can be determined if the chamber is placed in a strong magnetic field. If a stream of a particles, each carrying a charge q, moves initially in a straight line with velocity v and a magnetic field B is applied in a direction at right angles to the direction of motion, the panicles will be forced by the field to follow a circular path of radius R. The magnetic force Bqv is exactly balanced by the centrifugal force mv_2/R. Thus

$$Bqv = mv^2/R \text{ or } p(= mv) = BRq \quad ...(43)$$

The kinetic energy E of the particle can be easily calculated, if the rest mass energy m_0c_2 of the particle is known, by the relation

$$E = \sqrt{[p^2c^2 + (m_0c^2)^2]} - m_0c^2 \qquad ...(44)$$

The *disadvantage of the cloud chamber* lies in the fact that it needs a definite time to recover after an expansion and hence it is not possible to have a continuous record of events taking place in the chamber. A continuously sensitive, known as *diffusion cloud chamber* has been developed by A. Langsdorf in the United States in 1939. It consists of a vessel containing air or other gas, kept warm at the top and cold at the bottom. The liquid, usually methyl or ethyl alcohol, vaporises in the warm region, where the vapour pressure is high. The vapour diffuses downwards continuously through a region in which a vertical temperature gradient is maintained by cooling the bottom of the chamber. As the

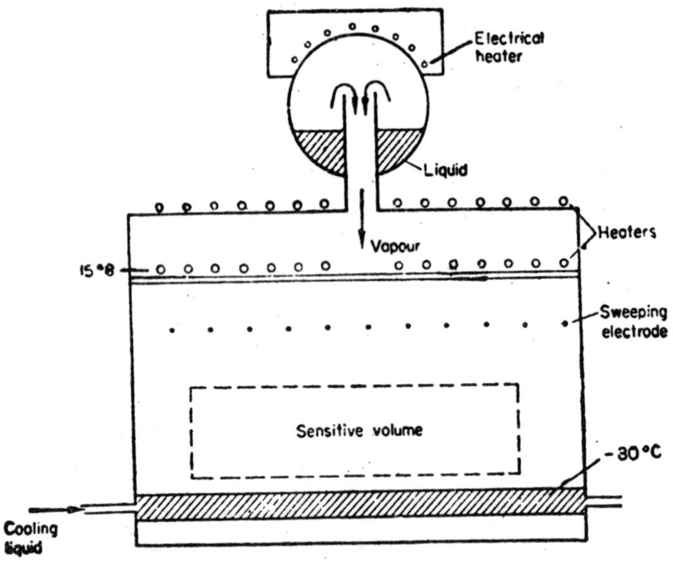

Fig. 2.25 : Schematic diagram of diffusion cloud chamber.

vapour diffuses into the colder regions of the gas, the saturation vapour pressure decreases rapidly and there is a volume near the cold base of the chamber where the supersaturated condition is maintained and thus eliminating the insensitive cycling time of the conventional cloud chamber. If some ions are present in this region, vapour forms droplets around them. In order to increase the frequency of occurrence of nuclear events, the pressure of the gas in the chamber is increased. *The main drawback of the diffusion chamber is that the sensitive region is no more than 7.5 cm deep.*

The diffusion cloud chamber is a continuously sensitive chamber, has many uses in the study of high energy particles obtained from accelerators. Its interesting application is the determination of the half-life of the free neutron. The discovery of the positron by electron positron pair production method, K^o meson by observing the decay in flight into pions, hyperon Λ^o by the observation of the decay in flight and that of muons in secondary cosmic radiation were made with the cloud chamber. The first reasonably accurate determination of the rest mass of the muon was made by W, B. Fretter with the help of cloud chamber.

2.10 BUBBLE CHAMBER

To detect very energetic particles the Cloud Chamber is not very suitable since the interaction of energetic particles cannot be completely observed in the chamber. To overcome this disadvantage it is necessary to use a chamber filled with a substance having a much greater stopping power. This has led to the development of the Bubble Chamber. In 1952, D.A. Glaser, at the University of Michigan, designed and made the first bubble chamber and this has now become one of the largest and at the same time one of the costiliest type of particle detector in use.

We know that normally the liquid boils with the evolution of bubbles of vapour at the boiling point. If the liquid is heated under a high pressure to a temperature well above its normal boiling point, a sudden release of pressure will leave the liquid in a super heated state. If an ionizing particle traverses the liquid with in a few milliseconds after the pressure is released, the ions left in the track of a particle act as condensation centres for the formation of vapour bubbles. The vapour bubbles grow at a rapid rate and attain a visible size in a time of the order of 10 to 100μ sec. Upon this principle chamber operates. The bubbles are like the droplets in a cloud chamber, visible under strong illumination. If nuclear reactions take place in the liquid of the bubble chamber, the sets of tracks diverge from the collision centre as in the cloud chamber. The tracks are photographed against a dark background.

The time sequence of events for a typical chamber is shown in Fig. 2.26. At the instant A, the high pressure P_1 (400 units) applied to the bubble chamber liquid is suddenly reduced to the lower pressure

P_1 (100 units). The pressure starts to rise due to the bubble formation. The liquid is recompressed at C and the chamber returns to its initial position at A'. The time interval between A and A' gives the cycling rate of the chamber. The expansion is timed so that the particle enters the chamber at X and the bubbles are illuminated and photographed at the moment Y.

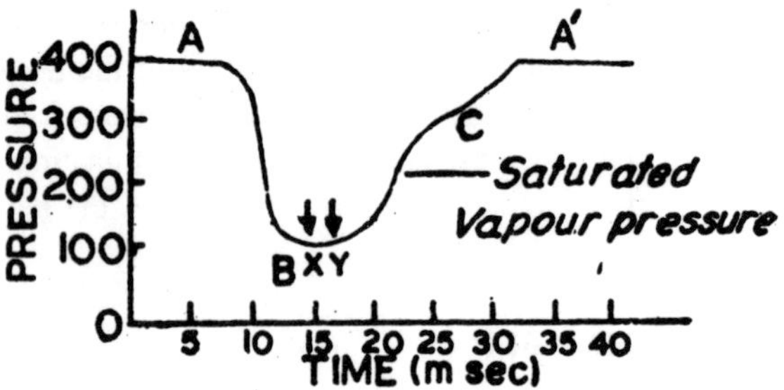

Fig. 2.26 : Time sequence of events in bubble chamber.

The liquids suitable as the working substance of a chamber are transparent and having a low surface tension and a high vapour pressure. The medium in first bubble chamber was ordinary ether, *i.e.*, diethyl ether, but liquid hydrogen, deuterium, helium, propane, xenon and other liquids have been utilized successfully for different purposes. In some experiments it is desired *to study the interaction of particles with protons*, the liquid in the chamber is then hydrogen. Propane (C_3H_8) is commonly used in the bubble chamber to avoid the need for low temperature (–246°C) operation with liquid hydrogen. It has almost 1.3 times as many hydrogen nuclei (*i.e.*, protons) per unit volume, but the carbon nuclei present are sometimes a drawback. For *observations on neutron interactions* liquid deuterium with a neutron fairly loosely bound to a proton if employed. Liquid helium chambers operating at 4°K and 1 atm. have been used for studying high energy interactions with the nucleons in the He^4 nucleus. A liquid containing heavy atoms is favoured where stopping power is the main objective, *e.g.*, *in studying the interactions of neutral particles and gamma rays.* Examples are the freon type compound, trifluoro-bromomethane (CF_3Br) and liquid xenon. The former can be used just above room temperature (30°C, 18 atm.) however there are three distinct complex nuclei present in this liquid. Liquid xenon operating at 14°C is ideal for studying interactions with high energy γ-quanta and neutral pions. For satisfactory bubble tracks a small amount of ethylene is to be added. The operating conditions for bubble chambers with different fluids are given in the following Table 2.2.

Table 2.2 : Operating conditions for Bubble Chamber fluids

Fluid	Temp. °C	Pressure (atmospheres)	Mean free path for 100 MeV γ-rays (m)	Density (Kg m^{-3})
Hydrogen	–246	5	27	60
Deuterium	–241	7	20	130
Helium	–269	1	18	130
Propane	58	21	2.2	430
Pentane	157	23	—	500
Xenon	–20	26	0.07	2300

Liquid hydrogen is an ideal substance for studying high energy interactions with protons, as it is a pure proton target having very small Coulomb scattering. L W. Alvarez at the Lawrence Radiation Laboratory, California University, thought that if the chamber was sufficiently large and the pressure of the compressed liquid was decreased rapidly, good tracks could be obtained in the interior in spite of the fact that bubbles were formed at the walls. This conjecture proved to be correct and a chamber six feet long and of cross-section 1½ feet, containing 520 litres of liquid hydrogen, was first operated in 1959 successfully. A schematic diagram of a hydrogen bubble chamber, operating at a temperature of 27°K and developed by Alvarez and his co-workers is illustrated in Fig. 2.27. The vessel was made of stainless steel with glass ports at the top for the viewing cameras. A box of thick glass walls was filled with liquid hydrogen and connected to the expansion pressure system. To maintain the chamber at constant temperature it was surrounded by liquid nitrogen and liquid hydrogen shields. From the side window W the high energy particles were allowed to enter the chamber. Liquid hydrogen was kept under pressure first but when a particle has passed through it the pressure was released so that it was super heated at a lower pressure. A sudden release of pressure from the expansion valve we followed by light flash and camera took the steroscopic view of the chamber.

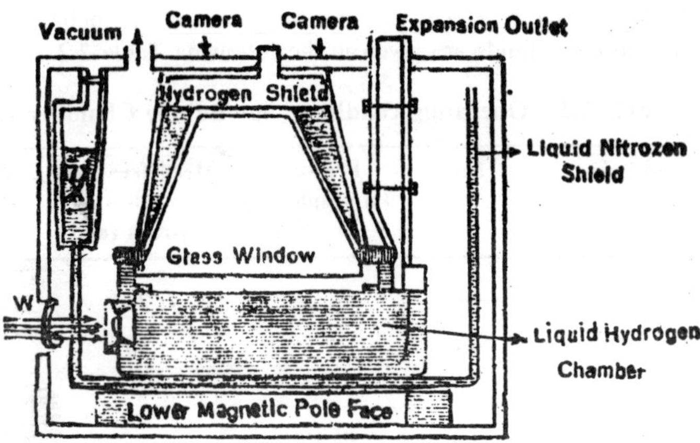

Fig. 2.27 : Schematic diagram of Alvarez bubble chamber.

This was the first of the so-called dirty bubble chambers. Subsequently, 80 inch hydrogen bubble chamber was built at the Brookhaven National Laboratory and similar one is in use at CERN (*Conseil Europeen Pourala Recherche Nucleaire*). Even larger bubble chambers, upto 14 feet in length, are being designed. Similar to the original cloud chamber bubble chamber is not continuously sensitive.

(i) A dirty bubble chamber is sensitive for only a few milliseconds after the pressure reaches its minimum.

(ii) The relatively slow expansion of the liquid does not permit the process to be initiated by an ionizing particle continuously.

(iii) The bubbles formed spontaneously at discontinuities and they interfere with the photography of the particle tracks.

(iv) Once the liquid has been expanded, it takes some time to recover sufficiently for another set of tracks to be recorded.

These drawbacks are overcome by using bubble chambers in connection with high energy particle accelerators, which function in a pulsed manner. The chamber can be made sensitive each time a pulse of particles emerges from the accelerator, and the large density and large volume of the chamber permit appreciable numbers of interesting collision processes even when the incident particles have energies of thousands of MeV. The timing is such that the particle beam from the accelerator enters the bubble chamber just prior to the pressure minimum. A few milliseconds later, powerful lights flash on and the tracks are recorded on a photographic film which moves forward automatically after each exposure. For accelerators, which emit particle pulses at more frequent intervals, smaller bubble chambers, with shorter recovery times, are employed.

A magnetic field is usually provided at the chamber in order to distinguish the sign of the charged particles and to measure their momenta from the radius of curvature of the bubble track. For particles of high momentum, the curvature is small and hence is difficult to measure unless a strong magnetic field is used. Thus, to analyse the collisions it is necessary to maintain the whole chamber in a strong uniform magnetic field (~20,000 gauss). To achieve this a magnet weighing several hundred tons is required with drastic cooling arrangement. Consequently, there has been much interest in the possibility of constructing superconducting magnets for use of bubble chambers. The problem in the construction of such superconducting magnets of appreciable size has been the difficulty of producing cable with satisfactory physical and mechanical properties at a reasonable cost. In recent years, certain metallic compounds (alloys) of niobium, *e.g.*, with tin, titanium, or zirconium, have been developed for fabrication into super conducting cables. The first successful operation of a bubble chamber, 10 inches in diameter, with a super conducting magnet was achieved at the Argonne National Laboratory in 1966.

The analysis is laborious and is greatly assisted by computing machines. A simplified account of the methods used is given below. The first step in the analysis, involving the selection of events of significance to the experimenter, is known as *scanning*. The magnified image of the film is inspected and the locations of interesting events are recorded. The next stage in track analysis, known as *tracking*, involves various degrees of automation. The tracking operation is greatly expedited by utilizing a flying spot digitizer. A small spot of light is moved rapidly in this system either mechanically or electronically, across and down the image of the bubble chamber event. Each time the moving spot of light crosses a track, co-ordinates of the point are expressed in digital form. The complete co-ordinates of the event of interest are next treated by a computer to yield what is called the *geometry* of the event. For the next stage, known as *kinematic analysis*, it is required to know the masses of the particles that produce the observed tracks. The kinematic analysis, made by the computer at each vertex of an event is based on the conservation laws of momentum and of kinetic energy.

Nevertheless, cloud and bubble chambers both are important detectors of radiation from a historical point of view since they have played such an important part in the study of fundamental particles and their interaction. They are used mainly in Nuclear Physics research. It has been estimated that a comparatively small bubble chamber is equivalent in stopping power to a cloud chamber some 140 ft. long.

2.11 SPARK CHAMBER

The spark chamber utilizes an incipient electrical discharge in a gas, like that used in a Geiger counter, to give track geometry information, like that provided by a bubble chamber. This has found increasing use in the field of high energy physics. The basic principle of operation is as follows:

A high voltage is maintained between two plates spaced in a vessel containing a gas but the electric field is not quite enough to permit the passage of a spark. Now if an ionizing particle enters the gas space, a spark will pass and it will tend to follow the path of the ion pairs produced by the ionizing particle. The passage of a spark through a gas that had been ionized by nuclear radiation was used by H. Greinacher, in 1935, for the detection of α-particles. F. Bella and C. Franzinetti in 1953, in Italy, confirmed the localization of the spark discharge and the first photographs were given in 1955 by P.G. Henning in Germany. The modern spark chamber stems from the work described by the British physicists T. E. Cranshaw and J. F. de Beer in 1957. In 1959, S. Fukui and S. Miyamoto in Japan found that with neon several particle tracks could be observed simultaneously in the spark chamber and with air only one particle track could be formed at a time. Spark chambers were used by Sree Kantan and coworkers at T. I. F. R. Bombay in 1960, to trigger high energy cosmic-ray events.

In its simplest form spark chamber consists of a series of large thin parallel metal plates (aluminium), several square feet in area, spaced about 1 cm. apart in a chamber filled with a neon gas at atmospheric pressure. All the plates are isolated from each other. The first, third, fifth, etc., plates are grounded and the second, fourth, sixth, etc., plates are connected to a high voltage direct current pulse generator, which gives them a high potential in short bursts of the order of a microsecond each.

This potential is just enough to cause sparks to occur between the plates in such regions as are ionized by a particle entering the chamber. The chamber is sensitive for about half microsecond preceding the application of the voltage.

Any charged particle passing through the chamber during this period produces along its path a visible and audible spark discharge, which can be located by photographing or by other methods. The photograph of all the individual spark reveals the sections of the trajectory of the particle between the plates. Ordinarily two cameras at right angles to one another are used to photograph.

In order to prepare the chamber for recording the path of another particle, any ion-pairs present between the plates must be removed, by applying a clearing field of about 200 volts per cm. continuously in the direction opposite to that of the high voltage spark field. Due to this field a spark can form only if the high voltage is attained within about a microsecond after the entry of an ionizing particle. In practice this delay time is about 0.2 microsecond. The recovery time depends on the gas, the design of the chamber and the energy of the spark. It is usually between one to 20 milliseconds, which is still very short in comparison with the recovery times of cloud and bubble chambers.

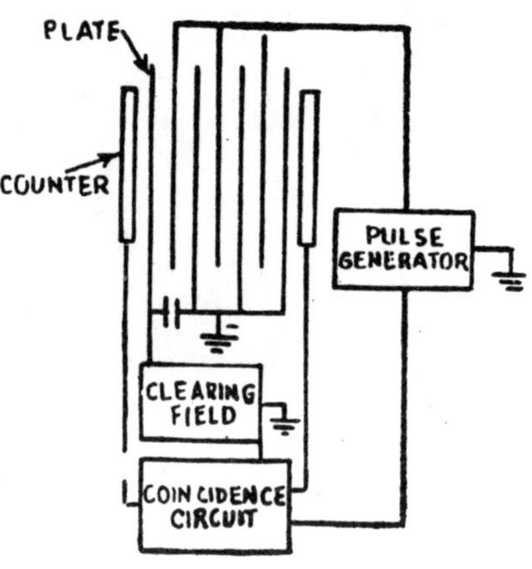

Fig. 2.28 : Schematic representation of spark chamber circuit.

The chamber can be operated in a magnetic field and the momentum and charge sign of the ionizing particle can then be determined from the curvature of the track. The number, thickness, spacing and material of the plates in a spark chamber are varied to suit the experimental requirements. The plates are commonly made of very thin aluminium to minimize energy absorption from the particle and are closely spaced to define the track more accurately. The sparks tend to be perpendicular to the plates in their passage between the adjacent plates. The photograph of spark chamber tracks and their interpretation is shown in Fig. 2.29. Sometimes it is desired to study the interactions of specific particles with matter. The plates are then thicker usually and are made of either carbon or a metal, such as iron, lead, tantalum or brass. The interactions take place almost within the solid plates as the density of nuclei in the gas between the plates is small.

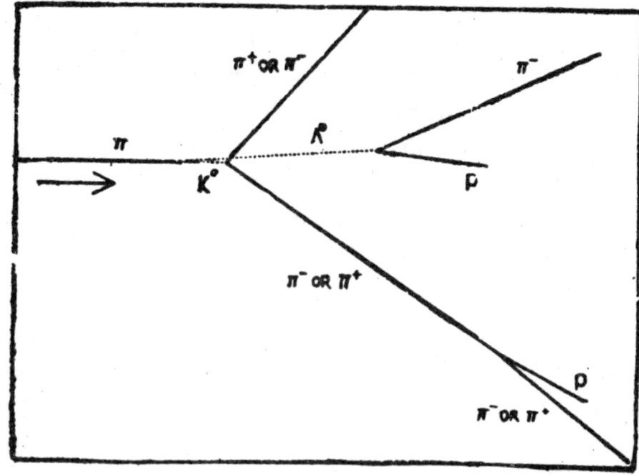

Fig. 2.29 : Photograph of spark chamber track (above) and their interpretation (below).

One of the main advantages of the spark chamber over the other chambers is that the triggering and removal of ions by the clearing field are comparatively simple. With the help of this chamber it is possible to study an individual event produced by one particle among the many passing through the chamber at a rate that may be as high as a million per second. The spark chamber event of interest is selected by an arrangement of counters, called a *Logic circuit.*

Analysis of spark chamber photographs can be carried out by ordinary calculations or by computer techniques. The preliminary scanning stage may not be required because spark chamber events of interest have been selected by the logic system. The tracking, geometry and kinematic analysis arc performed in the manner as with bubble chamber pictures.

There are *two main drawbacks* to a spark chamber as compared with a bubble chamber. First, in the bubble chamber the liquid may be chosen so as to restrict the interactions of the incoming particle to those of a special type. Second, the co-ordinates of tracks and of the points of interaction in a bubble chamber can be defined with much greater accuracy than is possible in a spark chamber. As a general rule, bubble chambers are preferred where good space resolution is desirable. On the other hand, spark chambers permit the selection of events of interest with a high repetition rate, *i.e., they have better time resolution.*

Since 1963, various "filmless" spark chambers have been developed in which photography of the spark discharge is avoided. In such devices, the co-ordinates of the spark tracks are obtained directly in a form suitable for a computer to determine the geometry of the event. The grounded plate of every pair is replaced by an array of closely spaced parallel wires. For the high voltage plates a similar arrangement of wires could be used but solid plates are generally employed since they are more convenient and easier to fabricate. The wires arc made from a magnetostrictive material (50% Fe and 50% Co alloy) in *one type of wire spark chamber*. A passage of a spark from a particular wire produces a local magnetic field and a deformation occurs at the affected point. The deformation travels then along the wire to suitable detectors at each end The ratio of the times between the firing of the spark and arrival of the deformation at the ends gives one co-ordinate, the other co-ordinate of the spark track is obtained by identifying the particular wire at which signals were received. The most promising wire spark chamber, referred to as a *spark-chamber hodoscope*, differs from the acoustic wire chamber in respect that the wires constituting the alternate grounded plates are alright angles to each other. At one end each wire is threaded through a ferrite core, such as is used in computer memory devices. A ferrite material has two stable states of magnetization and by a magnetic field (from an electric current) either one can be converted to the other. The magnetism of ferrite cores associated with the wires from which the spark passes is changed from one state to the other by the spark current or we can say that ferrite cores are nipped. By interrogating the cores in a manner similar to that employed in a computer the locations of the affected wires are determined and flipped cores are reset at the same time. Since successive planes of wires are at right angles, the information from the cores can give the co-ordinates of the spark track in three dimensions in digital form. It has a advantage of a very short recovery time. The amount of energy required to flip the core is very small and a feeble spark, which would not give off enough light to be photographed, is adequate for this purpose.

Sometimes it is desirable to make observations of interactions at very-very short intervals with high energy particles. *The scintillation counter hodoscope* can be used for this purpose. Although the accuracy of determining the co-ordinates of particle tracks decreases. It consists of an array of long thin rectangular strips (25 × 0.5 × 0.5 cm) of a plastic scintillator material. These strips are set side by side to produce a sensitive planer area. A photoelectric tube is associated with each strip and the passage of a charged particle through a given strip is indicated by the output signal from the corresponding photo-tube. This provides one co-ordinate of the particle track. For the determination of co-ordinates in the other two directions, it is necessary to build a hodoscope from several layers of scintillators arranged with the strips in successive layers at right angles. The output signals can be fed directly to an on-line computer for analysis. This type of hodoscope can record events at a much faster rate than wire spark chamber hodoscope can. It is because of the very short recovery time of a scintillation counter. The accuracy with which the track co-ordinates can be determined in a scintillation counter hodoscope is limited as it is dependent on the width of the scintillator strip.

2.12 NUCLEAR EMULSIONS

The discovery of radioactivity by Becquerel (1896) can be explained as:

As charged particles pass through a photographic emulsion of a photographic plate they interact with the tiny silver halide crystals producing a line of latent images along their path. In 1911, M Reinganum in Germany recognized that individual α-particles traversing a photographic emulsion produced separate tracks in a photographic emulsion as a series or closely spaced black specks of silver. Because of the relatively high density and stopping power of the photographic emulsion, the tracks produced by charged particles were very short compared with those obtained in a cloud chamber but do not differ greatly from bubble tracks. Miss Blau and Wambacher, in 1937, discovered 'Star' due to the interaction of high energy cosmic rays with the atomic nucleus in the emulsion.

The optical photographic emulsions is not suitable for quantitative work with nuclear radiations. The sensitivity is low and the tracks due to charged particles have non-clear range because the developed crystal grains are large and widely spaced. The composition of the emulsion was changed so as to make it more suitable for study of various ionizing particles, such as alpha particles, protons, mesons and even electrons The nuclear emulsions differ from the optical emulsions in that they have considerably higher silver halide content and smaller grain size In nuclear emulsion the thickness is greater than that of optical emulsion. The smaller the grain size, the more sensitive the emulsion to ionizing particles. Thus different commercially available emulsions, differing chiefly in grain size, can be used to discriminate between different particles. The

Table 2.3 : Composition in kg/m³ of Dry Nuclear Emulsions

↓ Element→ Type	H	C	N	O	S	Br	Ag	I
Ilford	49	300	73	200	11	1465	2025	57
Kodak Ltd.,	38	270	80	160	–	1440	1970	36
Eastman Kodak	43	340	110	170	–	1220	1700	54

least sensitive emulsions will show only fission fragment tracks, whereas the most sensitive ones reveal the tracks of singly charged minimum-ionizing particles. A series of nuclear emulsions has been developed for detecting particles of different specific ionization. D-l emulsion is the suitable type for detecting fission fragments and alpha particles. The C-2 Ilford emulsion is a useful general purpose emulsion, suitable for recoiling mesons, protons and heavy nuclei. The Ilford G-5 is used to record high energy electrons. The tracks produced in an emulsion are very short. They can be magnified and photographed. High quality microscopes are required for sinning the developed plates. The nuclear emulsion with their high silver halide concentration shrink during the fixation process unfortunately. A knowledge of this shrinkage factor is necessary when we want to determine the angle of dip of a track in the emulsion.

Table 2.4 : Nuclear and Optical Emulsions

Property	Optical Emulsion	Nuclear Emulsion
AgBr : Gelatin (mass)	4 : 53	80 : 20
AgBr : Gelatin (volume)	15 : 85	45 : 55
Grain size (micron)	1—3.5	0.1—0.6
Thickness (")	2—3	25—2000
Sensitivity to light	Very high	Poor
Response to α-particles	Dense blackening	Individual tracks
" " β "	Moderate "	Faint fog
" " γ "	Faint "	Almost none

Table 2.5 : Characteristics of Nuclear Emulsions

Ilford Emulsion type	D1	E-1(K1)	C_2(K2)	B-2	G-5
Mean grain diameter (mm)	0.12	0.14	0.10	0.21	0.18
Highest detectable velocity β	0.2	0.31	0.46	all	
Highest detectable energy of protons in MeV		20	50	120	all

Fast neutrons can be detected by the knock on protons in unloaded emulsions. Slow neutrons can be detected by using emulsions impregnated with lithium, boron or uranium. Gamma rays of energy above about 3 MeV can be detected through the track of the protons, released in photo-disintegration reaction, in emulsions loaded with deuterium compounds. Under favourable conditions it is possible to determine the charge, mass and incident K.E. of a charged particle brought to rest in a nuclear emulsion. The range of the particle, the grain density and its variation along the track and the small scattering angle due to collisions of the particle with nuclei in the emulsion are all relevant to these determinations. We shall confine ourselves here to discussion of the main relationships which have been established by experiments.

The number of grains deposited per micron dn/dx of track is closely related to the energy loss per unit of track dT/dx keV/micron. These two quantities are related by the empirical relation

$$\frac{dn}{dx} = a\left\{1 - e^{bz(dT/dx^{1/2} - c^{1/2}}\right\}, \quad ...(45)$$

where a, b and c are empirical constants and z is the integer characterizing the charge of the ionizing particle.

If m (the mass in units of proton mass) and z of a particle are known, its kinetic energy T (MeV) may be obtained from its total range R (microns) by an equation

$$T = Kz^{2n}\, m^{1-n}\, R^{n}, \quad ...(46)$$

where K and n are empirical constants.

The total number of silver grains deposited and the total particle range are related by an equation

$$N = K'\, z^{2n'}\, m^{1-n'}\, R^{n'}, \quad ...(47)$$

where K' and n' are another constants.

Fig. 2.30 gives the range of protons, deuterons, α-particles, pions and muons in Eastman Kodak NTA emulsions as a function of particle energy.

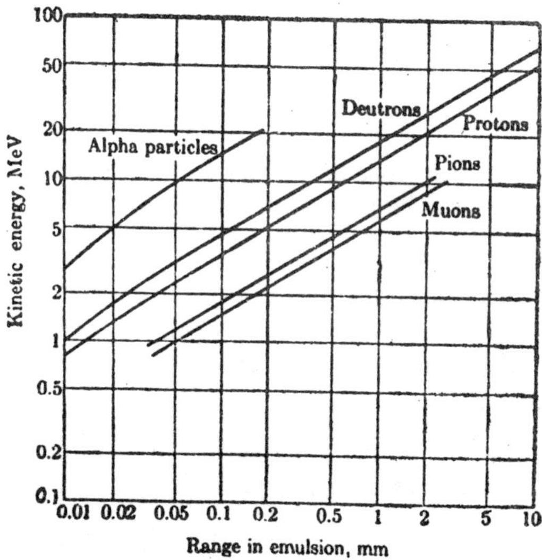

Fig. 2.30 : Range V/s energy for various particles in nuclear emulsions.

As compared with track chambers, like bubble and cloud chambers, photographic emulsions have following *drawback.* It is very difficult to determine the sign of the charge and the momentum of a particle from observations of the curvature of the path in a strong magnetic field. The actual path length in the emulsion is small compared to that in a cloud chamber and the multiple Coulomb scattering in the emulsion stimulates the curvature. By the use of very strong magnetic fields, the deflection of long (but not short) tracks in emulsions has been observed. This gives the sign of the charge, but the large amount of scattering makes the deflected tracks useless for the momentum calculations.

Some of the *advantages* of the nuclear emulsion detector are:

1. It produces a visible track following the complete trajectory of a charged particle. Particles of different ionizing powers produce tracks which appear quite different.
2. The emulsion is continuously sensitive and is consequently always available to record an event.
3. The energy of product particles can be determined from their range in the emulsion and the differential scattering cross section from counts of the number of tracks formed in each emulsion.
4. The emulsion is relatively light and cheap. This makes it better for high altitude cosmic ray experiments.
5. Due to high stopping power of emulsions, many short lived particles can be brought to rest in the emulsions, before they decay.
6. The measurement of the ranges of muons emitted from stopped pions, before they decay into electrons, enables the mean lifetime of the muon to be found.
7. They were widely employed in cosmic ray studies and led to the discovery of the π- and K-mesons. They are also employed in observations on hyperfragments.

The properties and applications of various nuclear detectors are listed in Table 2.6.

Table 2.6 : Common Nuclear Detectors

Detector	Detecting	Particles System	Application Detected
Ionization chamber	Air	α	Measurement of activity
Ionization chamber (with grid)	Argon + CO_2	α, fission fragments	Energy measurement, if particle stops in chamber
Proportional counters	Argon + CH_4	α, β, electrons	Measurement of energy.
Geiger-Müller counters	Argon + Cl_2	α, β, γ, cosmic-rays	Good for intensity determination (counts pulses).
Solid-state detectors (few microns to a few mm. in thickness)	Silicon doped with Li, P or Boron	α, β, p, d, t, fission products	Detection of charged particles. Measurement of energy.
Solid-slate detector (few mm. to 10 cm in thickness)	Germanium doped with Li	γ	Energy discrimination, but low efficiency.
Scintillation detector (NaI)	NaI (Thallium activated)	γ, high energy particles	Energy measurement.
Solid-organic; Scintillators	Anthracene stilbene	β^-, p	Energy selection for electrons, Detection of recoil protons.
Liquid-organic Scintillators	Toluene	β^-, soft X-rays	Detection of soft-rays from element dissolved.
Cerenkov counters	Glass, mica, cellophane	Electrons, X-rays, charged particles	Detection of high energy X-rays, To distinguish charged particles.
Cloud chamber	Alcohol +water	Charged particles	Determination of energy, Particle detection.
Bubble chamber	Propane, Xenon, liquid H_2	Charged particles	Detection of very energetic particles. To study particle interaction.
Spark chamber	Metallic plates	High energy charged particles	Localization of trajectories.
Photographic emulsions	Emulsions	Charged particles	Determir tion of energy and charge. To study particle interaction.

2.13 DIFFERENTIATING AND INTEGRATING CIRCUITS

Differentiating and integrating circuits are used for pulse forming and sharpening or for sorting out pulses. The need for a sharp pulse of voltage that repeats periodically at a high repetition rate is common in practical applications, *e.g.*, in electronic circuits including television, transmitters and receivers. Sharp pulses are used to initiate action of trigger circuits.

Consider the CR circuit shown in Fig. 2.30. For an initial potential across C zero, the input potential

$$e_i = \frac{1}{C}\int i\,dt + iR$$

or $$\frac{e_i}{R} = \frac{1}{RC}\int i\,dt + i \quad ...(48)$$

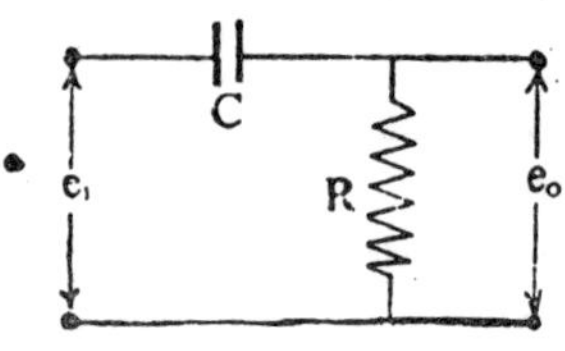

Fig. 2.31 : Differentiating Circuit.

If RC is small with respect to the time of integration, then the second term is small enough and can be neglected w.r. to the first. We thus have current

$$i = C\,de_i/dt.$$

Thus, the output potential $e_0 = CR\,de_i/dt$. ...(49)

Hence for a small value of RC, the output voltage is proportional to the derivative of the input signal. In this way we obtain the rate at which the voltage is changing with respect to time across the output. The differentiation of square wave by RC circuit is shown in Fig. 2.32. When e_i goes from zero to its positive value V_0, its rate of change is large and positive. It appears at once across the resistor, since the voltage across the capacitor cannot change instantaneously. As capacitor then charges, the potential across R decreases exponentially, $V = V_0 e^{-t/CR}$ and becomes zero. When input potential decreases from V_0 to zero, its rate of change is large and negative but equal in amount to the former positive

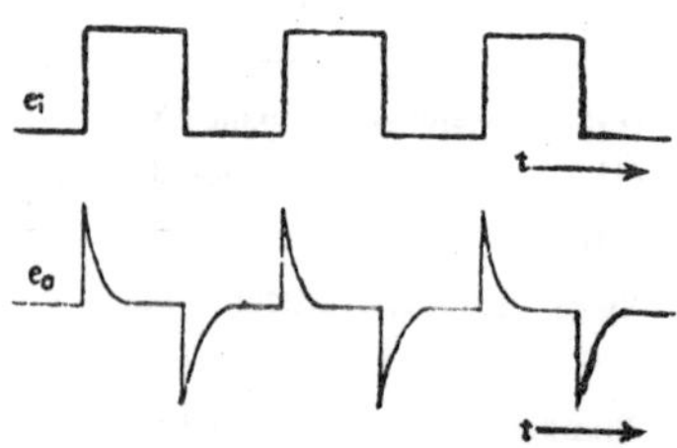

Fig. 2.32 : Differentiating Input (above) and out-put (lower).

rate of change so the negative value of e_0 at that instant represents this rate of change.

A sine-wave input becomes a cosine-wave at the output and a cosine-wave input becomes an inverted sine wave at the output. The rate of change with respect to time is zero at the maximum and minimum points of the sine and cosine waves.

The integration of a voltage wave may be done by the circuit shown in Fig. 2.33. In order to calculate the voltage change at the output over any time interval, the input volt-time area is simply divided by the time constant. The time constant *RC* has to be very large in comparison with the period of the input signal Mathematically it can be shown that for this circuit, output potential is same as obtained after integrating input potential. From eqn. (48), for large value of RC, we have

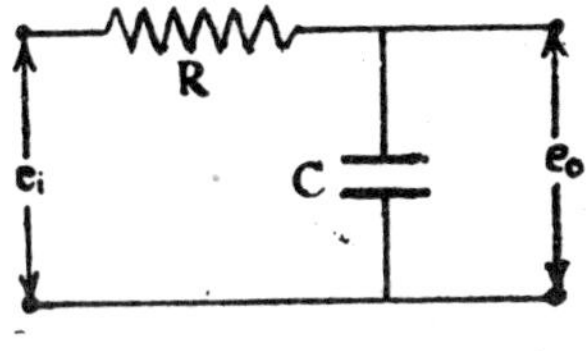

Fig. 2.33 : Integrating circuit.

$$i = e_j/R$$

$\therefore$ Output voltage

$$e_0 = \frac{1}{C}\int i dt = \frac{1}{RC}\int ei\, dt$$

The output of an integrating circuit is illustrated in Fig. 2.34 along with the input.

From above results we sec that the output of a differentiating circuit is independent of the d.c. level of the input, but the output of an integrating circuit must attain a d.c. level equal to that of the input. Thus, a true integration is attained for input signal having an average value of zero.

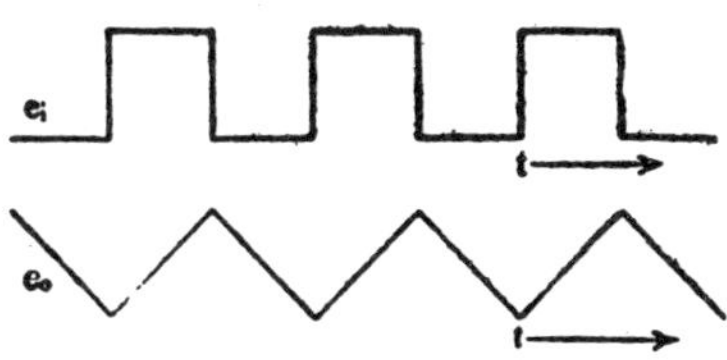

Fig. 2.34 : Integrating input (above) and output (lower).

2.14 PULSE AMPLIFIER

In nuclear radiation detectors, *e.g.*, in an ionisation chamber, the pulse height is very small and is of the order of micro volts. Since output voltages of 10 to 100 volts are often required for electronic pulse height discrimination, the gain requirement might be 10^5 to 10^6 for the amplifier. If we arc only interested in counting the ionization pulses then it is not necessary for the height of the output of the amplifier to be accurately proportional to the height of the counter pulse. In order to determine the energy of the ionizing particle, we require linear amplifier. To maintain proportionality between input pulse height and output, the stability of the amplifier is important. Linearity is obtained by proper choice of operating points for the amplification elements and by the use of negative feedback.

The pulse amplifier usually consists of three units in cascade. The first unit is the pre-amplifier which is attached to the nuclear detector by a very short cable This reduces distributed capacitance, which decreases rise time. Its gain is usually about 1 to 100. The output stage of the pre-amplifier is a cathode follower with its characteristic low output impedance. It is, therefore, practical to connect the output terminals of the pre-amplifier by a coaxial screened cable several feet long to the input terminals of the main amplifier, comprising cathode follower and negative feedback. The coupling network CR between pentode valves may be used as the short time constant differentiating network. The third stage of the unit is a triode connected pentode or a triode with a cathode load resistance shunted by a small trimming capacitor. The output signal

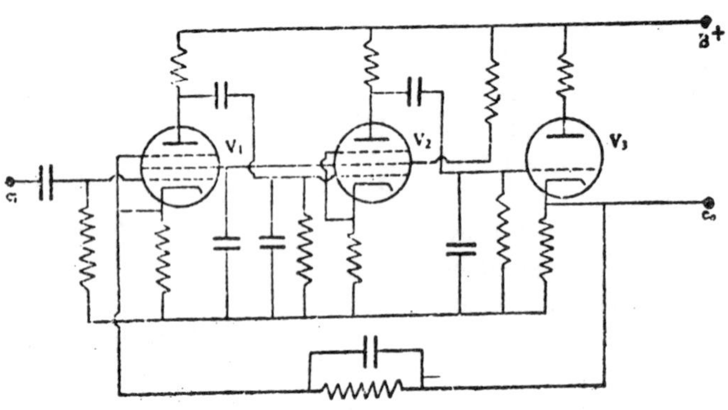

Fig. 2.35 : Linear Pulse Amplifier.

is feedback into the input stage through the direct connection between cathode resistances of pentode V_1 and triode V_3. The trimmer capacitor is adjusted so that no transient overshoot occurs with a fast rising input pulse to the amplifier.

2.15 THE TRIGGER (FLIP-FLOP) CIRCUIT

A closed loop regenerative system having two or more clearly defined stable or quasi-stable slates, each of which is maintained without recourse to external forcing, is known as *multivibrator*. Multivibrators are utilized in one of three basic configurations: Free running, monostable, or bistable. *Free running or a stable multivibrator* oscillates continuously to provide outputs used as clock pulses, or as sweep waveforms when a continuous time base is required. The *monostable or one-shot multivibrator* oscillates through one cycle to produce variable-width pulses for use as gate or timing pulses. The *bistable or flip-flop multivibrator* does not oscillate but is driven through each switchover action and is used in counting or toggle circuits when the stability characteristic in either of two states is important. It is a two stage device with one stable slate described as having one stage on or conducting and the other stage off or non-conducting. The second stable state is the reverse with the first state conducting stage off and the first stale non-conducting stage on.

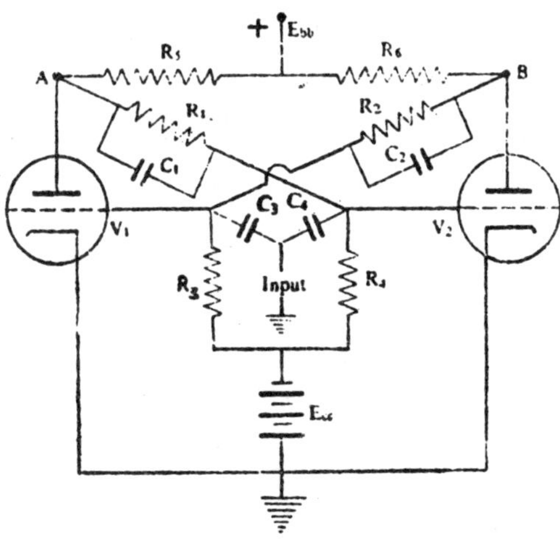

Fig. 2.36 : Eccles-Jordan Trigger circuit.

The-most famous circuit of this class is the *Eccles-Jordan Trigger circuit.* We assume that V_1 is conducting, it then follows that V_2 is not conducting because its grid bias is kept below cut off. For this purpose first assume that the bias battery E_{bb} is not present but that the bottom ends of the grid leak resistors are connected to the grounded cathodes. The voltage-divider action of R_1 and R_2 would cause the lowered plate potential at A to put a small +ve bias on the grid of V_2 As E_{cc} is larger than this +ve bias voltage, when it is put back in place it will make the grid negative with respect to the cathode.

If a positive trigger pulse is applied to the input, it will not alter the current in V_1 which is already conducting, but it will derive V_2 into conduction. This reduces the potential at B and at the grid of V_1, reducing the plate current of V_1 and raises the potential at A and the grid potential of valve V_2. This action further turns on valve V_2 and the process continues until V_1 is cut off and K_2 is conducting fully. The circuit will remain in this second stable condition until another +ve pulse conies along and reverses the action. Capacitors C_1 and C_2 are not needed to hold the grids of triodes negative between switching but are used to transfer instantaneously to opposite grids the changes in plate voltage of the triodes.

Negative pulses may also be used to trigger the circuit. They will not affect the cut off tube and will reduce the current in the conducting tube. The only requirement on the triggering pulses is that the signal must be strong enough and last long enough that the plate current in the on-coming tube exceeds that of the off going tube so that the regenerative action will continue the switching to completion.

2.16 VOLTAGE DISCRIMINATOR

It is often necessary to discriminate between pulses of different amplitudes. A voltage discriminator rejects all pulses below a predetermined height. It should have high stability and sensitivity to discriminate between input pulses differing in height by a fraction of one volt, should have short rise and recovery times, should generate an output pulse of standard height if the pre-determined level is exceeded. A circuit, which meets all these requirements, is *Schmitt trigger circuit.* It is a modified bistable multivibrator, *i.e., cathode coupled bistable multivibrator*, illustrated in Fig. 2.37. This circuit is also useful as a squaring circuit.

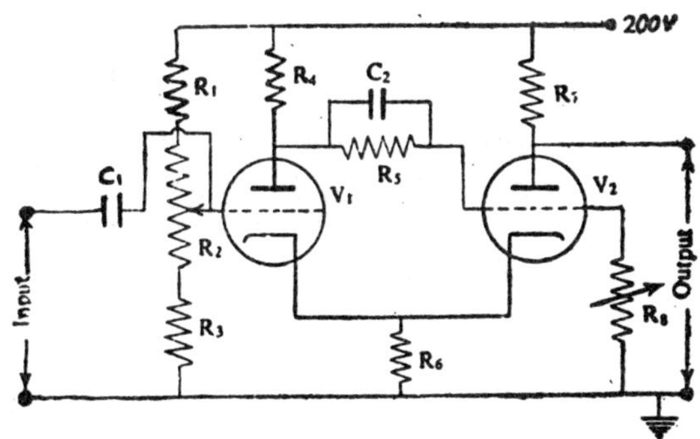

Fig. 2.37 : Voltage Discriminator.

In this circuit the direct coupling from the plate of V_1 to grid of V_2 is the same as for conventional bistable but reverse is eliminated. The common cathode resistor R_6 provides the other necessary coupling for regeneration between stages. When the input signal is below a preset voltage value, obtained by adjusting discriminator level potentiometer R_2, one tube conducts and the other is cut off. The moment the voltage exceeds the preset value, there is a rapid transition of states. The circuit remains in the new states as long as the input remains above a voltage level that is slightly lower than the initial triggering voltage. As soon as the voltage drops below this second level, there is another rapid transition of states back to the original states. Input signals below the preset voltage level do not provide an output signal.

The triggering level can be set by a potentiometer R_2 which gives an average value of E_{g1} of tube V_1. With E_{g1} sufficiently low, V_1 is cut off and R_8 is adjusted so that V_2 is heavily conducting. The plate current of V_2 causes a potential drop E_K across the common cathode resistor R_6. The difference $E_{g1} - E_K$ is now the grid bias of V_1. Since the transition of states occurs the moment the cut off voltage E_{co} is exceeded, hence the triggering level can be given by $E_{co} - (E_{g1} - E_K)$. At this point V_1 starts conducting. Its plate voltage and grid voltage of V_2 decrease, causing a decrease in plate current of V_2. The resulting drop of cathode voltage increases the plate current of V_1 and the regeneration process continues until V_2 is *off* and V_1 is on. The output from the plate of V_2 jumps to the plate supply value because of the transition.

2.17 SCALARS

The simplest type of operating or indicating device in a counting system is an electromechanical register with a suitable driving circuit. It consists basically of an electromagnet energised by the amplifier output. The energised electromagnet causes an armature to move which rotates a numbered drum through a pawl and ratched system. Each voltage pulse could momentarily close the switch contacts and advance a register one count. The drum carries ten numerals and on completion of one revolution it causes the adjacent drum to move forward. This type of registers are used when the pulse repetition frequency is not too high of the order of 5 per sec.

For most nuclear counting work, the register is preceded by a scaling circuit. The simplest one is the Eccles-Jordan scale of two. It has been seen in the flip-flop of Fig. 2.35 that at point B, a +ve pulse is obtained as output when V_2, cuts off and a –ve pulse appears when V_2 turns on. The second flip-flop circuit, which responds only to negative pulses, therefore, responds only to one pulse for every two pulses sent to the flip-flop circuit. The circuit is given the name of a scale of two circuit. Thus the count-down feature has a factor of 2^n, where n is the number of flip-flops.

Fig. 2.38 shows three flip-flops arranged to deliver only one pulse out of every eight pulses applied to the input. A small cold cathode neon indicating lamp is across the plate-load register of each of the odd numbered tubes V_1, V_3, V_5. Each will light up whenever its tube is conducting. When the tube is cut off, the potential difference across the neon lamp is well below the striking potential. The plates of V_2 and V_4 are capacitively coupled to the common plate lead of the tube that follows so that the signal will be transferred to the next flip-flop. A reset switch R is provided to put positive potential on the grids of V_2, V_4 and V_6, so that they will be on and odd numbered tubes will be off at the start of the count and ail indicator lamps are off.

The first input pulse (remembering that all input pulse are negative) reduces the potential at points A and B momentarily. Since V_1 is off there is no current between these points. The coupling capacitor sends this negative pulse to the grid of V_2, reducing its current. Thus the conduction is transferred to V_1, the light of lamp 1 goes on and potential of point C rises abruptly. This positive pulse at C is transferred to the second flip-flop, where nothing happens because V_4 is conducting already and V_2 has too much negative bias from the common cathode resistance.

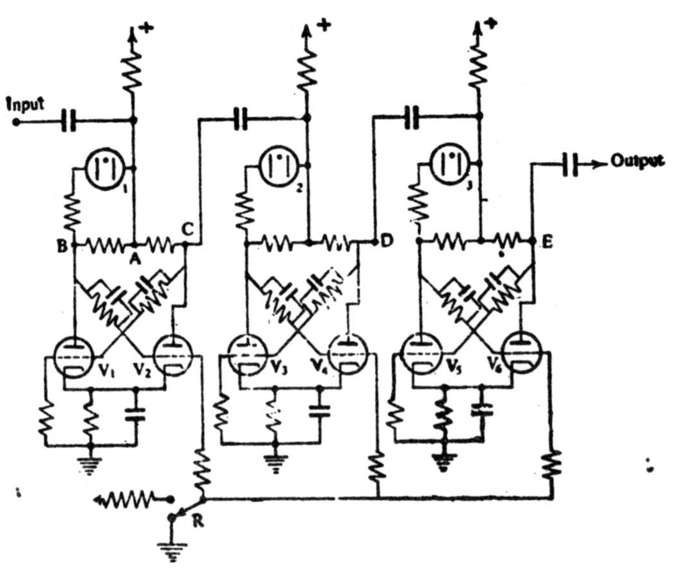

Fig. 2.38 : Three bistable multivibrators forming a scale of 8-circuit.

The second pulse switches conduction back to V_2 with a consequent drop in potential at C and triggering of the second flip- flop. V_3 now conducts and the light of lamp 2 goes on. The potential of point D rises hut causes no change in V_5 and V_6. The third input pulse turns off V_2, turns on V_1, and lamp 1, but the positive pulse sent forward from C does not do anything to V_3 or V_4. The fourth pulse at the input causes C to drop in potential and this negative pulse turns on V_5. Point D drops in potential and this turns on V_5 and the light of lamp 3 while V_6 goes off. The fifth pulse at the input causes C and D to rise in potential but no change occurs in V_3 or V_4. The sixth pulse at the input causes C to fall in potential and this turns off V_4 and rises potential at D. The state in the fast flip-flop, where V_5 on and V_6 off, does not change. The seventh pulse at the input causes C to rise in potential with no change beyond this point. V_4 is off. The eighth pulse at the input causes C to fall, turning V_4 on and causing D to drop in potential. This negative pulse turns off

V_5 and turns on V_6. The drop in potential at E is the negative pulse, the output needs to actuate the, mechanical counter. The increase in the register reading from start to finish must be multiplied by 8 to get the total number of input pulses.

Thus, we see that lamp 1 is on for one pulse, lamp 2 is on for two pulses and lamps 3 is on for four input pulses. With 7 pulses all three lamps arc lit and their indicated total value is

$$(1 \times 1) + (1 \times 2) + (1 \times 4) = 7$$

Scale of 10 counting is possible when feed back is used in the circuits. This is more convenient because our system of numbers is based on the factor 10.

2.18 DEKATRON

An alternate method of counting is provided by the multi-electrode cold cathode scaling tube known as dekatron. It is designed so a glowing spot moves successively around its circumferences with successive input pulses. It is a neon filled (13 cm. Hg pressure) tube containing an anode in the form of a flat disc of diameter ~1 cm. Surrounding the anode are thirty identical electrodes. They are symmetrically spaced and are about 0.05 cm. from the anode. These are grouped into three categories based on the operation principle of the device. Ten of the electrodes act as cathodes, A, to A, connected to a common point and K_0 the output cathode. Adjacent to each cathode is a pair of identical electrodes known as transfer electrodes or guides. The electrode nearest the cathode in a given direction is labelled guide 1 or 1G and the next electrode in the same direction is labelled guide 2 or 2G. The ten 1G electrodes are connected to a common terminal as are the ten 2G electrodes. The guide 1G and 2G electrodes are biased 20 V above the potential of the cathodes so that normally the discharge rests on one of the cathodes. Switch is normally closed but is opened for resetting to zero.

When the switch is open, K_0 is the most negative electrode and the gas discharge occurs between K_0 and the anode. With switch closed and a large negative input pulse applied to 1G, the slow discharge transfers from K_o to $1G_0$. The duration of pulse should be greater than 60 μ sec. to ensure transfer. A delayed negative pulse is applied to the 2G electrode so that $2G_0$ remains negative after lG_0 returns to its normal bias potential of +20 V and the glow discharge moves smoothly to $2G_0$. The subsequent removal of the negative pulse from $2G_0$ moves the discharge to the next closest and most negative electrode K_1 and thus lights up the number 1 on the face plate. A second pair of negative pulses moves the discharge to $1G_1$, $2G_1$ and then to K_2 where by the number 2 is illuminated. Thus we see that the net result of a single drive pulse is to transfer the discharge from one cathode to the adjacent cathode. Two guide electrodes are required to ensure that the discharge always transfers to the next digit

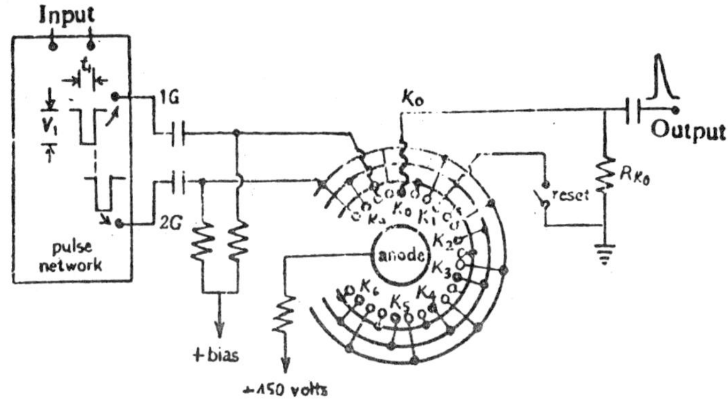

Fig. 2.39 : Simple Dekatron derive circuit.

cathode and does not jump back. The current for the discharge is supplied by the anode voltage source. The input pulses are used only to stimulate the movement of the discharge from one cathode to the next. Every tenth input pulse transfers the discharge from K_9 to K_0. A load resistor R_{K0} produces a positive output or carry pulse. The carry pulses are amplified to produce large negative drive pulses for the next Dekatron stage.

Direct readout of the count can be accomplished within each tube by placing numerals around the periphery, adjacent to the glow discharge positions. Other techniques include stacking discs with number shaped perforations to makeup the common anode and by forming each cathode into the shape of a number.

2.19 COINCIDENCE AND ANTI-COINCIDENCE CIRCUITS

We often need to tell when two pulses occur simultaneously. In practice we can only learn that they occurred within a time, the resolving time of the coincidence circuit, of each other. A variety of circuits may be used to establish the coincidence of pulses from independent detectors. One of the first coincidence circuits was introduced by Rossi. A schematic diagram of the Rossi circuit with two input channels is shown in Fig. 2.40.

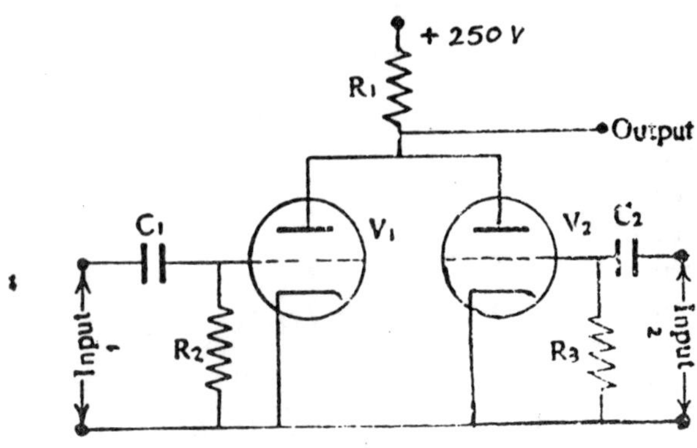

Fig. 2.40 : Rossi Coincidence Circuit.

Triodes are shown in the figure, but in practice pentodes are often preferred. The common anode load resistance R_1 is chosen large in comparison to the effective d.c. resistance of each triode. Normally tubes conduct small currents limited by the large value of R_1. Negative pulses of amplitudes, able to cut-off the anode current, from two counters are applied to inputs 1 and 2 respectively. If V_1 or V_2 is cut-off by a negative input pulse, the common rise in anode potential is only a few volts. If both valves are cut-off by the arrival of coincidence pulses at the input terminals, the common anode potential will rise to that of the H.T. (+250V) and a large positive coincidence output pulse is produced which can be discriminated easily from the small output pulses produced when one valve is cut-off

The anode potential rises at a rate dete mined by the time constant $T(= R_1C)$, where C is the common an ode to earth capacitance. If t is the duration of the simultaneous input pulses, the height of the coincident output pulse is where V_B is the battery voltage and V_C is the common anode potential. This height is well below the theoretical maximum value $(V_B - V_C)$ if t is small compared to T. In the normal Rossi circuit T > t and T the resolving time of the circuit is greater than t.

$$V_0 = (V_B - V_C)\,[1 - e^{-t/T}],$$

For higher resolution coincidence, the fast Rossi circuit is preferred. In this circuit anode load resistance R_1 is reduced so that the time constant R_1C is reduced to 1 μ sec. The germanium diode is biased by adjusting the potentiometer so that it does not conduct when one or both valves are conducting. Delayed-coincidence arrangements are frequently used for measuring the time interval between pairs of generically related pulses, such as the pulses which result from a single particle passing through two widely separated counters or the pulses arising from two particles which are emitted in a radiative cascade. To detect a delayed coincidence, the first pulse is to be converted into a artificial pulse having various controllable delay times. The delay time is adjusted until the maximum number of coincidences are recorded between the artificial pulses and the second pulses. This delay time gives the time interval between pairs of pulses.

It is sometimes important to impose upon an arrangement of counters the condition that one or more tubes shall not count while other tubes are counting. The detection of anti-coincidence is useful in differential analysers as well as in may other instruments. For this purpose the Rossi circuit to record coincidence is modified. The lower end of the grid leak resistor R, is connected to a negative bias supply, so that in the quiescent state F+ is cut-off. A negative pulse from counter 2 triggers a univibrator. The later circuit generates a positive rectangular pulse of duration T_2 which is of sufficient amplitude to turn on the anode current in V_2 A simultaneous negative pulse from counter 1 is delayed by a short time t_d, before triggering a univibrator. The latter circuit generating a negative rectangular pulse of width T_1, T_1, e_d is chosen to be shorter than T_2. so that V_2 is still conducting at the end of the negative pulse applied to grid of V_1. V_1 is cut-off by the negative pulse as V_2 is conducting the whole of this time, no large output pulse appears at the common anode.

Counter → Delay t_d → Univibrator J →

Anti-coincidence counter → Univibrator 2 →

Anti concidence state

SOLVED EXAMPLES

Example 1:

Estimate the number of ion pairs produced in a proportional counter by a 10 MeV proton if the counter size and pressure are large enough for all the proton's energy to he absorbed. If the gas multiplication factor is 10^3, how many coulombs flow in the counter when this proton is absorbed? If the pulse of current flows for about 10^{-3} sec and if the resistance is 10^4 ohms, estimate the height of voltage pulse.

Solution:

Energy required for the production of one ion pair = 35 eV, hence the no. of ion pairs produced by a proton of energy 10 MeV

$$= 10 \times 10^6/35 = 2.86 \times 10^5$$

After amplification the total number of ion pairs = nM

$$= 2.86 \times 10^5 \times 10^3 = 2.86 \times 10^8.$$

Charge carried by these ions $\Delta Q = 2.86 \times 10^8 \times 1.6 \times 10^{-19}$

$$= 4.57 \times 10^{-11} \text{ coulomb.}$$

Hence the current passes through the resistor ΔI

$$= 4.57 \times 10^{-11}/10^{-3} = 4.57 \times 10^{-8} \text{ amp.}$$

$\therefore$ The height of the pulse ΔV = I.R. $= 4.57 \times 10^{-8} \times 10^4$

$$\mathbf{= 4.57 \times 10^{-4} \text{ volt.}}$$

Example 2:

A slow neutron entering a U^{235} fission chamber releases fission fragments of a total energy of 200 MeV. If the capacity of the collector system is 25 $\mu\mu f$, calculate the resultant pulse height.

Solution:

Since the energy required for the production of one ion pair is 35 eV, hence the total number of ion pairs produced

$$N = 200 \times 10^6/35 = 57.14 \times 10^5$$

$\therefore$ Total charge produced = Ne $= 57.14 \times 10^5 \times 1.6 \times 10^{-19}$ coul.

$$= 91.42 \times 10^{-14} \text{ coul.}$$

Pulse height $\Delta V = \frac{\Delta q}{C} = \frac{91.42 \times 10^{-14}}{25 \times 10^{-12}}$ = **3.65 × 10⁻² volt.**

Example 3:

A.G.M. tube with a cathode 5.0 cm. in diameter and a wire diameter of 0.012 cm is filled with argon to a pressure such that the mean free path is 7.8 × 10^{-4} cm. Calculate the value of the voltage that must be applied to just produce an Avalanche.

Solution:

Electric field strength E at any radius r between two concentric cylinders, when a potential V is applied is given by

$$E = \frac{V}{r \log_e (b/a)}$$

Avalanche will just, start when the integral of this equation, taken over one mean free path is equal to the ionisation potential of the filling gas, *i.e.*,

$$E_{a\lambda} = \frac{V\lambda}{a \log_e (b/a)} = I \quad \text{or} \quad \frac{Ia \log_e (b/a)}{\lambda}$$

$$V = \frac{15.7 \times 0.006 \times 10^{-2} \times 2.3026 \log_{10} (0.025/0.00006)}{7.8 \times 10^{-6}}$$

= **730 volts.**

Example 4:

A Geiger-Müller counter has a plateau slope of 3% per 100 volts. If the operative point is at 1100 volts, what is the maximum permissible voltage fluctuation if the counting is not to be affected by more than 0.1?

Solution:

As the plateau slope $= \frac{n_2 - n_1}{n_{av}} \times \frac{100}{V_1 - V_1} \times 100\%$ per 100 volts

In this problem $(n_2 - n_1)/n_{av}$ = 0.1% = 0.1 × 10^{-2} and slope = 3

$\therefore V_2 - V_1 = 0.1 \times 10^{-2} \times 1100 \times 100/3$ = **3.3 volts.**

As operating voltage is 1100 volts, hence the voltage fluctuation is 3.3 volts in 1100 volts or 0.3%.

Example 5:

An organic-quenched G.M. tube operates at 1000 volts and has a wire diameter of 0.2 mm. The radius of the cathode is 2 cm, and the tube has a guaranteed life-time of 10^9 counts. What is the maximum radial field and how long will the counter last if it is used on the average, for 60 hours per week at 2000 counts per minute?

Solution:

The maximum radial field or the field along a radius at the wire surface

$$E_{maz} = \frac{V_0}{2.3026 \text{x} \log_{10} (b/a)} = \frac{1000}{2.3026 \times 10^{-4} \times \log_{10} 200}$$

$$= 1.8^9 \times 10^6 \text{ Volts/metre.}$$

If the lifetime of the tube is N years the total number of counts recorded will be equal to the guaranteed counts. Hence

$$N \times 50 \times 60 \times 60 \times 2000 = 10^9$$

or $$N = 2.77 \text{ years.}$$

Example 6:

An anthracene crystal and a 10 stage photomultiplier tube are to be used as a scintillation spectrometer. It is desired that a 10 keV β-particle incident on the scintillator produce a 2 mV pulse on the photomultiplier output circuit, which has a capacity of 120 μμf. What average electron multiplication per stage is required in the photomultiplier? Assuming a light collection efficiency unity and a photocathode efficiency 0.1.

Solution:

Anthracene gives the highest yield of photons, about 15 for each 1000 eV of energy dissipated in the crystal. Hence, the number of photons emitted in our problem is 150. Since the light collection efficiency is unity and the photocathode efficiency is 0.1, hence number of photoelectrons emitted is 15. If each dynode produces n secondary electrons, after passing through 10 stages the electrons are multiplied by a factor n^{10}

∴ No. of electrons in the output $= 15 \times n^{10}$,

Hence the output pulse $\Delta V = \frac{\Delta q}{C} = \frac{15 \times n^{10} \times 1.6 \times 10^{-19}}{120 \times 10^{-12}}$

As it is given to be 2mV, hence

$$n^{10} = \frac{120 \times 10^{-12} \times 2 \times 10^{-3}}{15 \times 1.6 \times 10^{-19}} = 10^5$$

or $\quad$ **n = 3.16 (Approx.).**

Example 7:

Calculate the capacitance of a silicon detector with the following characteristics : area 1.5 cm², dielectric constant 12, depletion layer 50 microns. What potential must be developed across this capacitance by the absorption of 4.5 MeV α-particle which produces one ion pair for each 3.5 eV expended?

Solution:

Silicon detector is equivalent to a parallel plate capacitor.

∴ Capacitance of the detector

$$= \frac{\varepsilon A}{d}$$

$$= \frac{12 \times 8.85 \times 10^{-12} \times 1.4 \times 10^4}{50 \times 10^{-6}}$$

= 318 pf.

No of ion pairs produced by α-particle

$= 4.5 \times 10^6/35 = 1.286 \times 10^6$.

∴ Charge associated with these ion pairs

$= 1.286 \times 10^6 \times 1.6 \times 10^{-19}$

$= 2.057 \times 10^{-13}$ coul.

Hence the potential applied across the capacitor

$V = q/C = 2.057 \times 10^{-13} \times 10^{-12} =$ **6.5×10^{-4} volt.**

Example 8:

Compute the approximate number of quanta of visible light emitted as Cerenkov radiation in the frequency range corresponding to *wavelengths in vacuum of λ = 4000 to 7000 A.U. given off by an electron of energy 20 MeV traversing 1 cm of Incite (μ 1.49).*

Solution:

Relativistic kinetic energy is given by $E = m_0c^2\ [(1 - \beta^2)^{1/2} - 1]$

$$\frac{1}{\sqrt{(1-\beta^2)}} - 1 = \frac{E}{m_0c^2} = -\frac{20 \times 1.6 \times 10^{-12}}{9.109 \times 10^{31} \times (3 \times 10^8)^2}$$

$\therefore (1 - \beta^2)^{-1/2} = 40$ or $\beta^2 = 0.9995$

Hence the number of quanta emitted as Cerenkov radiation per metre

$$N = \frac{2\pi z^2}{137}\left(\frac{1}{\lambda_2} - \frac{1}{\lambda_1}\right)\left(1 - \frac{1}{\beta^2 \mu^2}\right)$$

$$= \frac{2 \times 3.14}{137} \times \frac{3000 \times 10^{+10}}{4000 \times 7000} \times \left[1 - \frac{1}{0.9995 \times (1.49)^2}\right]$$

= **26980 quanta/metre = 270 quanta/cm (Approx.)**

EXERCISES

1. Scintillator with a background count rate of 260 cpm is used to assay samples with average 900 cpm. What will be the coefficient of variation and the standard deviation of an, "equal-count" assay of 10,000 counts each ? What time will be required for each complete determination.

(± 1.32 ±, %, 12.6 cpm, 47 min)

2. A G.M. tube with a cathode diameter of 4.0 cm, and a wire diameter of 0.016 cm is filled with argon and alcohol to a pressure such that the mean free path is 4.6×10^{-3} cm. Calculate the maximum radius at which secondary ions will be formed when 1200 volts is applied to the tube.

(0.087 cm from the wire)

3. A proportional counter has a gas magnification factor of 1000. Estimate the number of mean free paths in which this multiplication takes place and calculate the distance from the

wire within which this multiplication takes place assuming $\lambda = 2 \times 10^{-4}$ cm, for the electrons in the gas. **(10,0.00 2cm)**

4. It is required to operate a proportional counter with a maximum radial field of 10 mV/meter. What is the applied voltage required if the radii of the wire and tube are 20 urn and 10 mm respectively ? **(1.24 kV)**

5. A sample of uranium, emitting a-particles of energy 4.18 MeV, is placed near an ionization chamber. Assuming that only 10 particles per sec. enter the chamber, calculate the current produced. **(0.192 pA)**

6. The output circuit of a ratemeter consists of a resistance of 50,000 ohms shunted by a condenser. The output pulses from the ratemeter have a constant amplitude of 5.0 milliamperes, and a variable width. What pulse width is required to produce a voltage of 5.0 volts across the tank circuit at a count rate of 200 cpm? What pulse length will be required to obtain the same voltage at a count rate of 50,000 cpm? What capacitance will be required to obtain a circuit time constant of 2 sec ?

   (100 μ sec., 0.4 μ. sec., 40 pf)

7. A square wave whose peak to peak value is 1.0 volt extends 0.5 V with respect to earth. The duration of the positive section is 0.1 sec and of the negative section is 0.2 sec. If this waveform is impressed upon a R-C differentiating network of time constant 0.2 sec, what are the steady state maximum and minimum values of the output waveform. **(0.314, — 1.006 V)**

8. A. 0.1 MeV electrons is absorbed in an anthracene scintillator accompanied by the 10 stage photomultiplier. If it produces a 0.01 V pulse in the photomultiplier output circuit, which has a capacity of 160 μμf. What average electron multiplication per stage is required in the photomultiplier. Assume a photocathode efficiency of 0.07. **(3.1)**

9. The plateau of a G.M. counter working at 1 kV has a slope of 2.5% count rate per 100 volts. By how much can the working voltage be allowed to vary if the count rate is to be limited to 0.1%? **(0.4%)**

10. A silicon detector has an area of 3.0 cm^2 and a depletion layer of 60 microns when polarised with 50 volts. Calculate the time constant of the circuit when used with a 100 megaohm resistance, the time required to collect all of the holes and the amplitude of the voltage pulse produced by the absorption of a 3.78 MeV α-particles.

    (530 pf, 5.3 × 10^{-2} sec., 1.4 × 10^{-9} sec., 3.3 × 10^{-4} volt)

11. Calculate the resolving time of a pulse counter from the following schcdule of count rates: background 65 cpm, first source in position 8350 cpm, both sources in position 16832 cpm, second source in position 9416 cpm. **(5.5 × 10^{-6} min)**

12. The paralysis time of a G.M. tube is 400m sec. What is the true count rate for measured count rates of 100, 1,000 and 10,000 counts per min. **(100.07, 1007, 10715)**

3

Acceleration

3.1 INTRODUCTION

The term particle accelerator will be generalized to include devices which in themselves provide only large difference of potential, but when combined with ion tubes accelerate the ions to high energies. The accelerators have been important instruments in conducting research concerning particles such as mesons, anti-proton and antineutron. Other applications have been in studies designed to throw light, on the complex and possibly related problems of *nuclear structure, nuclear forces and strong and weak interactions*. At present the accelerator is the basic instrument in the rapidly growing branch of nuclear science known *high energy particle physics*, and has undergone a tremendously rapid development during past 40 years.

Charged particles obtained from ion source are accelerated only by electric fields in the direction of the field. Magnetic fields exert forces at right angles to the direction of particle motion and cannot change the energy, but is used to produce circular orbits of the otherwise extremely long particle paths. However, varying magnetic fields can accelerate the charged particles. The particle (*charged*) acted upon both by electric and magnetic fields or by any one only obeys the laws of electrodynamics.

3.2 ION SOURCES

The characteristics of an ion source are:

(a) The energy spread should be small.

(b) It should be designed such that final adjustments of its position can be made by external manipulations.

(c) It should be compact, rugged, dependable and have a long life.

(d) It should give a large, well focused ion output.

(e) efficiency of the positive ion source should be large, to simplify the problem of maintaining a vacuum and to reduce the gas consumption.

In most of the ion sources the pressure is between 10^{-1} to 10^{-3} mm Hg. Within an arc discharge, a high density of ionization exists. Positive ions go to the cathode and electrons move toward the anode. On account of their larger mass, the positive ions form positive space charge and radial electric field. The potential of the central region (*plasma*) becomes 5 to 10 volts positive relative to the walls, the positive ions are thus forced outward. Several types of sources are grouped according to the nature of the discharge phenomena and to the physical arrangements used to produce ions.

1. Hot Cathode Arc

Many modern ion source make use of a heated cathode to supply electrons by thermionic emission process. The emission can be obtained from tungsten or tantalum filaments or from oxide-coated cathodes. This arc can be operated at considerably lower potentials (200 volts). The design developed by Scott consisted of a cylindrical cathode of large area to give high emission of electrons, aligned along the axis of the arc to emerge parallel beam. The cathode was formed of Ni-strip, oxide coated. The anode was water cooled and very close to the cathode. The chief advantage of this source is in the high beam densities obtained because of the focusing action of the accelerating fields.

2. Cold Cathode Canal Ray Tube

This type of source was first used by Cockcroft and Walton in their earliest experiment. The cathode and anode were long coaxial cylinders with the anode inside, having a very small gap. A potential of about 20kV was required to maintain discharge currents between 10 and 100μa. Gas (at 10^{-2} mm Hg) was admitted through the base into the anode. The source was oil cooled at the base. The separated ions were measured by magnetic deflection. Beams of about μa of protons were normally obtained. Because of the large energy spread, relatively small ion currents, and the rapid erosion of the electrodes, this type of source becomes of historical importance only.

3. Spark Discharge Source

A wide range of atomic and molecular ions were successfully produced by this method. In it high voltage sparks were developed between electrodes of the chosen material. The accelerators such as the linac and the proton synchrotron require short pulses of particles, hence can make use of such ion sources. The discharge is usually maintained by a radio-frequency oscillator and one of the electrodes either is made of the material to be analyzed or is packed with the material. The ions thus produced are formed into a beam by the collimating system. In the ion source used at CRL, the electrodes of titanium metal were heated and cooled in a hydrogen or deuterium until the metal was saturated with hydrogen. An electrical delay circuit was used to develop a voltage pulse of 10-20 kV across the discharge chamber. The disadvantages are relatively short life and inclusion in the beam of a large metallic ion component.

4. Magnetic Ion Source

It is the advanced form of hot cathode arc source in which the axial magnetic field concentrates the discharge in the region near the exit hole. Magnetic field collimates tightly the electron beam and keeps +ve ions from diffusing to the walls of the chamber and being lost and hence increases the yield. Very efficient magnetic ion source called *Duoplasmatron*, was devised originally by M. von Ardenne. According to the inventor this ion source can give 500 μa of proton through a hole of 1 mm diameter. The output of the source is practically 100% H^+ with very few percentage of H_2^+ and neutral atoms. In this hot cathode arc source, the anode structure forms the pole pieces of an electromagnet that produces an intense axial magnetic field in the discharge region. This field coils up the orbits of the electrons, produced by thermionic emission from a hot cathode, therefore, it enhances the probability of an electron's colliding with a gas molecule before striking the anode. This ion source was involved in the U.S.S.R. and in East Germany. It is now available commercially from the high Voltage Engineering Corporation of Berlington, Massachusetts and also from VEB Dresten (East Germany).

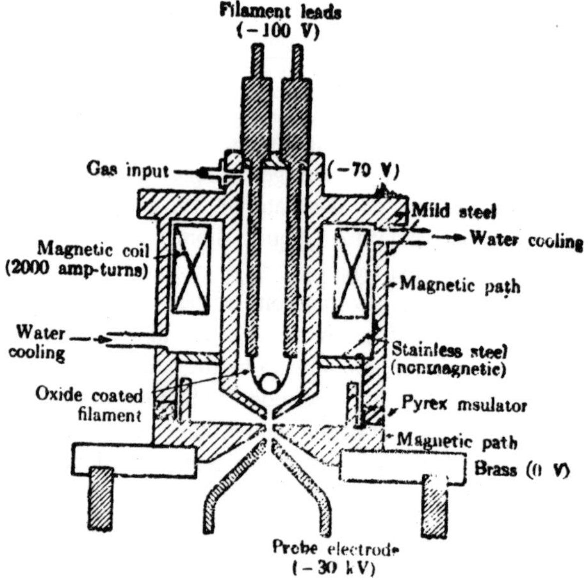

Fig. 3.1 : Duoplasmatron ion source.

5. Capillary Arc

The ionization density can be increased by concentrating the discharge in a constriction between cathode and anode. This constriction, or *capillary*, not only increases current density but makes it possible to choose the wall surface to improve the atomic ion concentration. Positive ions are ejected radially through the plasma sheath and can emerge through a small aperture in the wall of the capillary with only 5-10 volts variation in energy.

The chief advantages are: The increased beam intensities and increased atom ion percentage.

6. Electron Oscillation (P.I.G.) Ion Source

One of the most recently developed high intensity ion sources utilizes electron oscillations to increase the ionization density in the discharge. In the ion source devised by Gow and Foster, the discharge occurs between a cylindrical anode and two disc-shaped secondary emission cathodes at the ends of the anode cylinder. Hydrogen gas pressure of ~0.02mm Hg, an axial magnetic field of ~1000 gauss and pd of a few hundred volts between electrodes provide the necessary conditions for the discharge. The primary advantage of this source over the hot cathode type is in the large ratio of ion current to electron current in the discharge. It has been very successful when applied to pulsed operation. With a pulse time of 0.001 or smaller, the heating in the discharge can be readily controlled. Its modified form is a cyclotron ion source, in which instead of ions being pulled out through an aperture in one of the cathode, they emerge through a slit. Thus very high intensities can be obtained.

Here we see that these sources produced positive ions and electrons only. But for machines like tendem accelerator, negative ion sources are needed. Some of the electrons released in the gas break down in a source for positive ions attach themselves to gas molecules and produce negative ions which could be extracted by reversing the extraction potential. In most of the machines negative ions are not produced directly in the ion source because the intensity of ion beam is very low, but arc formed by charge exchange from a beam of positive ions of about 50 keV passing through an *adder gas*.

7. Radio-frequency Discharge

Very efficient ion sources have been produced that operate on the principle of electrodeless gas breakdown in a radiofrequency field. An axial magnetic field is often used to enhance the ion yield. The gas is enclosed in a Pyrex tube and the radiofrequency field inside the tube is produced by the two ring-shaped electrodes outside the tube. To produce a breakdown in the gas, about 60 watts of radiofrequency power at 100 Mc/sec is needed at a pressure of about 2×10^{-2} mm Hg. The ions thus produced are extracted from the source by a superimposed dc-field between the extraction canal and the metallic electrode at the top source. The chief advantages are:

(i) The cathode life is not a limitation,

(ii) The electrons circulate in orbits through the gas, multiplying the collision probability and hence increasing the atomic-ion concentration. Its application to the production of positive ions has been investigated by several workers. The ion source at CERN laboratory yields currents of the order of 100ma in pulses 10μ sec long. It requires ~7 kw of rf-power at 139mc/see.

3.3 ELECTROSTATIC ACCELERATORS

An electrostatic accelerator is one in which a high voltage is built up between two terminals and charged particles are accelerated in their passage across the space. After traversing a potential difference of V volts, a particle with charge Z times that of the electron, will have a kinetic energy ZV electron volts. There exist quite a few generator producing potential differences in the million volt range. Two of them, the Cockcroft-Walton and the Van de Graaff generators have been in use for many years as particle accelerators.

1. Cockcroft-Walton Accelerator

Cockcroft and Walton, in 1932, at the Cavendish laboratory, Cambridge (England), employee an electrostatic accelerator baser" upon the *voltage multiplier principle.* In Fig. 3.2, a number of condensers C_1, C_2, C_3, C_4, etc. of equal capacitance are arranged together with rectifiers R_1, R_2, R_3, R_4, etc. These rectifiers were of the vacuum tube type in the original machine but now are made of selenium, act effectively as switches. These rectifiers allow to pass electricity in one direction, indicated by the arrow.

In the following argument we shall assume that there are no current losses across any of the components The transformer T applies an a.c. voltage of maximum value V, may range from 25kV to 100 kV at the secondary output between two columns of condensers. The point B is connected to earth, so its potential will have a fixed value, say zero. The potential of A oscillates between –V and +V. For the first negative half cycle B goes positive with respect to A and the rectifier R_1 conducts and C_1 is charged with the charge to V. During the next half cycle R_1 no longer conducts leaving the point C isolated at a potential of V. Once charged the condenser will maintain the potential difference V between C and A, hence during the positive half-cycle, the point C will reach the voltage 2V with respect to earth. As in this cycle R_2 conducts, hence the charge accumulated on C_1 is now shared with C_2. On repeating the first half cycle the condenser C_1 is recharged upto V. In this half cycle C_2 retains its charge but is increased by sharing with C_1 again during the fourth half cycle. After few cycles an equilibrium is reached in which there is no current through either R_1 or R_2 at any time and the potential of D is now equal to the maximum' potential of C with respect to B *i.e.*, 2V If we now add the condensers C_3 and C_1 through the rectifiers R_3 and R_4 we can repeat the whole argument and the potential appearing finally at F is 2V with respect to D and 4V with respect to B. Thus, by using such arrangements of two capacitors and two rectifiers, also known as voltage *doublers*, the voltage V can be multiplied upto any multiple of V. It, because of the arrangement of several such doublers, is known as a *voltage multiplier*, sometimes also known as a *cascade rectifier*.

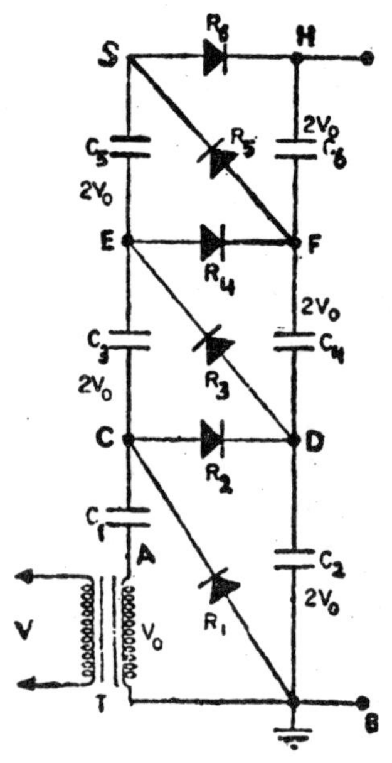

Fig. 3.2 : Voltage multiplier circuit.

For a circuit of N-capacitors and N rectifiers, with peak input voltage V_0, frequency f, capacitance of each capacitor C and load current 7, the output voltage V is given by

$$V = NV_0 - \frac{I}{12fC}\left(N^3 \frac{9}{4}N^2 + \frac{1}{2}N\right) \qquad ...(1)$$

The ripple voltage is given by

$$\pm V = IN\,(N + 2)/16fC \qquad ...(2)$$

Hence to have V greater we require higher values of f and C. These values also reduce ripple voltage. If C is very large, the size of the apparatus will become very large. Now-a-days frequency f is increased by using radio frequency source to obtain high voltage without increasing C to a very large value.

The high voltage between the terminals of the machine is applied to the charged particles to be accelerated by means of an evacuated accelerating tube. Starting with V = 100 kV, Cockcroft and Walton obtained an output of nearly 400 kV in the their first accelerator. In later forms of the apparatus particle energies upto 4 MeV have been realized. The Cockcroft add Walton machine has the advantage of simplicity of design and construction. The maximum energies which can be obtained are low compared with those from other accelerators, but it provides fairly large ion currents at constant voltage. It is very useful for experimental work at moderate particle energies.

2. Van de Graaff Accelerator

For many years electrostatic generators, such as the Wimshurst machine have been used to produce high voltages. In 1931, a new type of machine was invented by R J. Van de Graaff in the United States. It is an important instrument for producing high voltages in the million volts region. It B design is based on the principle that *if a charged conductor is brought into internal contact with a hollow second conductor, all of its charge transfers to the hollow conductor no matter how high the potential of the latter may be.* The potential of the hollow conductor can be raised to any desired value by successive adding the charges to it.

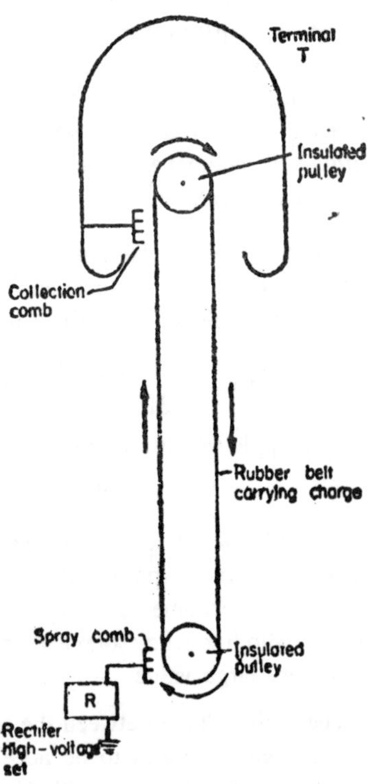

Fig. 3.3 : Schematic diagram of a Van de Graaff generator.

Its schematic diagram is shown in Fig. 3.3. It consists of a belt, made of paper, silk, rayon or other flexible non-conducting material, which Is run by means of a matter at high speed over two pulleys. The positive pole of dc-power supply at a few kilovolts is connected to a pointed comb like conductor. Because of the large electrostatic field in the air near the spray points, positive and negative ions are formed and a spray of positive ions is repelled from the points. These positive ions attach themselves to the surface of a moving belt and are carried mechanically by the belt to the terminal where they are removed by another corona discharge to another set of needle points. This set of needle points is connected to hollow hemispherical conductor mounted on insulating supports. The upper pulley is within this large metal hemisphere. The charge is thus transferred to this hemisphere and increases its potential.

In the later forms of the generator, a pointed conductor, also connected to the sphere, is placed at the lop on the down side of the belt. A discharge of negative electricity thus takes place from this pointed conductor, so that electrons are collected and carried downward by the belt. This negative charge ultimately passed to the dc-source. Although it does not alter the potential, it increases the current which the apparatus can supply.

If C is the capacitance of the terminal to the ground, the terminal will be raised to a potential V by a charge Q (= CV). The rate of increase of potential

$$\frac{dV}{dt} = \frac{1}{C}\frac{dQ}{dt} = \frac{i}{C} \qquad ...(3)$$

The capacity of a spherical shell, at a long distance from ground is given by the relation $C = 4\pi\epsilon_0 R = 1.11 \times 10_{10}10$ R farads. If the terminal of radius r_1 is enclosed within a grounded concentric shell of radius r_2, the capacity

$$C = 1.11 \times 10^{-10}\ r_1 r_2/(r_1 - r_1) \text{ farads} \qquad ...(4)$$

Another geometry is a cylindrical terminal with spherical end caps. The capacitance per unit length of coaxial cylinders of radii, r_1 and r_2 is

$$\frac{C}{l} = \frac{2\pi\varepsilon_0}{\log_e(r_2/r_1)} = \frac{0.55\times10^{-10}}{\log_e(r_2/r_1)} \text{farads.} \qquad ...(5)$$

Above relations show that for a spherical capacitor of radius 1m, (C = 100 pf, for a spherical shell of inner radius 1m and outer radius 2m, C = 220pf, for a 1 m long cylindrical section with 1m radius caps enclosed in a 2m cylinder, C = 300pf. Due to this higher; value of C for the same size, the cylindrical terminals are now preferred.

Maximum charging currents used now-a-days are a few μa. For C = 500pf, i = 1μa, we have dV\dt = 2MV/sec. The power required to charge the terminal is the product of i and V. The resistor is sometimes placed between the insulated upper pulley and the terminal to control the upper stray potential and to make the charge densities equal on the upward and downward runs.

The charging current returns to ground from the high voltage terminal is the sum of following currents:

1. Current in the resistive potential divider or leakage down the column is proportional to the terminal potential. For 5MV terminal voltage it is equal to

$$5 \times 10^6/5 \times 10^{12} = 1\mu a.$$

2. Positive-ion-current, determined by ion source conditions rather than by total voltage, is generally constant. It is the beam of +ve ions being accelerated down the accelerator tube, and is of the order of 100μa.
3. Secondary electron current, produced by the stray ions striking the tube wall, increases rapidly above an apparent threshold. These secondary electrons are accelerated up the column to produce X-rays on striking the ion source.
4. The corona current, caused by the break down of insulation (either gas or solid) due to excessive fields at the surface of the terminal, is zero at low voltages and rises rapidly above a threshold value. The Paschon's Law gives the condition of corona discharge as

V_{min} = Pressure of the gas of the medium × gap × specific breakdown potential K.

These components vary with voltage as shown in Fig. 3.4. The maximum available charging current determines the equilibrium

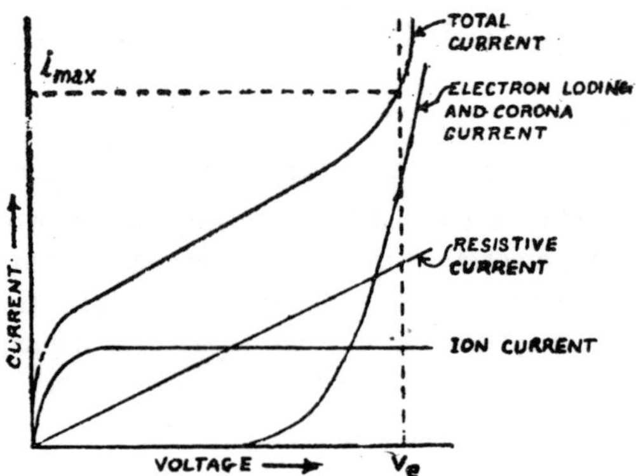

Fig. 3.4 : Charging current v/s voltage.

potential V_e of the terminal. The extremely rapid change of corona current with voltage tends to stabilize voltage at the equilibrium value.

Electrical breakdown in compressed gases depends on the following factors:

1. Distribution of potential along the insulator.
2. Shape, material and surface conditions of the insulating support.
3. Composition and pressure of the gas.
4. Electrode material, shape and surface contaminations.
5. Radius of curvature, area and surface smoothness of the terminal.

The breakdown potential a radius of curvature of the terminal. The type of discharge depends on the shape of the electrodes, gap, length, pressure and external circuit. The presence of oxide films, oil, fingerprints and adsorbed gases changes the sparking limit. Dust particles or surface roughness produces local fields. The relative humidity increases the sparking potential in air. The contamination problem is solved by enclosing the terminal in a pressure housing and by filtering and purifying the gas. The maximum field is 800 kV/m. for aluminium terminals and is 1200kV/m for stainless steel terminals. The aluminium is commonly used as it can be spun to the desired spherical shapes and has a low density. Stainless steel is more difficult to form.

The break down potential depends strongly on gas composition, particularly at high pressures. Air, CO_2 and N_2 were used in early generators. The gaseous compounds, containing Cl_2, F_2 or other electronegative gases have better insulating strength. Several compounds, *e.g.*, CCl_4, freon (CCl_2, F_2) and SF_6 are studied and SF_e was found the best gas. The terminal' potential increases smoothly with the air pressure. We do not want to go beyond 20atm, as there is a break down. The ignition temperature becomes lower at high pressures and the belt may burn.

Solid insulators are used to support the terminals in both vertical and horizontal type accelerators. In all modern generators the supporting insulators are surrounded by closely spaced equi-potential rings. The corona rings along the column give a uniform potential gradient and eliminate corona, as the weak point at the terminal connection is shielded from high fields Trump and Andrias suggested that the cylinders of Incite, texlolite and isotantite were better, particularly for high gas pressures.

The Charge carrying belt should have high mechanical strength, high surface and volume resistivity, high dielectric strength and be fire resistant. It should not absorb moisture when opened to the atmosphere. The endless woven cotton belts are dried and are used at the MIT lab. Instead of corona discharge for transferring the charge to the belt, one can acquire the charge by induction. For this purpose we require conducting material on the inside surface of the belt, which will acquire charge by induction. It has a disadvantage that the belt material losses due to friction with the pulley. However, coating is done on the belt to check this frictional effect. The current delivered by a belt of width w and speed v is given by

$$i = \sigma \; w \; v \text{ amp.}$$

where σ is the charge density, having its maximum value

$$\sigma_m = 2\varepsilon_o E_m \simeq 5.3 \times 10^{-5} \text{ coul./m}^2.$$

The accelerating tube is the evacuated tube of insulating material, commonly porcelain or glass cylinders several inches long and of large

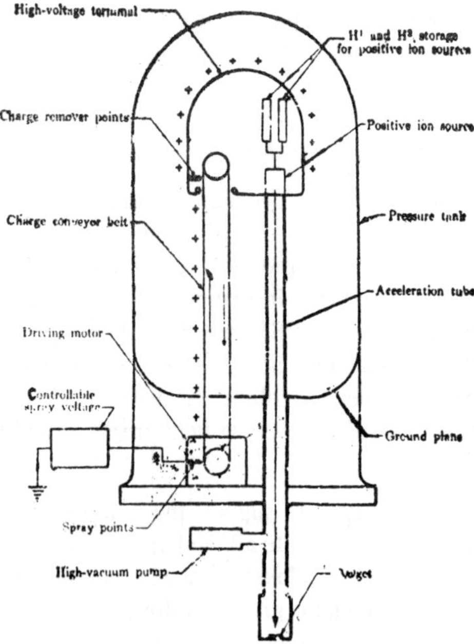

Fig. 3.5 : Van de Graaff accelerator.

diameter with vacuum tight seals to metal plate electrodes between sections. These electrodes are connected to corresponding equipotential rings in the column to maintain a uniform distribution of potential along the tube. Vacuum pumps for the accelerator tubes have speeds compatible with the tube diameter.

A Van de Graaff accelerator consists of three main parts t the pressurized tank, the generator and the accelerator proper. Ion source is placed inside the hollow hemisphere and the ions are accelerated down the tube on to the target located at the low voltage end of the tube. This tube is usually made up of sections of glass, porcelain or other insulating material joined end to end by vacuum tight scale. To avoid the possibility of a spark or other discharge passing from one end to other when the potential is applied, the tube must be of enough length.

The maximum particle energies, can be realized with this type of accelerator, are relatively small in comparison with most other types of machine. Its advantages over all other types of accelerators are the sensitivity and control of the voltage.

Tandem Accelerators

In the conventional single stage accelerator positive ions are produced inside the high voltage terminal and accelerated to ground potential. In the case of two stage operation, negative ions arc produced at ground potential and accelerated towards the positive terminal. There the electrons are stripped off and the remaining positive ions proceed again to ground potential. In this way same voltage is used twice. The operation of two stage tandem accelerator is illustrated schematically in Fig 3.6. Positive ions are produced in the ion source and accelerated to 50keV approximately. In passing through charge exchange canal containing gas at low pressure, approximately 1% of the positive ions are converted into negative ions by pick up of two electrons. The negative ion beam is then separated from unwanted neutral and positive particles by means of a magnet. The negative ions are attracted toward the positive terminal and gain an energy of V eV. In the high voltage terminal the negative ions traverse a second gas stripper canal or a thin carbon foil target. Their energy is already much higher than that required to remove the most tightly bound electrons in an atom; therefore, all the ions emerging from the stripper will have lost both electrons. These positive ions will gain again V eV energy in passing from positive terminal to ground potential. This implies that second stage is extremely powerful when heavy ions

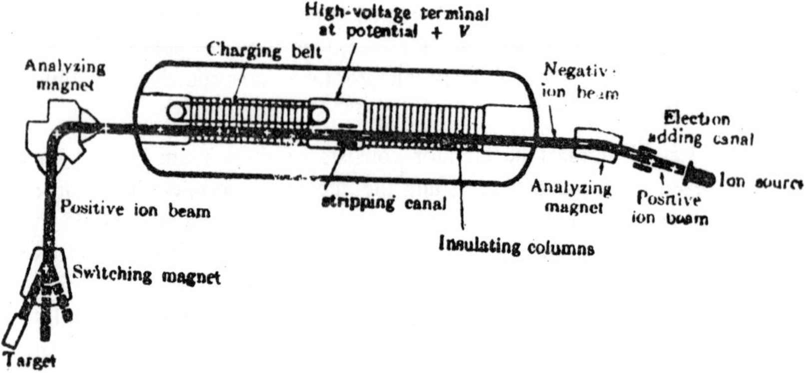

Fig. 3.6 : Two Stage Tandem accelerator.

are accelerated. Take for example oxygen nuclei (z = 8) and a terminal voltage 5 million volts. They are singly charged in the first stage and gain energy 5 MeV. But in the second stage they have eight positive charges and acquiring thus 40 MeV. Thus the total energy becomes 45MeV.

3. Insulating Core Transformers

Transformers are very well known to us. They consist of two coils, called primary and secondary. But increasing the number of turns in secondary, output ac potential can be increased easily. The transformers consisting iron core cannot deliver voltages in the million volt range. The limitation is due to insulation required between the secondary coil

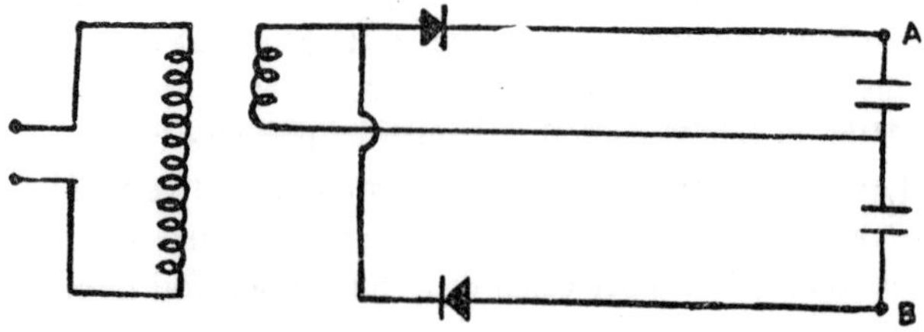

Fig. 3.7 : Insulating core transformer (Equivalent circuit).

(high potential) and the core, which is not far from ground potential. In the insulating core transformer the iron core has not been dispensed with, but is made out of sections which are electrically insulated from one another. The segments form a central column surrounded by an outer shell of magnetic material, where the primary coil is wound. Each section in the central core is surrounded by a secondary winding, where an ac voltage of about 30 kV is induced. The output from each secondary is rectified by means of the voltage doubling circuit. By connecting in series all the dc power supplies thus formed, we can obtain high voltage. The first insulating core transformer, designed in 1960, was successfully operated at one million volts and 25 milliamperes with a power conversion efficiency of about 90%.

3.4 LINEAR ACCELERATORS

In a linear accelerator (often abbreviated as linac) the energy of a charged particle is increased steadily or in stops as it travels in a straight line. There are two principal devices which make use of direct acceleration in a straight tube.

1. Drift Tube Accelerators

The idea of accelerating positive ions by the aid of an alternating radio-frequency field was first suggested by Ising in 1925. By such a method D.H. Sloan and Lawrence in 1931 produced a beam of mercury ions emerging with energy 1.25 MeV, using initially a 30 mega cycles per second radio frequency supply of only 40 kilo volts. The principle is shown in Fig. 3.8. The linear accelerator consists of a number of cylindrical electrodes of increasing length arranged in a straight line.

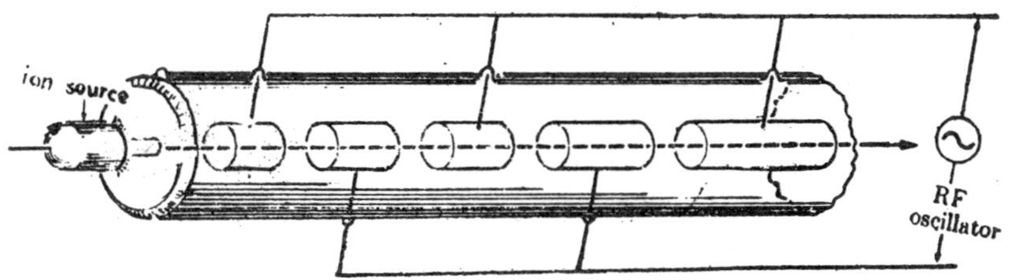

Fig. 3.8 : Schematic diagram of a linear accelerator.

Alternate cylinders are connected together, the cylinders numbered 1, 3, 5, etc., being joined to one terminal and the remaining cylinders to the second terminal of the high frequency power supply. At any instant, therefore, alternate electrode carried opposite electrical potentials. All the electrodes which are at positive potential in a particular half cycle becomes negative in the next half cycle.

Suppose positive ions from the ion source move from left to right along the common axis of the cylindrical electrodes. While passing through first electrode, the ions receive no acceleration, since they are within a uniform potential. If the first is positive and the second is negative, the positively charged particles will be accelerated in the gap between these electrodes. The positive ions then enter the second electrode and travel through it at a constant but higher speed. The length of the second electrode is such that just as the ions reach the gap between it and the third electrode the potential of these electrodes are reversed. Now, the second becomes positive and third negative, thus the positive ions are accelerated in this gap.

The length of a cylindrical electrode plus a gap or the separation between two consecutive gaps is the distance traversed by the particles during one half cycle of the applied field of frequency f and is given by

$$L_n = v_n T_2/2 = v_n/2f \qquad ...(6)$$

where v_n is the velocity of the particle in this n^{th} electrode. If V is the peak value of the potential difference between two terminals, then the energy acquired by the ions in passing one gap will be qV. If there are n such gaps, the total kinetic energy acquired by the ions will be qV_n. The velocity of the ions in the nth electrode is, thus, given by the relation

$$\frac{1}{2}Mv_n^2 = nqV + K, \qquad ...(7)$$

where M is the mass of the particle and q its charge. The constant K allows for the fact that the injection energy is greater than zero and for the fact that the first gap has only half the accelerating voltage V if the oscillator is balanced with respect to ground. By combining eqns. (6) and (7) we obtain the length in the nth electrode

$$L_n = \frac{1}{2f}\left[\frac{2(nqV + K)}{M}\right]^{1/2} \qquad ...(8)$$

This relation indicates that in order to get a large energy the peak voltage of the oscillator should be large and number of cylinders should be as large as possible. The length of the cylinders can be reduced by increasing the frequency of the oscillator.

Relativistically, for particles of high energy ($v \rightarrow c$) the distance between gaps is constant, *i.e.*,

$$L = c/2f = \lambda/2. \quad ...(9)$$

Hence, each energy increment is mainly represented by an increase in mass rather than velocity. The cylinders then approach uniformly in size and spacing and the distance between two adjacent gaps is a half wave length of the wave emitted by the electric oscillator.

Radial Focusing

The ion beam during its passage through the gaps is not only accelerated but also focussed radially due to the curved nature of the electric lines of force in the gap. The Fig. 3.9 shows the gap between two coaxial cylinders when a difference in potentials exists between them. In the first half of the gap, any oncoming ion with a velocity Vi will receive an increment of velocity ΔV forward and ΔV_c toward the axis of the cylinder. During the latter half of the gap it will receive an equal increment in its forward motion but the radial component of force is directed outwards. This latter velocity increment V'_c is not as great

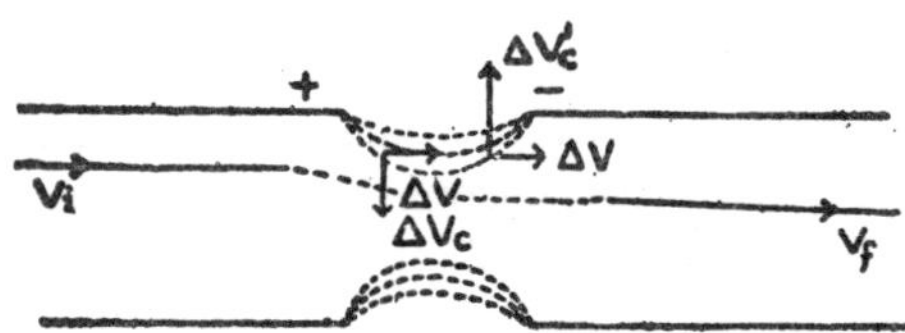

Fig. 3.9 : Focusing due to electric field.

as that toward the axis, because ion is travelling faster and hence is acted upon by a force for a shorter time. Thus, there is a net displacement of ions towards the axis of the cylinders. The defect of this instrument is that of its larger size and it is very difficult to maintain vacuum in large chamber.

Phase Focusing

In the derivation of relation (8), we have assumed a constant potential difference V applied between consecutive cylinders. As alternating potential Vo sin wt is applied by the radio-frequency source, hence the ion should cross the gaps at times O, T, 2T,...., T = 2π/ω or the ions should approach in same phase at successive gaps. Particles which arrive at each gap at the instant corresponding to the point indicated by P in the figure will always receive the appropriate impulse. A particle which

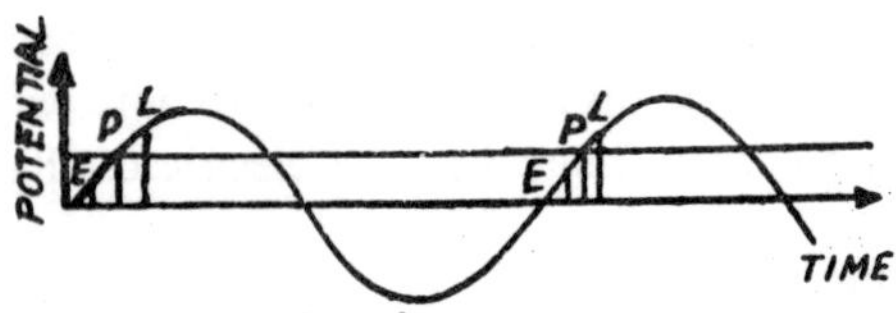

Fig. 3.10 : Principle of phase stability.

has more energy than that called for at a particular gap will necessarily have greater velocity and will have phase such as E. It will thus be accelerated less than normal and will arrive at the next gap with a phase closer to P. An ion that arrives at a gap with less than the characteristic energy will arrive late, phase L, It will be accelerated more than the normal and will gradually shift its phase toward. Thus all ions within a certain phase range will be trapped and will oscillate about the point of *stable phase* (point P).

2. Wave Guide Accelerators

Modern accelerators make use of waveguides to establish the electric field. A waveguide is a pipe of conducting material in which an oscillating electromagnetic field is established. It electromagnetic waves with frequencies around 10^4 or 10^5 megacycles per sec are fed into one end of a waveguide, the waves and the associated oscillating electric and magnetic fields travel along the length of the guide. It may at first appear surprising that the rate of propagation of the wave-front in a waveguide, the *phase velocity*, in a simple waveguide, is usually greater than the velocity of light. The reason is that the electromagnetic waves approach the walls of the waveguide at an angle so that they are reflected back and forth and travel along the guide in a zig-zag path. The interference of these waves leads to a new wave pattern in which the wavelength is greater than in free space. The phase velocity v_p is equal to the product of this wavelength and the normal frequency which is constant and is given by the relation

$$v_p = c[1 - (\lambda/\lambda_c)^2]^{-1/2} \quad ...(10)$$

where λ is the wavelength of the applied r-f in free space and λ_c is the cut off wavelength which depends on the type of wave and the dimensions of the guide. $\lambda_c = 2.615$ a for the circular guide of radius a and for TM_{01} (or E_{01}), *transverse magnetic field pattern* waves. If the electromagnetic wave is completely absorbed at the end of the waveguide, the result is a travelling wave. But if there is an electrical conductor at the end, part of the wave will be reflected back along the guide, and the combination will be a standing wave. The radiofrequency signal is produced by master oscillator and amplified at each feeding station by klystrons. The speed of the electromagnetic wave is synchronised with the speed of the particle so that the particle gains energy from the electric component of the wave.

Linear accelerators have been developed for both electrons and protons. The design considerations of proton and electron linear accelerators are basically different for the fundamental reason that the less massive electron travels most of the time with a velocity very close to that of light. On the other hand, the protons gain speed along the total light path. The chief drawback of linear accelerator is its very high power consumption. A linear accelerator of length l excited by a field whose free space wavelength is λ and accelerating particles to a final energy of T eV, requires a total power

$$P = CT^2\lambda^{1/2}/l \text{ watts,} \quad ...(11)$$

where $C \simeq 3 \times 10^{-8}$ for proton machines and $\simeq 10^{-7}$, for electron machines.

The ratio of the square of the maximum electric field on the axis to the power dissipated per meter of length is called the *shunt impedance*. It is given by

$$Z = 10^{-6} E_z^2/P \text{ megohms/m.} \quad ...(12)$$

Electron Linear Accelerator

If a bunch of electrons injected along the axis of a guide carrying an E_{01} travelling wave it may be possible for the electrons to be accelerated by the axial electric field of the wave and thereby reach high energies.

In order for an electron to be accelerated in a wave guide system, the phase velocity must be essentially equal to that of the accelerating electron. The phase velocity can be reduced by placing metal discs with holes in the centre, at intervals along the guide. The resulting system is known as an *iris loaded* or *disc loaded wave guide*. The beam of electrons to be accelerated passes through these holes. Near the injection end of the loaded guide the electron velocity may be of the order of 1/2c (for 79 keV) and the phase velocity of the wave must be adjusted to this value. After travelling a short distance the electron velocity is very close to c and the phase velocity is made equal to c and remains same over the remaining length of the accelerator.

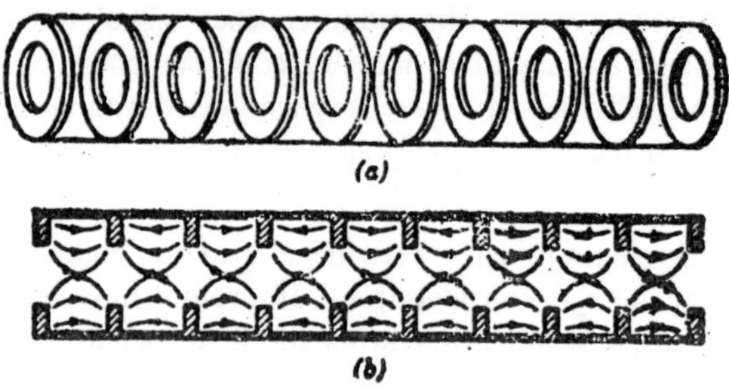

Fig. 3.11 : (a) A disc loaded wave guide. (b) The electric field configuration.

The tendency for the electron bunch to diverge under the influence of the transverse field components can be compensated in the initial stage of acceleration by an axial external dc magnetic field. As the electron reaches an energy of a few MeV, the defocusing effects become ineffective.

The radii of the hole in the iris a and the guide radius b are not independent. For any choice of a there exists a value of b. As a is increased b also increases rather slowly. For relatively small values of a, b is represented by the relation

$$b/\lambda = 0.383 \; [1 + 20 \; (a/\lambda)^3]. \qquad ...(13)$$

The flow of energy is strongly influenced by the size of the apertures. The energy distribution along the wave guide follows from following relations:

Group velocity v_g = energy flow/energy density.

Energy dissipated $= \frac{d}{dz}$ (rate of energy flow)

Q-factor of the waveguide $= \frac{\omega(\text{energy stored})}{\text{rate or energy dissipation}}$

or Rate of energy dissipation $\propto e^{(\omega z/vgQ)}$.

Since the energy dissipation is proportional to the square of the electric field E_z, hence

$$E_z = E_{zo}\ e^{-(\omega z/2vgQ)}.$$

If the distance between two feedpoints is l the total energy gain of a particle reaching from one feed point to the next will be

$$e\int_0^l E_z dz = \frac{2ev_g QE_{z0}}{\omega}[1-e^{(wl/2v_g Q)}] \qquad ...(14)$$

The first travelling wave accelerator was constructed by D.W. Fry and his associates in England in 1946 In the following year Hansen and his colleagues completed a similar device to produce 6MeV electrons. Electrons with energies up to 1BeV have been obtained with the 220ft Stanford linear accelerator constructed at Stanford University. This accelerator is divided into 10 sections, each of which is fed with high frequency power from a separate klystron amplifier and producing a travelling wave in the tube having a frequency of 3×10^9 cycles/sec. The accelerator is operated at the rate of 60 pulses/sec, each pulse lasting one microsecond and produces about 5×10^{10} electrons per pulse. These electrons have a narrow energy spread of 0.5 per cent about the mean energy BeV. A linear electron accelerator, 2 miles long, was constructed at Stanford Linear Accelerator Centre (SLAG) and was completed in 1966. It consists of 960 sections of copper tubing, each 10 feet long and 4 inches external diameter. With 240 klystrons in operation, the maximum electron energy is 20 BeV. It is planned eventually to increase the number of klystrons to 960, the electron energy should then be 40BeV.

Proton and Heavy ion Linear Accelerators

Above technique cannot be applied to proton acceleration. The main difference arises because of the much lower injection velocity of protons (4 MeV proton has a velocity of ~0.1c). Disc loaded-wave guide is impracticable for phase velocities below about 0.4 c.

To accelerate protons a long cylindrical resonant cavity is excited in a TM_{010} mode with an infinite phase velocity. The electric field lines are parallel to the axis of the cavity and terminate on the end walls of the cavity. Since an ion takes many periods of the rf-field to travel from one end of the cavity to the other, some device must be introduced to shield it from the field while the field direction is such as to decelerate rather than to accelerate. It is done by introducing a series of tubes along the axis through which the beam passes. These tubes are known as drift tubes, as ions experience no forces while they travel through beam and cross the gap between two *drift tubes* when the electric field is increasing with time. Their length increases along the accelerator as the ion velocity increases. The tubes decrease in diameter to keep the resonance frequency uniform along the length of the accelerator.

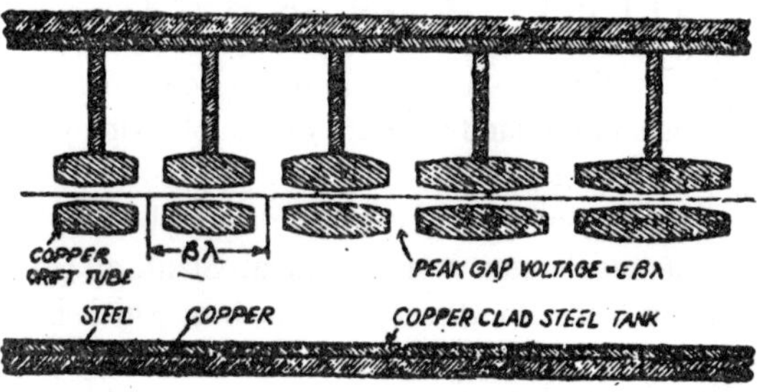

Fig. 3.12 : Drift tube linear accelerator.

One of the difficult problems was rf-defocusing of the beam. At velocities low compared with c a phase stable particle will be lost very rapidly by the radial components of rf-field. The radial field at the entrance to the drift tube is removed by placing a grid just infront of the drift tube. However, this focusing grid or foil in-front of each drift tube intercepts or scatters some of the protons and results in the considerable loss in beam intensity. The use of electric quadrupole lenses along the axis of the accelerator is an alternative approach to the focusing problem.

In 1965, the highest energy attained with a proton linear accelerator was 70 MeV. Several linear accelerators have been built follower energies. At Brookbaven a design study has been made for a 500 MeV linac and at Los Alamos design study has been made for an 800 MeV linac.

3.5. LOW ENERGY CIRCULAR ACCELERATORS

Instead of building up a high potential, as is done in electrostatic accelerators, a generator of relatively low voltage is used repeatedly to add energy to the ions in cyclic accelerators. In these accelerators the length of path can be increased without increasing the length of the apparatus. Examples are the cyclotron and betatron.

1. Cyclotron (Fixed Frequency)

Lawrence realized that, instead of using cylinders of gradually increasing lengths, the particles could be accelerated by using one pair of electrodes only if the particles were compelled to cross the same accelerating gap repeatedly by means of a magnetic field. This instrument was named as *magnetic resonance accelerator*. By 1936, the name *cyclotron* has come into general use. In 1939, Professor Lawrence was awarded the Nobel Prize in Physics in recognition of this achievement in the conception and development of the cyclotron.

In its simplest form, the cyclotron consists of two flat, semicircular metal boxes, called dees because of their shape. These hollow chambers have their diametric edges parallel and slightly separated from each other. A radio-frequency alternating potential of the order of megacycles per second is applied between the dees, which act as electrodes. These dees are surrounded by a closed vessel, containing gas like hydrogen, helium, deuterium at low pressure and the whole apparatus is placed between the poles of a strong electromagnet which provides a magnetic field perpendicular to the plan of the dees. The positively charged particles arc produced at the center between two dees, as shown by S in Fig. 11.13.

Suppose that at any particular instant the alternating potential in the direction which makes D_1 positive and D_2, negative. A positive ion starting from the source S will be attracted by the dee D_2. Since there is a uniform magnetic field B acting at right angles to the plane of the dees, the ion of charge q and mass M will move in a circular path of radius $r = M_v/Bq$. In the interior of the dee, the speed of the ion remains constant. After it has traversed half a cycle, the ion comes to the edge of D_2. If in the meantime, the potential difference between D_1 and D_2 has changed direction so that D_2 is now positive and D_1 negative, the positive ion will receive an additional acceleration while going across the gap between the dees, and then travel in a circular path of larger

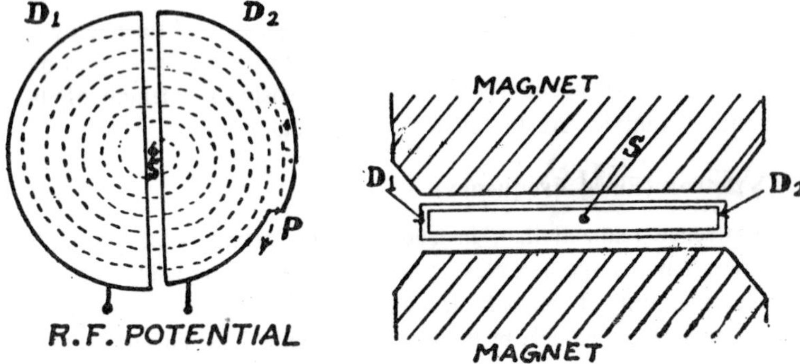

Fig. 3.13 : Cyclotron (Magnetic Resonance Accelerator).

radius inside D_1 under the influence of magnetic field. After traversing a half cycle in D_1, it will reach the edge of D_1 and receive an additional acceleration between the gaps because in the meantime the potential difference between dees has changed. The ion will continue travelling in a semi-circle of increasing radii, each time it goes from D_1 to D_2 and from D_2 to D_1. The time taken by the charged particle to traverse the semi- circular path in the dee is given by $t = \pi r/v = \pi M/Bq$ and hence the time for a complete circular path

$$\tau = 2t = 2\pi M/Bq \qquad ...(15)$$

Thus, for any given value of M/q, τ is determined by the magnetic field intensity B. By adjusting B the time can be made the same as that required to change the potentials. The frequency of the oscillations required to keep the ion in phase is given by the relation

$$f = l/\tau = Bq/2\pi M. \qquad ...(16)$$

If f and B are adjusted to keep the charged ion always in phase, each time the ion crosses the gap it receives an additional energy and at the same time it describes a flat spiral of increasing radius. Eventually, the ion reaches the periphery of the dee, where it can be brought out of the chamber by means of a deflecting plate charged to a high negative potential. This attractive force draws the ion out of its spiral path and thus can be used easily. If R is the radius of the dee, the kinetic energy of the ion emerging from fie cyclotron is thus given by

$$T = \frac{1}{2} M(BqR/M)^2 = q^2B^2R^2/2M = 2\pi^2R^2f^2M. \qquad ...(17)$$

This relation indicates that the maximum energy attained by the ion is limited by the radius R of the dees, magnetic field B or the frequency of the alternating potential f. It is also clear that the maximum energy acquired by the charged particle in a particular cyclotron is independent of the alternating voltage. It can be explained by the fact that when the voltage is small the ion makes a large number of turns before reaching the periphery, but when the voltage is high the number of turns is small. The total energy remains same in both the cases, provided B and R are unchanged.

Above equations are also true for relativistic velocities, providing $M = M_0(1 - v^2/c^2)^{-1/2}$. The quantity Br, often called *rigidity of a beam* of ions is thus given as

$$Br = \frac{M_v}{q} = \frac{W_v}{qc^2} = \frac{1}{qc}(W^2 - W_0^2)^{1/2} \qquad ...(18)$$

If the ion has Ze charge, and T and W_0 are in MeV then above relation gives

$$Br = \frac{1}{300Z}(T^2 + 2TW_0)^{1/2} \text{ weber/meter} \qquad ...(19)$$

If there are accelerations and the average potential difference between D's is V, the final energy can also be written as

$$T = NqV. \qquad ...(20)$$

Eliminating T from eqns. (17) and (20), we get

$$R = \frac{1}{B}\left(\frac{2MV}{q}\right)^{1/2} N^{1/2} \qquad ...(21)$$

This shows that the radii of successive paths increase as $N^{1/2}$, getting closer together as N becomes large.

Electric and Magnetic Focusing

The electric vertical forces are analogus to those already discussed in the linear accelerator. If the particles cross the gap when the field is rising the focusing field they experience when they approach the gap is over compensated by the stronger defocusing field that exists after the gap is passed. If the field is decreasing as they cross the gap, a net focusing field experienced. If the ions are to be kept in the vicinity of the median plane there must be forces returning them to this plane, whenever they leave it. These forces may result by shaping the magnetic field so that the vertical component B_0 decreases with increasing distance from the centre of the dees and by adjusting electric field between the dees. The magnetic field in a circular magnet tends naturally to decrease near the edge of the pole tips and the lines of force curve slightly outwards, as shown in Fig. 3.14. The field is still vertical in the median plane, but above and below it there are radial components of the field.

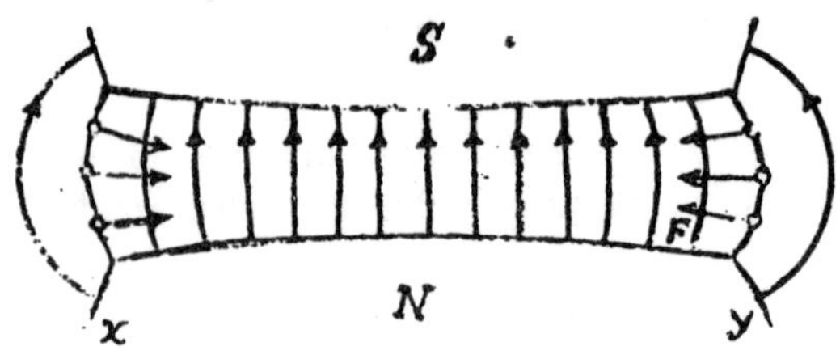

Fig. 3.14 : Vertical focusing in the cyclotron.

Being normal to the velocity, this radial component acts upon a particle with a force E (= Bqv) directed towards the median plane An ion that starts out with a slight vertical component of velocity will, because of these vertical forces, execute harmonic motion about the n-icdian plane. Without such restoring forces, it would be likely to drift towards the inside of the dees and be lost. The magnetic field varies as

$$B = B_0 (r_0/r)^n, \qquad (22)$$

where B_0 is the field at some fixed radius r_0, the index n is obtained by differentiation- It is zero at the centre, rises almost linearly upto the point where fringing effects start, and increases rapidly to exit-slit location. The frequency of the axial oscillations is given by

$$f_z = f_0 n^{1/2} \qquad ...(23)$$

where f_n is the cyclotron resonance frequency. The radial oscillations also occur, with a frequency

$$f_r = f_0(1 - n)^{1/2} \qquad ...(24)$$

Let us discuss the main components of a cyclotron:

(a) *Vacuum Chambers:* The vacuum inside the machine is obtained with the help of diffusion pumps. Oil-diffusion pumps of large diameter which have two or three pumping stages and automatic fractionation of the oil are best suited to this purpose. The baffles are used to prevent back streaming of the oil vapour, so that the oil deposition within the vacuum chamber is very small.

(b) *Deflector:* The deflector is used to pull the ion beam out of its circular path and direct it in a linear path. The deflector system may be divided as dc, rf, and dc-plus rf-types. In the dc-type the dc-electric field is applied between two curved plates forming a cylindrical condenser. The inner one is called the *septum* and the outer the *deflector*. The blade of the deflector is made of Cu with oil cooling arrangement and the septum is made of very thin tungsten plate. As a chosen maximum radius R a thin septum is inserted having a larger radius of curvature $R + \Delta R$, which splits the beam and allows a fraction of ions to pass into a channel behind the septum. If V_d is the deflector voltage and d is the spacing, then we have

$$\frac{mv^2}{R} = Bqv$$

and
$$\frac{mv^2}{R+\Delta R} = Bqv\frac{V_d q}{d}$$

$$\therefore \quad \frac{V_d}{d} = \frac{2T}{qR}\frac{\Delta R}{R+\Delta R} \qquad ...(25)$$

This equation determines the maximum permissible value of radius.

(c) *Oscillator:* To produce a high frequency potential difference of 10 to 100 kv in the frequency range between 10 to 20 megacycles/sec between the dees, resonant circuits are required. The resonant circuit is equivalent to a pair of quarter wave coaxial transmission lines with the dees supported on the ends of the inner conductors. The frequency is controlled by *Cu-shorting bars* or *spiders* reaching from stem to outer cylinder or in shielded pair structure from stem to stem.

(d) *Magnets:* The main magnet of a cyclotron has a basic structure of two cylindrical tapered poles, to rectangular yokes and two rectangular uprights. The purpose of the taper is to keep flux density approximately constant. The pole base has diameter greater than that of D's to keep the flux density in the base of the pole (about 52" for 42" cyclotron). A stabilized current is supplied to the magnet coils.

To tune up a cyclotron to preserve resonance and maintain beam intensity, shimming coils are used. These are flat-wound coils of many concentric turns with individual leads brought out radially from the inner turns.

(e) *Dees:* The dees are made of Cu, preferably the oxygen free high conductivity variety, and are supported in cantilever fashion by *dee stems* of 3" diameter. Few cyclotrons have only one dee, in which case the stem is located so as to be ⊥ to the open edge of the dee. This yield a symmetrical voltage pattern across the edge. To obtain a vertically symmetrical field pattern, a *dummy dee* (consists of a Cu frame) faces the active electrode.

(f) *Ion Source:* Recent sources are of the hooded arc variety. A hollow cylinder or chimney is closed at the top but with a hole in its side at mid-plane height, facing one dee. The cathode is formed of heavy tungsten or tantalum rod, formed as a short stemmed 'V' or 'U' clamped in terminating lugs and using large diameter water cooled leads. The chimney of the ion source may be moved in x-y-z direction to place the hole to any desired position so as to maximise the output current. The filament may be replaced without the loss of vacuum in the main chamber through air locks.

Cyclotrons are usually described in terms of the diameter of the pole faces of magnet. The first machine constructed by Lawrence and Livingston in 1932 used a magnet with pole faces 11 inches in diameter and produced 1.2 MeV protons. Consequently, cyclotrons with larger and larger magnets were built by Lawrence and his associates at Berkeley and later by others at several laboratories in the United States and aboard. Among them are several of about 60" pole-face diameter that accelerate deuterons to about 20 MeV and helium ions to about 40 MeV. The relativistic mass increase has limited the energies achieved with standard cyclotrons to about 25MeV for deuterons and 50 MeV for alpha particles. Although the cyclotron is capable of producing beams of energetic ions that are more intense than those from the electrostatic accelerators, the particle energies are much less constant. Hence, cyclotrons have found their widest use for studies requiring particles of high energy but where essential to know the exact value of energy. These high energy particles are used for the study of nuclear reactions, for the production of protons neutrons and radionuclides.

2. The Betatron

The cyclotron principle is inapplicable for the acceleration of electrons to high energies because of the large relativistic increase of mass at low energies. The voltage multiplier and the Van de Graaff electrostatic generator can both be used to accelerate electrons but the energies are limited to a few MeV. High energies X-rays, used in biological and medical research as well as in atomic research, can be secured by bombarding a target with high energy electrons. The first important machine for producing high energy electrons was the *betatron*. The first *magnetic induction accelerator* or *betatron* was constructed by D. W. Kerst at the University of Illinois in 1940, to accelerate electrons to an energy of 2.3 MeV. Several other betatrons have been constructed, the largest being the one at the University of Illinois, completed in 1950. It yields electrons of 300 MeV energy.

The action of the betatron depends on the same fundamental principle as that of the transformer in which an alternating current applied to a primary coil induces an alternating current usually with higher or low voltage in the secondary coil. In the betatron secondary coil is replaced by a doughnut shaped vacuum chamber. Electrons produced in the doughnut from a hot filament, are given a preliminary acceleration by the application of potential difference of 20 to 70 kV. When an alternating magnetic field is applied parallel to the axis of the tube, two effects are produced: *an electromotive force is produced in the electron orbit by the changing magnetic flux that gives an additional energy to the electrons; a radial force is produced by the action of magnetic field whose direction is perpendicular to the electron velocity which keeps the electron moving in a circular path.* Instead of spiralling, as in the cyclotron, conditions are arranged such that the increasing magnetic field keeps the electrons in a circular orbit of constant radius.

Using Faraday's Law of induction, the work done on an electron of charge e in one revolution $W = e d\Phi/dt$. If F is the tangential force acting on the electron, the work done W can be expressed as $W = 2\pi RF$.

$$\therefore \qquad F = \frac{e\, d\Phi}{2\pi R\, dt} \qquad ...(26)$$

For an electron moving in a circular orbit

$$Bev = Mv^2/R \quad \text{or} \quad Mv = BeR \qquad ...(27)$$

This relation indicates that the magnetic field B at the orbit must increase as the electron energy increases, otherwise radius R will not be constant. If R is kept constant, then

$$\frac{d}{dt}(Mv) = eR\frac{dB}{dt} \qquad ...(28)$$

From Newton's Second Law, the rate of change of momentum is equal to the force. Hence on combining eqns (26) and (28), we get

$$\frac{d\Phi}{dt} = 2\pi R^2 \frac{dB}{dt} = 2\frac{d}{dt}(\pi R^2 B)$$

This relation shows that the rate of change flux Φ within the orbit of radius R is always twice what it would have been if the magnetic field intensity were uniform throughout the orbit. This relation is known as the betatron condition or flux condition. It is also called as "2-1" rule of the betatron, this can be written as

$$dB_C/dt = 2\ dB_G/dt, \qquad ...(29)$$

where B_C is the space-averaged field strength in the core and B_G is the guiding field at the orbit. Since the induced potential is determined by the rate of change of flux, the iron core is laminated as in a transformer and alternating potential at 60 or 180 cycles is used to produce varying magnetic field. The electrons are injected at an instant when the magnetic field is just rising from its zero value in the first quarter cycle. The increasing magnetic field induces a potential within the doughnut which increases the energy of the electrons. When the field strength passes its maximum value and starts to decrease, the direction of the induced electromagnetic force is changed and the electrons start to slow down. This effect is avoided by removing the electrons from their stable orbit by passing a pulse of current through an auxiliary coil when the field reaches its positive maximum. These high energy electrons leave the path tangentially to strike a tared which then emits X-rays. The emission from the machine consists, therefore, of a succession of short pulses of X-radiations, each occurring at the peek of the a.c. maximum. Only the first quarter of each a.c. cycle is usefully employed in acceleration.

The energy of electrons can be estimated from the average Induced emf and the total number of revolutions made by the electrons. If be flux variation is given by the relation $\Phi = F_0 \sin \omega t$.

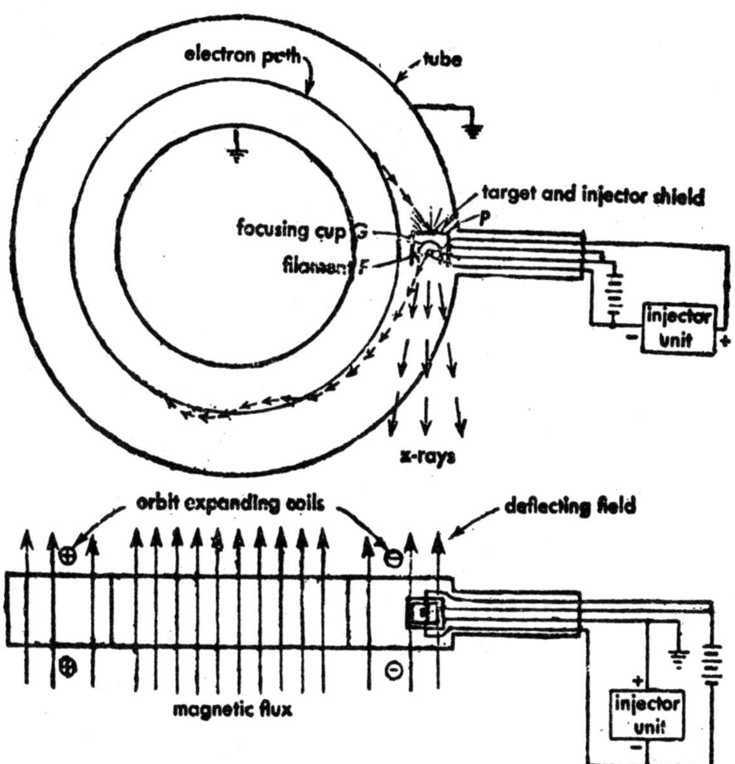

Fig. 3.15 : Schematic diagram of a betatron: (above) cross-section of betatron and (lower) plane view.

The time during which acceleration takes place will be $\pi/2\omega$. The energy gained by the electron per turn when the flux changes will be

$$eV = e\, d\Phi/dt = ew\Phi_0 \sin \omega t \qquad ...(30)$$

Thus, during the acceleration period, its average value will be $(2/\pi)\, e\omega\phi_0$. For most of the time electrons travel with a velocity close to the velocity of light and, therefore, the total distance travelled during the acceleration process will be $c\pi/2\omega$. If R is the radius of the orbit, the number of revolutions N will be $c/4\omega$ R, hence the total energy gained by electrons will be the product of the number of revolutions and the mean energy per turn.

$$\therefore \qquad E = e\, c\phi_0/2\pi R. \qquad ...(31)$$

The energy of electrons can also be estimated by the help of relativistic equation for energy

$$K.E. = pc = B_0 eRc, \qquad ...(32)$$

where B_0 is the magnetic field at the end of the acceleration cycle at the orbit. The energy obtainable is, therefore, limited by the radius and the peak strength of the magnet at the orbit. Using rest energy W_0 (= m_0c^2) of the electron. The kinetic energy T is given by

$$B_0 = \frac{mv}{eR} = \frac{(T^2 + 2TW_0)^{1/2}}{ceR} \qquad ...(33)$$

This relation shows that the high energy electrons can be retained in reasonably small orbits by readily available magnetic field.

Orbital Stability: Stability of the betatron orbit is achieved by making the para-axial field B: at the orbit decreasing with radius according to the relation

$$B_z = B_0\,(R/r)^n = B_0\,(r/R)^{-n}, \qquad ...(34)$$

where n is a constant, R the radius of *equilibrium orbit* and By is the value of Bz at equilibrium orbit. Since r makes small variation from R, hence can be written as r = R + x, Maxwell's equation for this case curl B = 0 gives

$$\partial Bz/\partial r = \partial B_r/\partial z$$

$$\therefore \qquad B_r = -\, nzB_0/R \qquad ...(35)$$

The *radial equation of motion* is

$$\frac{d}{dt}(m\dot{r}) = \frac{mv^2}{r} - B_z ev \qquad ...(36)$$

For radius r, eqn. (34) gives

$$B_z = B_0 \left(\frac{R + x}{R}\right)^{-n} = B_0 \left(1 - \frac{nx}{R}\right) \qquad ...(37)$$

Hence eqn. (36) becomes

$$\frac{d}{dt}(m\dot{x}) = \frac{mv^2}{R + x} - B_0 \left(1 - \frac{nx}{R}\right) ev$$

or

$$m\ddot{x} = \frac{mv^2}{R}\left(1 - \frac{x}{R}\right) - ev\, B_0 \left(1 - \frac{nx}{R}\right) \qquad ...(38)$$

For equilibrium orbit x and their derivatives are zero, hence we get $mv^2/R = evB_0$ and thus eqn. (38) becomes

$$m\ddot{x} = \frac{mv^2}{R}\left(1 - \frac{x}{R} - 1 + \frac{nx}{R}\right) - \frac{mv^2}{R^2}\,x\,(-1 + n)$$

or $$\ddot{x} + \omega^2 (1 - n)\, x = 0, \qquad ...(39)$$

where $\omega = v/R$ is the angular velocity in the equilibrium orbit. Above relation shows that the frequency of radial oscillations

$$f_r, = \omega\,(1 - n)^{1/2}/2\pi \;\; (n < 1) \qquad ...(40)$$

The *vertical equation of motion* for a small axial displacement z out of the median plane is written as

$$\frac{d}{dt}(m\dot{z}) = B_r\, ev = ev\left(\frac{-nB_0 z}{R}\right)$$

or $$\ddot{z} + \omega^2 nz = 0 \qquad ...(41)$$

Thus, the frequency of vertical oscillations is

$$fz = \omega n^{1/2}/2\pi \qquad ...(42)$$

For equations (39) and (41) to describe oscillations, the coefficient of x and z must be positive, *i.e.*, $1 > n > 0$. It shows that the field must decrease from the centre to the periphery, or in other words that the lines of force must be concave towards the axis of the machine. This decrease must not be too rapid, as $n < 1$. These oscillations are named as *betatron oscillations*, and are naturally damped as the field grows, the oscillation amplitude is proportional to $B_0^{-1/2}$. These principles are utilized in the designing of the magnet. The same magnet structure is used to provide the accelerating flux across the short central gap and the guiding field in the wider gap which surrounds the flux gap. The gap lengths are chosen to satisfy relation (29) and the guide field gap is tapered to given an n value between 0 and 1.

Most of the modern betatrons arc operated from a 60 cycle/sec a.c. source. If the length of the circular path within the doughnut of a betatron is supposed to be 3 metre and the energy gained by the electron in each revolution 400 eV, the electrons will then have a total energy of 167 MeV. In practice the energy of electrons can be varied from 10 to 100 MeV by applying the orbit shifting magnetic field at different times during the quarter cycle in which the field is increasing. The betatron, constructed at the General Electric Research Laboratories, produces

100 MeV electrons. The diameter of the pole face is 76 inches and that of stable orbit is 66 inches. The magnet weighs 130 tons and the maximum magnetic field at the orbit is 4000 gauss. Electrons, injected at energies of 30-70 keV, travel around the doughnut about 2.5 × 106 times, gaining about 400 eV of energy at per revolution. University of Illinois has a large machine with doughnut diameter 97 inches and magnet weighing 350 tons giving 340 MeV electrons.

Since the operation of the betatron is unaffected by the increasing mass of the electron, as it gains energy it might be thought that there would be no upper limit of energy obtainable by means of this device. In fact, there is, not because of mass increase but because of radiation loss, a limit of energy where the radiation off-sets the gain. Because of the centripetal force, the circulating electrons are expected to emit radiations. Using classical arguments, Schwinger in 1949 gave a relation for the total energy radiated by an electron. According to him the total energy radiated by an electron of energy W in one revolution is

$$U = \frac{1}{3\epsilon_0}\left(\frac{e^2}{R}\right)\left(\frac{W}{m_0c^2}\right)^4 = 88.5\frac{W^4}{R} \quad ...(43)$$

where U is in keV, W in geV (giga electron volt, in Europe which is known as beva electron volt in U.S.A.) and R in metres. This relation indicates that the energy radiated per revolution increases with the fourth power of the electron energy and thus becomes very serious at very high energies. For the 300 MeV betatron the total loss is about 13% of the total energy. If no compensating means are included the orbit will shrink about 20% and the beam will loss by collision with the inside wall of the vacuum chamber. Electrons can however be accelerated to higher energies without appreciable radiation losses by means of a linear accelerator.

3.6 PRINCIPLE OF PHASE STABILITY

In a phase stable accelerator, particles will be accelerated at a series of gaps by an alternating electric field. *The gap separation and the frequency and strength of the field are adjusted so that a particle of a specified energy arriving at a particular gap at a specified equilibrium phase of the accelerating field will arrive at the next gap at the same phase.* This principle is most important in some accelerators, known as *synchronous or phase stable accelerators.* These are most easily understood by analyzing the electric field patterns as travelling waves. If x represents distance travelled by the particles, the significant field component can be represented as

$$E_x = E \sin\left[\int \omega \, dt - \int \frac{\omega}{v} dx + \phi_0\right] \quad ...(44)$$

where ϕ_0 is the equilibrium phase, v the wave velocity and $E = V/x_0$. Here V is the voltage across the accelerating gap and x_0 the distance between two consecutive gaps. If the particle is travelling on the equilibrium orbit at the equilibrium phase ϕ_0, we have to adjust ω and v such that bracket term of eqn. (44) reduces to ϕ_0. If the particle is not at the equilibrium phase but as at a phase $\phi_0 + \Delta\phi$. Associated with the phase error will be a momentum error Δp from the equilibrium momentum p_0 and position error Δx from the position of the equilibrium particle. Since a phase change of 2π results in a position change Δx of one wavelength, hence

$$\Delta\phi = -\omega\Delta\, x/v \quad ...(45)$$

and

$$\frac{\Delta p}{p_0} = \frac{1}{p_0}\left[\frac{m_0\dot{x}}{(1-\dot{x}^2/c^2)} - \frac{m_0 v}{(1-v^2/c^2)^{1/2}}\right] = \frac{\Delta\dot{x}}{v(1-v^2/c^2)} \quad ...(46)$$

Δp also appears in the equation of motion

$$\frac{d}{dt}(p_0 + \Delta p) = qE \sin(\phi_0 + \Delta\phi)$$

$$\therefore \quad \Delta\dot{p} = qE\, \Delta\phi \cos\phi_0 \text{ (for small } \Delta\phi). \quad ...(47)$$

The combination of equations (45), (46) and (47) gives

$$\frac{d}{dt}\left[\frac{p_0}{v(1 - v^2/c^2}\frac{d}{dt}\left(\frac{v\Delta\phi}{\omega}\right)\right] + qE\,\Delta\phi\cos\phi_0 = 0 \quad ...(48)$$

For a *linear accelerator* ω is kept constant and we get

$$\frac{d}{dt}\left[\frac{1}{(1 - v^2/c^2)^{3/2}}\frac{d}{dt}\left(\frac{\Delta\phi v}{c}\right)\right] + \frac{\omega qE}{m_0 c}\Delta\phi\cos\phi_0 = 0 \quad ...(49)$$

In the case of particles travelling on a *circular or spiral orbit,* the momentum error results not only in an azimuthal position error but also in a radius error Δr from the equilibrium radius R. We shall consider acceleration by the hth harmonic of the revolution frequency. Thus, eqn. (45) becomes

$$\Delta\phi = h\omega\, \Delta x/v = h\omega r\, \Delta\theta/v = -h\Delta \qquad ...(50)$$

$$\therefore \quad \Delta\phi = h\Delta\dot{\theta} = -h\, [\Delta(r\dot{\theta}) - \dot{\theta}\Delta r]/r \qquad ...(51)$$

or
$$\Delta\dot{\phi} \simeq \frac{hv}{R}\left[\frac{Dr}{R} - \frac{D\dot{x}}{v}\right] \qquad ...(52)$$

Since $p = Bqr$ and $B \propto r^{-1}$ for stability of betatron oscillation

$$\therefore \quad \Delta p/p_0 = (1 - n)\, \Delta r/R \qquad ...(53)$$

Substituting in eqn. (52) for Δr from eqn. (53) and $\Delta\dot{x}$ from eqn. (46), we get

$$\Delta\dot{\phi} = \frac{hv}{R}\frac{\Delta p}{p_0}\left[\frac{1}{1-n} - \left(1 - \frac{v^2}{c^2}\right)\right] \qquad ...(54)$$

In a circular accelerator the equation of motion is

$$\frac{1}{r}\frac{d}{dt}(rp) = \dot{p} + \frac{\dot{r}\,p}{r} = qE\sin(\phi_0 + \Delta\phi) + qE' + qr\, B_z$$

As $\quad q\dot{r}\, B_z = rp/r$, hence

$$\dot{p} = qE\sin(\phi_0 = \Delta\phi) + qE'. \qquad ...(55)$$

In this eqn. E' represents other accelerating or decelerating forces such as betatron acceleration term

$$E_b' = \frac{1}{2\pi r}\int \frac{\partial B}{\partial t}\, 2\pi r\, dr \qquad ...(56)$$

and radiation deceleration term

$$E_r' = 9.60 \times 10^{-10}\, (1 - v^2/c^2)^{-2}\, r^{-2} \text{ volt/metre.}$$

E in eqn. (55) is $l/2\pi r$ times the voltage V, which is the sum of the peak voltages applied across all accelerating gaps. Thus eqn. (55) becomes

$$rp = \frac{qV}{2\pi}\sin(\phi_0 + \Delta\phi) + q\int \frac{\partial B}{\partial t} r\, dr \qquad ...(57)$$

In terms of momentum, radius and phase errors thus becomes

$$\dot{p}_0\, \Delta r + R\Delta\dot{p} = (qV/2\pi)\, \Delta\phi\cos\phi_0 + q(\partial B/\partial t)\, R\, \Delta r \qquad ...(58)$$

Since $p_0 = BqR$ hence first and last terms of this eqn cancel each other, and we are left with

$$\Delta\dot{p} = (qV/2\pi R)\ \Delta\phi \cos \phi_0 \qquad ...(59)$$

Combining this eqn with eqn. (54), we get the phase equation for the *synchrotron*. Using $\gamma = \left[\dfrac{1 - v^2}{c^2}\right]^{1-1/2}$ we have

$$\frac{d}{dt}\left[\frac{\gamma\Delta\dot{\phi}}{1/(1-n) - 1/\gamma^2}\right] = \frac{hqV}{2\pi}\frac{\cos\phi_0}{R^2 m_0}\delta\phi \qquad ...(60)$$

In the *synchrocyclotron* B is constant and radiation losses are negligible, hence E' in eqn (55) is zero. The harmonic number h is usually unity. n is very small can be neglected and R is not constant, hence eqn. (54) becomes

$$\Delta\dot{\phi} = \frac{v}{R}\frac{\Delta p}{p_0}\left(1 - \frac{1}{\gamma^2}\right) = \frac{qB}{m_0^2 c}\frac{\beta}{\gamma^2}\Delta p \qquad ...(61)$$

Equation (57) takes the form

$$r\dot{p} = p\dot{p}/Bq = (qV/\pi) \sin \phi \qquad ...(62)$$

$$\therefore \quad \frac{1}{m_0 c}\frac{d}{dt}(p\Delta p) = \frac{d}{dt}(\beta\gamma\ \Delta p) = \frac{q^2 BV}{\pi m_0 c}\Delta\phi \cos\phi_0 \qquad ...(63)$$

Combining eqns (61) and (63) we get the second order differential eqn in $\Delta\phi$ and is known as the *phase equation or the synchrocyclotron.*

$$\frac{d}{dt}(\gamma^2\ \Delta\phi) = \frac{(q/m_0)^3\ VB^2}{\pi c^2}\Delta\phi \cos\phi_0 \qquad ...(64)$$

This again describes a stable damped oscillation provided only that $\cos\phi_0$ is negative. Since $\sin\phi_0$ must be positive to permit acceleration, hence ϕ_0 must lie between $\pi/2$ and π.

3.7 HIGH ENERGY CIRCULAR ACCELERATORS

The acceleration of nuclear particles to relativistic energies was only made possible by the advent of the *principle of phase stability*. When applied in conjunction with a magnetic guiding field, this principle has resulted in accelerators whose cost per MeV is even lower than that of smaller machines. According to the way in which this phase stability is achieved, we shall have to study the *synchrocyclotron,* the electron *synchrotron* and the *proton synchrotron*. All these machines are also called *synchrotrons*.

1. The Synchrocyclotron (Frequency Modulated Cyclotron)

The relativistic limitation on energy for fixed frequency cyclotrons has restricted the useful size of the magnets. This limitation can be removed and the ions can be accelerated indefinitely if the applied frequency is varied to match exactly the ion revolution frequency. Equation (64) is applicable in this situation. The instantaneous augular frequency of the phase oscillation is $(q/m_0)^3\ VB^2 \cos \phi_0/\pi c^2\gamma^3$.

If V = 10 kV and the particle energy is 300 MeV, the frequency of phase oscillation is about 1/1000 of the rotation frequency of the ions. The eqn. (64) shows that the oscillation is damped as energy increases so that the acceleration can be carried on to whatever limit is set by the greatest radius at which magnetic field is usable.

This upper limit is set by the fact that the field falls of rapidly with radius towards the edge of the magnet pole. The limit is set in practice by the passage of n through a value of 0.2 at which the radial betatron oscillation frequency is exactly twice the vertical betatron oscillation frequency

$$\frac{f_r}{f_z} = \frac{(1-n)^{1/2} f_0}{n^{1/2} f_0} = \frac{(1-0.2)^{1/2}}{0.2^{1/2}} = \frac{2}{1} \qquad ...(65)$$

At this point a resonance takes place between two oscillations and the vertical oscillations amplitude increases until the beam is lost. It is found generally about one magnet gap width in from the edge of the magnet pole.

In synchrocyclotron light +ve ions (p, d, α) are accelerated to energies significant relative to the rest energy of the particle. The ions traverse circular orbits with increasing radii as energy increases. The ions pass many times (2 times in an one revolution) through the rf electric field of a large D-shaped hollow electrode and experience an acceleration on each traversal of the accelerating gap. For each value of panicle energy in the magnetic field there is a particular orbit radius and a specific frequency of revolution.

$$r = \frac{mv}{qB} = \frac{[T(T + 2\ m_0c^2]^{1/2}}{cqB} \qquad ...(66)$$

At every instant the synchronous ion and the oscillator have the same frequency, given by

$$f = \frac{v}{2\pi r} = \frac{qB}{2\pi m} = \frac{qB}{2\pi m_0}\frac{m_0c^2}{(m_0n_0 + T)} = \frac{f_0}{1 + T/m_0c^2}$$

where $f_0 = qB/2\pi m_0$ is the non-relativistic cyclotron frequency.

The decrease in ion revolution frequency is caused primarily by the increasing T but is also affected by the slight decrease in B with increasing radius required for orbit stability. Above relation also shows that the fractional change in oscillator frequency is proportional to $\frac{T}{M_0c^2}$ electrons, because of the small rest mass, cannot be readily accelerated in the synchrocyclotron. The change in frequency is ordinarily made with a rotating multibladed capacitor.

The *principle of phase stable orbits* can be explained by reference of Fig. 3.16, which represents the variation of the radio-frequency oscillating potential with time. Particles which arrive at each gap at the instant corresponding to the point indicated by P in the figure will always

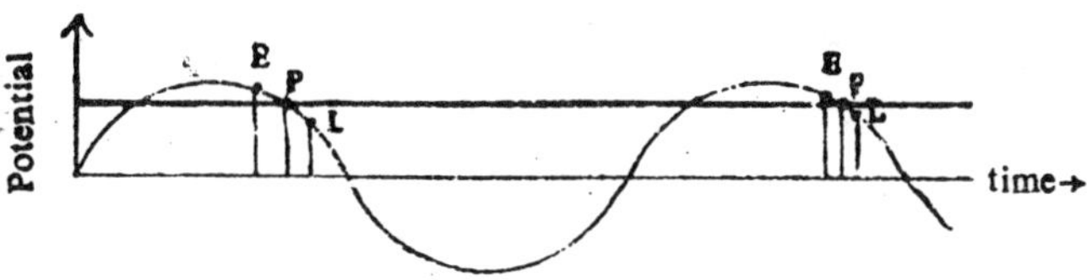

Fig. 3.16 : Principle of phase stability.

receive the appropriate impulse. A particle which has more energy than that culled for at a particular gap will necessarily traverse an orbit of larger radius which requires a time longer than the r-f-period and will arrive later as indicated by the point L. The potential will then have decreased and this particle will receive a smaller energy impulse. The increase in mass is hence small. It is clear that the time T required to traverse the path will be less and its next arrival will be closer to the point indicated by P (prompt). After traversing several times, the particle will be brought to the stable phase and the resonant energy. On the other hand, if particle has smaller energy, it will arrive earlier as indicated by the point E. This particle will receive greater energy and mass than normal, since the potential is now higher than at P- It will thus take a slightly longer time to return to the point where it again receives impulse. This point will be nearer to the point P. In this way after several revolutions the particle will be brought to the stable phase. *The revolution of the charged particles is thus automatically synchronized with the changing frequency of the accelerating potential.* It should be noted that phase oscillations occur if phase lies between $\pi/2$ and π.

Dee Voltage

If the final energy is 500 MeV, obtained in time 0.01 sec by a source of frequency 20 Me/sec. The average energy gained per revolution = $(500 \times 10^6/0.01)/20 \times 10^6 = 2500$eV/rev. If the synchronous phase angle is $\phi_0 = 150^o$, so that $\sin \phi_o = 0.5$. As there are two accelarations per turn, a single dee with a peak potential of 2500volts is adequate, which is very small in comparison of 100 kilovolts required for cyclotrons. The *D-shaped electrode* is similar to the electrodes used in the standard cyclotron. The *ion source* is quite -similar to that used for the standard cyclotron. The *magnetic field* is produced by a solid core magnet or large pole-face area. It must be steady and approximately uniform over the pole face, decreasing slightly with increasing, radius to provide focusing. The weight of the magnet varies as T^3. Exciting coils for the magnet have also been developed. Some designers use oil-bath cooling and others prefer dry-wound coils. The *radio frequency oscillator* for giving power to the D depends for its frequency variation on the mechanical modulation of the resonant D circuit by a variable capacitor. In half wave and three quarter wave circuits, the capacitor is located at the outer end of the D stem and can have its own separate vacuum system. For maximum capacitance, it must have close spacing of the blades, smooth rounded corners, highly polished surfaces and fast pumps for a good vacuum. In quarter wave circuits, the capacitor operates in the fringing magnetic field at the back of the D and is of multiple parallel plate design. *The deflection* of resonant ions to produce an emergent beam is complicated as the change in orbit radius per turn is very small. The deflector must pull the beam through a large radial distance. The magnetic deflector known as regenerator is used as it acts on all particles of the beam. The emergent beam is focused to small diameter and passed through channels in a shield.

As soon as the phase stability was recognized, frequency modulation was applied to cyclotrons, first in Berkeley, then in many other laboratories. The first frequency modulated cyclotron was the 184 inch machine at the university of California at Berkeley which produced 190MeV deuterons and 380 MeV α-particles. It consists of only a single dee placed inside a vacuum chamber between the poles of a very large electromagnet. The pole faces are specially shaped to provide a field which decreases almost linearly from the centre (15 kilo gauss) out to the position of maximum orbital radius. The field thus produced will focus the particles in the median plane. The oscillating potential is applied between the dee and the ground plane. The radio-frequency (peak voltage 15 kV) is supplied by a tube oscillator circuit modulated by being coupled to a variable capacitor. By means of this capacitor the frequency of the oscillating potential is decreased to compensate for the gain in the effective mass of the particle as its speed increases. For deuterons and α-particles the frequency was modulated from 11.5 mega cycles/sec at the instant of injection to 9.8 mega-cycles/sec when the particles reached the periphery of the dee Protons with energies of about 350 MeV were obtained with an oscillator frequency modulated from 23 to 15.6 mega cycles/sec. In 1957, protons of 720 MeV energy were produced by increasing the strength of the magnetic field from 15-28kilo gauss. Another large synchrocyclotron with a magnet 236 inches in diameter weighing 7200 tons was constructed at Dubna in U.S.S.R. in 1954. This produces protons of energy 680 MeV. A 600 MeV

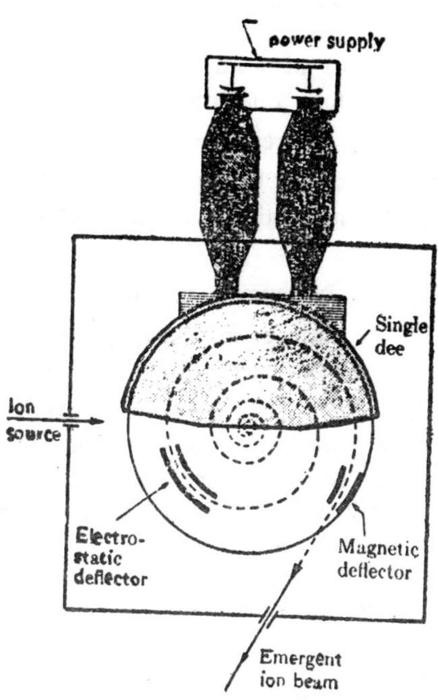

Fig. 3.17 : Schematic diagram of the 184 inch synchrocyclotron.

synchrocyclotron with magnet 196 inches in diameter weighing 2500 tons was completed in 1958 at the CERN laboratory in Geneva, Switzerland. Designs for even larger machines seen practical. But the excessive cost of solid core magnets for higher energies (weight proportional to T^3) has transferred interest to the proton synchrotron.

The main differences between the operation of the ordinary cyclotron and synchrocyclotron may be summarized as: In the latter case;

1. The magnetic field decreases towards the edge of the magnet pole.
2. The frequency is modulated.
3. The particles may turn around 10^5 times before reaching the maximum energy, instead of 10^2 times in the former case.
4. The ions from slugs that circulate in the machine and come out in spurts lasting about 50msec and repeating roughly 100 times per sec. The synchrocyclotron has been most valuable in providing information about the production and properties of mesons.

2. The Electron Synchrotron

When studying the betatron we found that maximum energy attainable was limited mainly because of the very low energy gain per turn. The gain was compensated by the losses due to electromagnetic radiation at high energies. We have seen in the last sub-article that electrons cannot be accelerated by synchrocyclotron because of their small mass. By adapting the betatron to use the synchrotron principle, Goward and Barnes, in 1946, in England, gave a machine which is known as electron synchrotron. In the synchrocyclotron the nature of particles was such that their acceleration entailed a decrease of their frequency of revolution in the presence of a particularly constant magnetic field. Conversely, since relativistic electrons (energy $\geq$ 2 MeV) revolve on a fixed orbit with constant frequency (as they have reached the velocity of light v = 0.98c at 2 MeV), the synchronous energy can only increase with time in the electron synchrotron if the magnetic field is also made to increase so as to satisfy equation.

$$\omega_s = eBc^2/W_s, \qquad \text{...(67)}$$

where the subscript A' denotes synchronous particles.

The electron synchrotron has, like the betatron, the doughnut in an ac-magnetic field. The magnet can be a ring of C-shaped units, not filling the hole in the doughnut as the case with the betatron. The weight of the magnet is decreased in this way. In some synchrotrons the magnetic field is produced by coils only, no iron being present. In the central gap some flax bars serve as the central core of the magnet to start up the machine us a betatron. These bars are made of high permeability metal and do not have to be large; they short the magnetic field at low inductions but become saturated at high inductions and the transition from betatron action to synchrotron action can be made smoothly. Part of the interior of the doughnut is coated with copper or silver to give a resonance cavity. A small break in the coating separates it into two parts. A high frequency electric field from a radio-frequency oscillator is applied across this gap at the proper time in the magnetic cycle. When the accelerator is on, the electron is accelerated each time it crosses through the resonator.

The electron synchrotron accelerates electron in an orbit of constant radius by means of a radiofrequency electric field applied across a gap. A ring shaped magnet provides the magnetic field over the doughnut shaped vacuum chamber. The pole faces are accurately shaped to provide a field which decreases with increasing radius, with n~0.6, to supply focusing forces for the electrons, similar to the betron cash. The maximum energy of the electrons depends on orbit radius and on to maximum magnetic field.

$$T = ceBR = 300 \text{ BR MeV.} \quad \text{...(68)}$$

Synchronous acceleration starts normally when the electrons reach the velocity of light. At this constant velocity frequency

$$f = c/2\pi R = 47.8 \times 10^6 \text{ cps} \quad \text{...(69)}$$

The rf-accelerating field must have a frequency equal to this electron frequency or to some harmonic of it. The volts per turn requirement comes from the rate of magnetic field and other constants of motion. We can obtain it by the relation

$$V_e = 2\pi R^2 \, dB/dt. \quad \text{...(70)}$$

During acceleration the electrons are kept in step with the accelerating field by the phase stability. The equilibrium electrons continue to return to accelerating point at the correct phase; other electrons in the phase stable range oscillate in phase around.this equilibrium phase. The phase oscillations are accompanied by oscillations in energy and so in orbit radius. Hence, the electrons are accelerated in a bunch as are the protons in synchrocyclotron.

By a little manipulation of eqn. (60) it can be shown that for electron energies higher than 2 or 3 Mev, we have

$$\frac{\text{Frequency of phase oscillation}}{\text{Frequency of revolution}} = \left[\frac{he\ V \cos\phi_0}{2\pi m_0 c^2 (1-n)}\right] \qquad ...(71)$$

Thus, for V = 1500 volts, h = 1, n = 0.6 and energy 3MeV, the ratio is about 0.014.

Electrons are injected into the doughnut after a preliminary acceleration in an electrostatic field to 50-100 keV. When these are

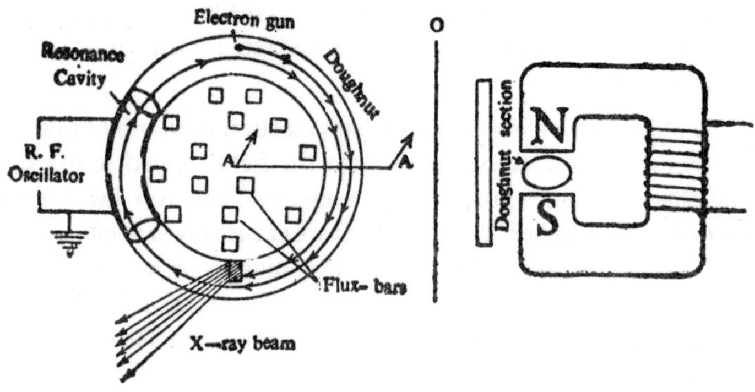

Fig 3.18 : Electron Synchrotron, (left) Plan with magnet removed (right) vertical section showing annular magnet.

injected at the beginning of the magnetic cycle, the device operates as a betatron. The electrons travel in a circular paths and increase their energy as the field increases, until the electrons reach some 2 MeV at which point the bars are magnetically saturated and are no longer able to induce an e.m.f. This saturation causes the betatron mechanism to stop. At this point the synchrotron mechanism begins to operate. If the potential applied to the resonator operates at the proper frequency, the electrons are all kept in phase and receive increments of energy at each revolution, as they pass through the cavity. The high frequency field remains on while the magnetic field is increasing and is automatically cut off when the electrons have acquired their maximum energy or if desired at an earlier time in the cycle. As the magnetic field still has a little increase to go before reaching its maximum, the radius of the beam of constant velocity (~c) diminishes a little so that the beam spirals in slightly to strike a target projecting from the inner edge of the doughnut which gives off short wavelength X-rays. The rays emerge in pulses as in the betatron.

Several methods of pre-acceleration have been used in the electron synchrotron. Van de Graaff generator or a linear accelerator used as an injector in many of the recently constructed electron synchrotrons. This type of injector has the advantage of providing a well collimated highly mono-energetic beam of electrons. Electrons are injected at 80 keV, accelerated to 7 MeV by betatron action and finally reach energies of 330 MeV in the machine which is at the Massachusetts Institute of Technology. This machine has a 50 ton magnet and radio frequency of 46.5 mega cycles. The orbit radius is 40 inches.

Higher-energy (1500 MeV) electron synchrotron consists principally of four quadrants in each of which a magnetic field is created by electromagnets with C-shaped cross-sections. The quadrants are separated by short straight sections where there is no magnetic field. All these sections are exploited to connect pumps and instruments to the accelerator tube and two of them are used for injection and acceleration of the electron beam.

3. The Proton Synchrotron

We know that 1500 MeV is an upper limit for betatrons and electron synchrotrons because of the energy loss by radiation of the high energy electrons revolving in an orbit. With protons, however, the situation is different as the rate of loss of energy by a charged particle revolving in an orbit is not only proportional to E^4/R but also to $1/M_0^4$. Since the rest mass of the proton is nearly two thousand times that of an electron, the proton energy can hence be raised to extremely high values.

The principles on which the proton synchrotron is designed and operated are basically the same as those of the electron synchrotron. Its name does not imply that its principles of operation apply exclusively to protons (they are equally valid for deuterons, α-particles or other ions), but results from the fact that protons, being elementary particles, are preferred for investigations in elementary particles physics. Protons revolve in an orbit of constant radius in a doughnut shaped vacuum chamber. A ring shaped magnet surrounding the doughnut is required. In the synchrocyclotron, however, the protons follow a spiral path and the magnetic field must cover the whole area in which the particles spiral around and solid core magnet is required. Because of the large reduction in the weight of the magnet a proton synchrotron is much cheaper to construct than a synchrocyclotron.

Protons are accelerated to a few MeV and then injected into the doughnut, tube, when the magnetic field is small. A magnetic field which increases with time as the proton gains energy is applied normal to the chamber, as in the betatron. Unlike the electrons, which approach the velocity of light at relatively low energies (v = 0.98 c at 2 MeV) and so have an essentially constant frequency of rotation during acceleration, protons do not reach a speed of 0.98c until their energy is 4 BeV. A velocity and frequency of rotation of protons in an orbit of constant radius increase during the acceleration. The frequency increase is rapid at first but slows down as the proton velocity approaches c. The frequency of the oscillating potential must be increased, as in the synchrocyclotron to synchronize with the rotation frequency of protons and to keep the protons in a constant average orbit radius. This is the main difference between electron synchrotron and proton synchrotron. In most of the present machines the frequency modulation is accomplished electronically rather than by rotating condensers, as in the frequency modulated cyclotrons. The same type of phase focusing exists to bunch the particles about an equilibrium phase of the accelerating field as for the electron synchrotron.

The largest and most costly component of a proton synchrotron is its magnet. The first step in design is a decision on magnet dimensions and properties. It determines characteristics of the other components. The choice of the magnet structure and maximum field determines the orbit radius for a chosen particle energy. The K.E. of the particle T is given by

$$T(T + 2M_0c^2) = q^2c^2B^2R^2 \qquad \text{...(72)}$$

When T and M_0c^2 are expressed in BeV units, B in weber/m^2 and q = 2e. Then

$$R = 3.33[T(T + 2M_0c^2)]^{1/2}/Bz \text{ metres.} \qquad \text{...(73)}$$

The design involved the use of parallel bundles of iron laminations arranged in a radial array around a circle, leaving wedge shaped spaces between the bundles at the orbit. To focus particles in orbit, the magnetic field must decrease with increasing radius across the pole face. In both the *cosmotron* and *bevatron* the n value was chosen as 0.6, and was obtained by doping the pole faces so the gap on the outside of the orbit is longer than on the inside.

Magnetic field increases with time during the pulse, the rate of increase is determined by the characteristics of the power supply and the magnet windings. The average value of dB/dt is given by the maximum field and the rise time. The rate of increase of particle energy with lime is directly related to the rate of magnetic field increase. By differentiating eqn. (72), we get

$$\frac{dT}{dt} = \frac{q^2c^2R^2B}{T + M_0c^2}\frac{dB}{dt} \qquad ...(74)$$

The ion revolution frequencies at injection and at high energy determine the range of frequencies required in the radiofrequency system. The relativistic relation for the frequency of revolution in circular orbit is

$$f_c = \frac{c^2qB}{2\pi(T + M_0c^2)} \qquad ...(75)$$

In the orbit with for quadrants of radius R and straight sections of length L the orbital frequency is decreased by the path length ratio.

$$\frac{f_0}{f_c} = \frac{2\pi R}{[2\pi R + 4L]} \qquad ...(76)$$

The energy increase per turn can be obtained by dividing relation (74) by f_0 given in eqns (75) and (76).

$$\Delta T/\text{turn} = qR(2\pi R + 4L)dB/dt. \qquad ...(77)$$

The variation of frequency of revolution fo for cosmotron with the magnetic field shows a range of frequencies from 0.4 megacycle/sec. at 4 MeV injection (B = 3.11 gauss) to 4.2 mega-cycles/see. at the maximum energy of 3BeV.

f_0f_c = 0.824 and for the average value of dB/dt = 14 kilo gauss/sec. ΔT/turn = 880eV/turn.

Injecting particles into a synchrotron is not very easy. A high intensity, pulsed beam of panicles is needed, with small energy spread and delivered in a well-collimated beam of small cross- section. It is pulsed at the start of each magnet cycle for a short time during which the particles come in stable orbits. A particle directed into the synchrotron field on a path 11 to its equilibrium orbit but r > R, will oscillate in the radial plane about the equilibrium orbit with a frequency

$$f_r = \left(1+\frac{L}{\pi R}\right)(1-n)^{1/2} f_0 \qquad ...(78)$$

Angular divergence also introduces vertical oscillations about the median plane with a frequency

$$f_z = \left(1+\frac{L}{\pi R}\right)(1-n)^{1/2} f_0 \qquad ...(79)$$

To obtain a fine *vacuum*, pump stations are distributed around the periphery. Cosmotron chamber requires 12 pumping stations, 3 in each quadrant. A target is permanently mounted just inside the inner wall of one of the straight section vacuum chambers.

The first proposal of a proton accelerator using a ring magnet, in which both magnetic field and frequency of the accelerating electric field were varied, was made in 1943 by Prof. M.L. Oliphant of Birmingham University. Design studies of proton synchrotrons in the U.S.A. were started in 1947 and the first machine was in operation in 1952 at the Brookhaven National Laboratory, This machine produced protons of energy 2.3 BeV, although weight of the magnet was 1650 tons. The energy was later increased to 3 BeV. The Brookhaven machine is called *cosmotron* because it makes possible the study of some of the nuclear reactions that occur with particles with energies comparable to those of the primary cosmic ray particles. In 1954, 6.4 BeV protons were obtained in the machine at the Radiation Laboratory of the University of California. This machine is called *bevatron* (billion-electron-volt accelerator). A 10BeV machine was completed at Dubna near Moscow in 1957 and is known as *synchrophasotron*. Several other conventional proton synchrotrons nave been built in Saclay, France (Max. energy 2.5 BeV, magnet weight 1080 tons); Harwell, U. K. (Max. energy 7.0 BeV, magnet weight 7000 tons); Princeton, U.S.A. (Max. energy 3.0 BeV, magnet weight 350 tons) ; Argonne U. S. A. (Max. energy 12.5 BeV, magnet weight 4000 tons); Canberra, Australia (Max. energy 10.6 BeV, magnet weight 0 tons due to air core).

The Brookhaven cosmotron is not circular but in four quad- rants each of 30ft. radius and are joined by four straight sections each of 10ft. long. The magnetic field is applied over four quadrant and straight sections are kept free from magnetic field. The magnet, with a maximum diameter of 75 ft is a steel ring about 88 ft in cross-section and weighs about 2000 tons. Protons of about 4 MeV energy, derived from a 4 MeV Van de Graaff machine, an injected into one of the straight portions of the tube when the magnetic field has reached a value of about 300 gauss. The maximum magnetic field is 14000 gauss when the peak current through the magnet is 7000 amps. The electrostatic inflector plates,

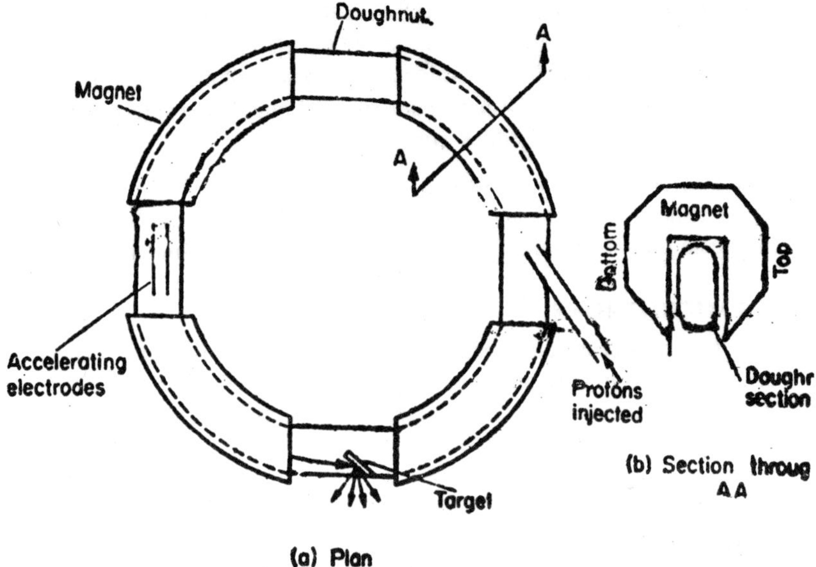

Fig. 3.19 : Schematic diagram of cosmotron.

located in this straight section, deflect the protons entering at a 30° angle to the equilibrium orbit, by this angle. The magnetic field is increased in proportion to the momentum to keep the protons in the synchrotron moving in a circular path of constant radius. As a result of the increasing speed of the particles as they gain energy in successive revolutions, the time required for the protons to make a turn of constant radius decreases, hence the frequency of revolutions increases correspondingly. To maintain the condition of phase stability, the frequency of the radio frequency potential is increased in the same manner. It is applied to one of the straight sections. This straight section is made of a non-metallic ferromagnetic material, called *ferrite*. This ferrite tube consists of one turn of wire supplied with high frequency current from an oscillator, the secondary is the proton beam itself and tube forms the core of a high frequency transformer. The proton receives an average energy of about 1000eV per revolution and thus makes about 2.3×10^6 revolutions per sec. The frequency of the oscillations is increased fairly rapidly at first, as the protons gain speed and then slowly. The frequency becomes almost constant when the speed attained is within a few per cent of the velocity of light. During the acceleration period the radio frequency changes from about 0.37 mega-cycles/sec to about 4 megacycles/sec. In the fourth straight section are placed the targets and arrangements for extracting the beams. The energy of the protons striking the target can be varied at will upto the maximum energy limit. The whole machine is surrounded with 8 ft. thick concrete shields.

Studies of the nuclear reactions, possible with cosmotron, are essential to an understanding of nuclear structure. The Bevatron has been used to create proton-anti-proton pairs as well as neutron anti-neutron pairs.

3.8 MICROTRONS

The microtron, also called an *electron cyclotron*, is an electron accelerator, originally proposed by Veksler, since relativistic electrons cannot be accelerated by fixed frequency cyclotrons. In a microtron a small radio frequency accelerating cavity is located near the edge of a uniform magnetic field. The source of electrons is a hot filament in the entrance hole. The electrons are drawn from it by the intense electric field. They then travel on a circular orbit in the magnetic field and return to the resonator in approximately an integral number of radio frequency cycles so that they are again accelerated. Thus, the electron-path is a sequence of circles of ever-increasing radius, all tangent at the resonator. The time period of electron of energy W is

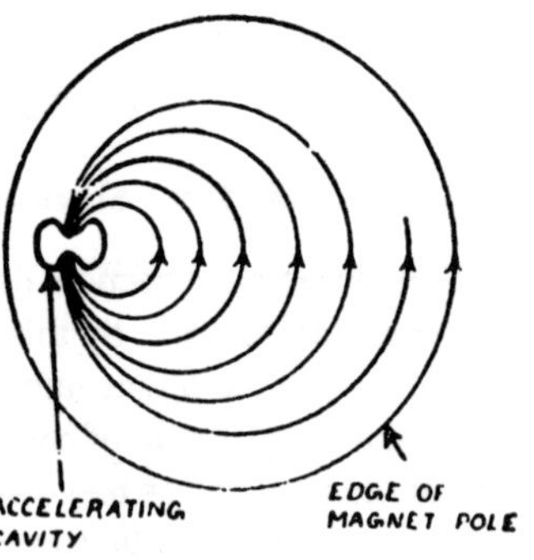

Fig. 3.20 : Orbits in a microtron.

$$\tau = \frac{2\pi}{\omega} = \frac{2\pi}{\frac{Be}{m}} = \frac{2\pi mc^2}{Bec^2} = \frac{2\pi W}{Bec^2} \quad ...(80)$$

$\therefore$ Change in period $\Delta\tau = \left(\frac{2\pi}{Bec^2}\right)\Delta W$ ...(81)

Thus, we see that the change in period depends on the change in energy, quite independent of total energy. The microtron operates on this principle.

Resonance Conditions

If the energy gained per turn is expressed as a fraction $\in$ of the rest mass energy W_0, *i.e.*, $\Delta W = \in W_0$. Electron started at rest gains $\in W_0$ in crossing the gap and hence has total energy $W_1 = W_0 = \in W_0$. Thus, eqn. (80) becomes

$$\tau = 2\pi W_0 (1 + \in) Bec^2 \quad ...(82)$$

It must be equal to some integral multiple μ of the period τ_{rf} of rf oscillator, *i.e.*,

$$2\pi W_0 (1 + 2\in)/Bec^2 = \mu\tau_{rf} \quad ...(83)$$

After a second acceleration electron acquires same energy increment $\in W_0$, hence its time period

$$\tau_2 = 2\pi W_0 (1 + 2\in)/Bec^2 \quad ...(84)$$

$$\therefore \quad \tau_2 - \tau_1 = (2\pi W_0/Bec^2)\in \quad ...(85)$$

This difference in periods must be integral multiple of oscillator frequency τ, *i.e.*, $\tau_2 - \tau_1 = v\tau rf$. Hence

$$(2\pi W_0/Bec^2) \in = v\tau rf \quad ...(86)$$

After the second rotation, the electron reaches the gap at the original phase and receives the same energy increment. Eqns. (83) and (86) give

$$\in = v/(\mu - v) \quad ...(87)$$

$$\therefore \quad \tau rf = \lambda/c = 2\pi W_0/Bec^2 /(\mu - v)$$

$$\text{or} \quad B\lambda = 2\pi W_0/ec\ (\mu - v) \quad ...(88)$$

Above relations show that m must exceed v. The difference in diameters between two consecutive orbits is

$$D_{n+1} - D_n = v\lambda/\pi \quad ...(89)$$

The output kinetic energy in N turns $T = N\in W_0$ ...(90)

If the electrons are injected into the resonator with an initial energy kW_0, above relations can be replaced as

$$\tau = 2\pi W_0 (1 + k + \varepsilon)/Bec^2 = \mu\rho rf$$

$$\in = (1 + k)v/(\mu - v)$$

$$B\lambda = 2\pi W_0 (1 + k)/ec (\mu - v) \qquad ...(91)$$

$$T = kW_0 + NeW_0.$$

The electrons in the microtron experience a variety of phase stability as found in the synchrotron. The microtron has the advantage over the betatron that its accelerated beam is rather easily extracted for use outside the machine.

3.9 THE ALTERNATING GRADIENT (STRONG FOCUSING) ACCELERATOR

Although the energies which could be reached in the proton synchrotron are much greater, in theory, than several BeV, practical limits are set by the size of the machine and by its cost. By the use of the *strong focusing principle*, the straying of the accelerated particles from the ideal circular path can be markedly reduced in both radial and axial directions. The size of the doughnut can be decreased correspondingly and high particle energies can be attained without the need for large and heavy magnets. A suggested method for achieving strong focusing was published in 1952 by E. D. Courant, MS. Livingston and H.S. Snyder of Brookhaven National Laboratory, based on the use of magnetic fields with alternating gradients. In this system a synchrotron is modified such that the particles in going round their orbiting vacuum chamber pass alternately through strong focusing and defocusing magnetic lenses. The result of this is a very strong net focusing.

The principle of alternate-gradient focusing can best be understood in terms of a very simple optical analogue. If two optical lenses, one convex and one concave, both with the same numerical focal length, are placed as shown in Fig. 3.21. In any lens, whether optical or ion optical, the angle of deflection experienced by a given ray is proportional to the distance from the central ray. The angle of deflection is, therefore, larger in the defocusing lens than in the focusing lens, and the net focusing effect always results. It can also be proved by using lens formula. If the lenses are separated by a distance d, the overall focal length F is

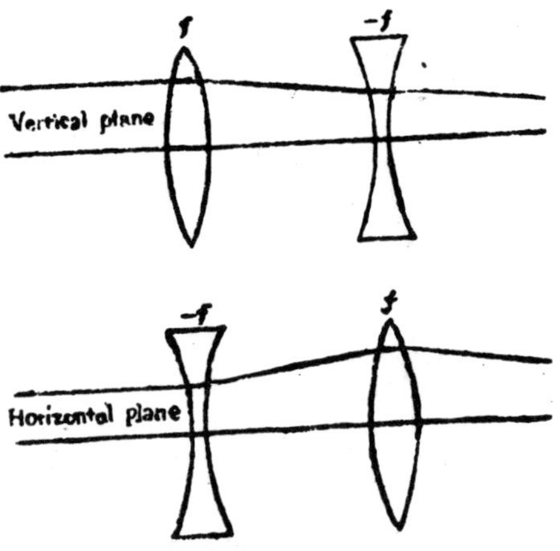

Fig. 3.21 : Optical analogue of both sections for each plane.

$$\frac{1}{F} = \frac{1}{f_1} + \frac{1}{f_2} - \frac{d}{f_1 f_2} \qquad ...(92)$$

If $f_1 = f$, $f_2 = -f$, then $F = f^2/d$. which is a +ve quantity. Thus, we see that *the combination of two lenses one covergent and other divergent is always he a convergent lens and is independent of the order of the lenses provided separation remains same.* The beam of particles is similar to a pencil of light, for which the sector with field index $n\left(= \frac{-dB}{dr}\frac{r}{B}\right) > 0$ behaves as a convergent lens in the vertical plane and as a divergent lens in the horizontal plane. The sector with field index $n < 0$ behaves as a convergent lens in the horizontal plane and divergent lens in the vertical plane. In this way there is a succession of convergent and divergent lenses in both planes thus, providing the vertical and radial focusing. The action of the alternating gradient magnets on the particles in the proton synchrotron is the same as the action on the light ray by a series of lenses which are alternatively focusing and defocusing.

If a beam of charged particles is moving along the z-axis across a region of transverse field E_x, given as $E_x = -G_x$, where G is a constant, the beam will be focused in the xz plane. To have div E = 0, E_v must have the form $E_v = Gy$ and the beam will be equally strongly defocused in the y-plane. If the beam now passes) through a second region of reversed gradients, the net effect will be focusing in both planes. In the case of magnetic fields G = particle velocity v × magnetic field gradient B'.

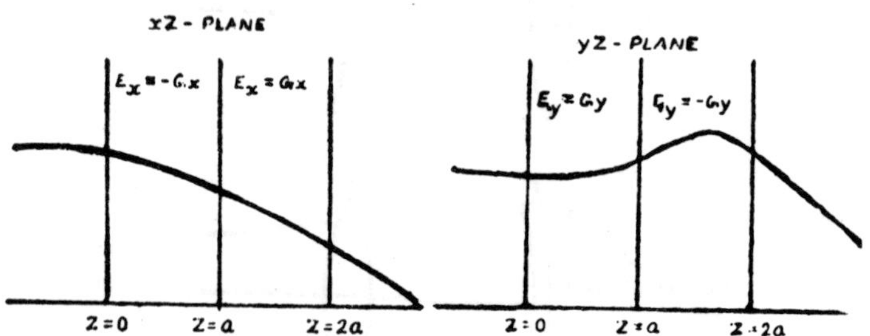

Fig. 3.22 : Focusing in a two-element AG lens.

In the xz-plane the eqn. of motion in the first region is

$$m\ddot{x} = -qGx \qquad \text{...(93)}$$

Its solution is

$$X = X_0 \cos (qGt^2/m)^{1/2} + \dot{x}_0 (qG/m)^{-1/2} \sin (qGt^2/m)^{1/2}$$

$$= X_0 \cos (qGz^2/mv^2)^{1/2} + \dot{x}_0 (qG/m)^{-1/2} \sin (qGz^2/mv^2)^{1/2}$$

At z = a, we have

$$x = x_0 \cos ka + x_0 (1/kv) \sin k a$$

$$x/v = -kx_0 \sin ka + x_0 (1/v) \cos ka, \qquad \text{...(94)}$$

where $k = (qG/mv^2)^{1/2}$

For the second region eqn. (94) becomes

$$x = x_0 \cosh ka + x_0 (1/kv) \sinh ka$$

$$x/v = k\, x_0 \sinh k a + \dot{x}_0 (1/v) \cosh ka \qquad \text{...(95)}$$

Using matrix method, the displacement and velocity of the particle after passing through these two field regions can be written as

$$\begin{vmatrix} x \\ \dot{x}/v \end{vmatrix} = \begin{vmatrix} \beta_2 & \beta_1/k \\ k\beta_1 & \beta_2 \end{vmatrix} \begin{vmatrix} \alpha_2 & \alpha_1/k \\ -k\alpha_1 & \alpha_2 \end{vmatrix} \begin{vmatrix} x_0 \\ \dot{x}_0/v \end{vmatrix}$$

$$= \begin{vmatrix} (\beta_2\alpha_2 - \beta_1\alpha_1) & +(\beta_2\alpha_1 - \beta_1\alpha_2)/k \\ k(\beta_1\alpha_2) - \beta_2\alpha_1) + & (\beta_2\alpha_2 - \beta_1\alpha_1) \end{vmatrix} \begin{vmatrix} x_0 \\ \dot{x}_0/v \end{vmatrix} \quad ...(96)$$

Similarly in the y-z plane or the defocusing plane

$$\begin{vmatrix} y \\ \dot{y}/v \end{vmatrix} = \begin{vmatrix} (\beta_2\alpha_2 + \beta_1\alpha_1) & +(\beta_2\alpha_1 + \beta_1\alpha_2)/k \\ k(\beta_1\alpha_2) - \beta_2\alpha_1) + & (\beta_2\alpha_2 - \beta_1\alpha_1) \end{vmatrix} \begin{vmatrix} y_0 \\ \dot{y}_0/v \end{vmatrix} \quad ...(97)$$

where we have assumed α_1 = sin ka, α_2 = cos ka, β_1 = sinh ka and β_2 cosh ka. After simplifying above equations, assuming ka very small, we see that two velocities ($\dot{x}$ and y) are equal for the same initial displacements and both are directed toward the axis.

Above treatment is modified to the lens element of an AG focusing system, consisting of the half focusing, full defocusing and the half focusing magnets. The transformation matrix is thus given by

$$\begin{vmatrix} \cos\frac{ka}{2} & \frac{1}{k}\sin\frac{ka}{2} \\ -k\sin\frac{ka}{2} & \cos\frac{ka}{2} \end{vmatrix} \begin{vmatrix} \cosh ka & \frac{1}{k}\sinh ka \\ k\sinh ka & \cosh ka \end{vmatrix} \begin{vmatrix} \cos\frac{ka}{2} & \frac{1}{k}\sin\frac{ka}{2} \\ -k\sin\frac{ka}{2} & \cos\frac{ka}{2} \end{vmatrix}$$

$$\begin{vmatrix} \cosh ka \cos ka & (\sinh ka + \cosh ka \sin ka)/k \\ k(\sinh ka - \cosh ka \sin ka) & \cosh ka \cos ka \end{vmatrix} \quad ...(98)$$

This matrix looks like the matrix in eqn (94), if we put

$$\cos\mu = \cosh ka \cos ka. \quad ...(99)$$

For a circular synchrotron composed of 2N magnets, which have the same strength of field at the midpoints of their radial apertures, the radial gradients are alternate in direction and having field indices n and –n. The shift in phase μ_z, of the axial betatron oscillation and μ_x for radial motion are thus given as

$$\cos\mu_z = \cos\mu_z = \cos(\pi n^{1/2}/N)\cosh(\pi n^{1/2}/N) \quad ...(100)$$

$$\left[\because ka = \left(\frac{qdB/dr}{mv}\right)^{1/2}\frac{2\pi r}{2N} = \left(\frac{qnB/r}{mv}\right)^{1/2}\frac{\pi r}{N} = \frac{\pi n^{1/2}}{N}\right]$$

cos μ_z and cos μ_x both lie in between +1 and–1. As cosh $(\pi n^{1/2}/N)$ never vanishes, hence both the above quantities will became zero if

$$n/N^2 = 1/4 \qquad ...(101)$$

Large values of n mean powerful forces and reduced amplitudes of betatron oscillations, so that magnets with apertures only 1 and 2 inches on a side appeared possible eqn (100) can represent the stability diagram for the AG synchrotron.

Phase Stability

The charged particle can acquire the synchronous energy increment if passes the gap at the appropriate phase. The ions which do not reach the gap at exactly the correct moment will gain too much or too little energy and their momenta will differ from the ideal value. The momentum compactness α is defined as: consider two particles, one with momentum p and orbit length L and the other with momentum p + dp and orbit length L + dL. The momentum compactness

$$\alpha = \frac{\frac{dp}{p}}{\frac{dL}{L}} = \frac{\frac{dp}{p}}{\frac{dR}{R}} \qquad (\text{As } L = 2\pi R) \qquad ...(102)$$

It depends on the machine parameters. For an AG accelerator it can be given as

$$\alpha = \frac{n^{3/2}\pi}{4N}\left(1 + \frac{L_s}{L_m}\right)\left(\coth\frac{n^{1/2}\pi}{2N} - \cot\frac{n^{1/2}\pi}{2N} + \frac{L_s n^{1/2}\pi}{L_m N}\right),$$

where L_s is the length of the straight sections and L_m that of magnets.

As a special case, consider that there are no straight sections ($L_s = 0$) and $\cos \mu_z = \cos \mu_x = 0$ for which $n/N^2 = 1/4$. Under these conditions

$$\alpha = \frac{n\pi}{8}\left[\coth\left(\frac{\pi}{4}\right) - \cot\left(\frac{\pi}{4}\right)\right] = \frac{n\pi}{8}[1.525 - 1] = \frac{n}{4.85} \qquad ...(103)$$

The first two AG proton synchrotrons to go into operation (in 1961) were the 28 BeV machine at CERN (European Council for Nuclear Research, Geneva, Switzerland) and the 33 BeV machine at Brookhaven National Laboratory. The Brookhaven AGS. is essentially circular, with a diameter of 842.9' (about one half mile in circumference). The magnet, weighing about 4000 tons, consists of 240 sectors, 36" × 39" in cross-section, and has a maximum field strength of 13000 gauss. The magnet bends the protons into a path of just this diameter and applies strong focusing forces which always tend to bring the protons back toward their orbit within the vacuum chamber. The vacuum pipe is only 7" wide and $2\frac{3}{4}$" high and is maintained at a pressure of less than 10^{-5} mm of Hg. The pole pieces of each successive pair of magnets are shaped so that the magnetic field alternatively increases and decreases in the radial direction. This alternating magnetic gradient causes the circulating proton beam alternately to focus and defocus in vertical and horizontal plane and yields a tightly focused beam after many traversals of the magnets. The magnets are arranged in 12 super-periods or identical groups of 20 magnets each. Between super periods and in the middle of each superperiod is a 10' straight section, that can be used either for one of the 12 radio frequency accelerating stations, for injection, for targets or for other beam ejection means. The variable frequency increases from 1.4-4.5 mega-cycles per sec in each pulse. Protons of 50 MeV energy arc injected into one of the straight sections. These protons are obtained from a hydrogen gas by stripping the negatively charged electrons from the gas molecules, leaving as positively charged protons in the ion source, and are accelerated by Cockcroft-Walton generator to 750 keV and then directed into the linear accelerator, of a long cylindrical tank about 3' in diameter and 110' long containing 124 drift tubes along its axis. These protons, circulating around the ring, are accelerated by electric fields produced in acceleration stations. A high frequency voltage is impressed across two gaps at these stations. If the frequency applied is correct, the same proton will always be accelerated at each gap. The protons gain about 90 keV of energy per revolution. To reach about 30 BeV 325000 revolutions around the ring are required. The proton beam thus obtained can be directed at appropriate target substances and the resulting reactions can be studied by means of the emitted radiations.

The 70 BeV proton AGS designed at Serpukhov near Moscow, U.S.S.R. in 1967 has 120 magnet units separated by straight sections of three different lengths, namely, about 15, 8.5 and 4.25ft. The maximum field strength used is 12000 gauss. The doughnut chamber is 6.5" wide and 4.4" high and is of diameter 1500'. The 100 MeV energy protons, injected from a linear accelerator, are accelerated by means of 53 accelerating stations distributed around the chamber. The energy gain is 350 keV per revolution. The maximum beam intensity is of the order of 1012 protons per pulse and the repetition rate is 5-10 pulses per min.

One of the World's most powerful accelerators is the 200 BeV A.G. synchrotron at the Lawrence Radiation Laboratory. It was ready in 1974. The machine is a system of four accelerators, feeding one another in series. The first, a Cockcroft-Walton generator injects 750 keV protons into a 200 MeV linear accelerator of length 160 metres and thus serves as a pre-injector into an 8 BeV injector alternating gradient synchrotron, which in turn delivers high energy protons into the main synchrotron. The latter is a strong focusing machine with 528 magnets in the ring about 1500 metres in diameter. The machine operates at a pulse rate of 30 per min producing 3×10^{13} protons per pulse at 200 BeV energy. A 300 BeV alternating gradient proton synchrotron is now available at CERN. It would use the same method of injection but the main accelerator would be larger. It is improbable that construction of the 600-1000 BeV accelerator started in 1972 in the United States will be completed upto the end of 1982.

3.10 AVF CYCLOTRONS

In the synchrocyclotron the frequency is varied with the radius in the same way as the mass increases with the radius. Another way is to increase the magnetic field with the radius as particle mass increases, keeping the frequency of the accelerating voltage as fixed. This requirement conflicts with the condition of axial stability of the particle orbit. For this Thomson proposed the use of alternately high and low regions of magnetic field around the orbit. Machines using such magnets arc, therefore, called AVF (*Azimuthally Varying Field*). As azimuthally varying fields are produced by *sector or spiral magnets*. On account of focusing process these cyclotrons are named as *Sector Focused (SF)* or *Spiral Focused Cyclotrons*.

Principle

The magnetic field varies with azimuth, regions where it is high are called *hills* (H) and where it is low as *valleys* (V). The radial velocity Vr interacts with the field Be to produce an axial force known as Thomas force. In the spiral shaped magnets, the convex side works as a focusing lens and concave side as defocusing lens. The net effect is focusing and the force as Kerst force. Laslett suggested that the particle entering the focusing surface of the hill would reach the defocusing surface normally or very close to the normal, hence the net effect would be focusing.

When the focusing always takes place at each hill, the focusing is known Kerst and Lasslet focusing.

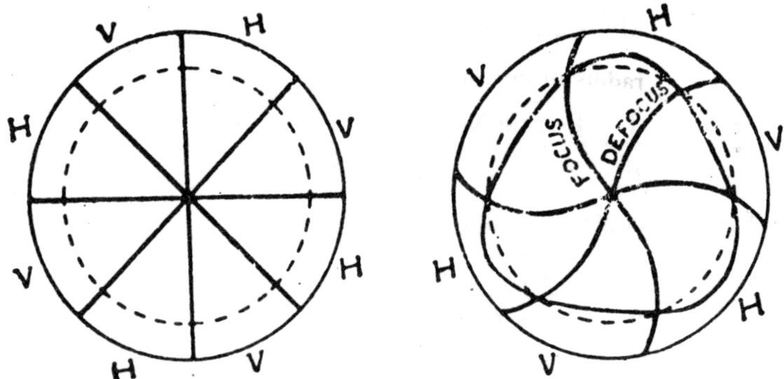

Fig. 3.23 : Thomas three sector field (left), Kerst and Laslett spiral shaped field (right).

Flutter

The change in field strength between the hills and valleys is defined in terms of a quantity known as flutter amplitude

$$B = B_{av} (1 - f \sin N\theta) \quad ...(105)$$

$$\therefore \quad B_H = B_{av} (1 + f), \quad N\theta = 3\pi/2 \text{ or } \theta = 3\pi/2N$$

$$B_V = B_{av} (1 - f), \quad N\theta = \pi/2 \text{ or } \theta = \pi/2N.$$

Thus for first harmonic (N = 1). B_H corresponds to $3\pi/2$ and B_V to $\pi/2$. Another quantity known as *flutter function*

$$F = [(B^2)_{av} - (B_{av})^2]/(B_{av})^2 \quad ...(106)$$

Eqn (105) gives $B^2 = (B_{av})^2 (1 + f^2 \sin^2 N\theta - 2f \sin N\theta)$.

$$\therefore \quad (B^2)_{av} = (B_{av})^2 [1 + 1/2 f^2],$$

or $$F = 1/2 f^2 \quad ...(107)$$

Flutter amplitude $$f = \frac{B_H - B_v}{2B_{av}}. \quad ...(108)$$

Thomas Focusing

For the arrangement of n hills or valleys, where the z-componetn of the magnetic field vary as

$$B_z = (B_z)_{av} (1 - f \sin B\theta) \quad ...(109)$$

For an assumed constant momentum of a particle of mass M moving with a velocity v, we get

$$Mv = qB_z, \ \rho = (B_z)_{av} r_0 \quad ...(110)$$

$\therefore$ radius of curvature $\rho = r_0 (B_z)_{av}/B_z$

or
$$\frac{1}{\rho} = \frac{1 - f \sin N\theta}{r_0} \quad ...(111)$$

For a particular curvature

$$\frac{1}{\rho} = \frac{r^2 + 2r'^2 - rr''}{(r^2 + r'^2)^{3/2}}$$

Since r changes very slowly with θ, hence $r' = dr/d\theta$ is small and r'^2 can be neglected.

$$\therefore \quad \frac{1}{\rho} \simeq \frac{r^2 - rr''}{r^3} \simeq \frac{r_0 - r''}{r_0^2} \quad ...(112)$$

Solving eqns. (111) and (112), we get

$$r'' = r_0 f \sin N\theta, \ r' = (r_0 f/N) \cos N\theta$$

Here constant of integration is zero as $r' = 0$ when $\theta = \pi/2$

$$\therefore v_r = \frac{dr}{dt} = \frac{dr}{d\theta} \cdot \frac{d\theta}{dt} = -\omega \frac{dr}{d\theta} = \frac{r_0}{N} \omega f \cos N\theta$$

Let us find B_θ, which can be obtained by the first term of Taylor's Expansion.

$$B_\theta = [B_\theta]_{z=0} + z\left[\frac{\partial B_\theta}{\partial z}\right]_{z=0} + ... = z\left[\frac{\partial B_\theta}{\partial z}\right]_{z=0}$$

$$= z\left[\frac{\partial B_z}{r_0 \partial \theta}\right] \qquad \text{(As curl B = 0)}$$

$$= \frac{z}{r_0} fN \cdot (B_z)_{av} \cos N\theta$$

$\therefore$ Thomas Force $F_z = qv_r B_\theta = -zN\omega^2 f^2 \cos^2 N\theta$

Average vertical force $(F_z)_{av} = -1/2 \, M\omega^2 f^2 z$

$$\therefore \qquad M\frac{d^2z}{dt^2} + \frac{M\omega^2 f^2}{2} z = 0 \qquad \text{...(113)}$$

and Thomas Frequency $\omega_{ih} = \omega f/\sqrt{2}$

$$\text{The ratio } \nu = \frac{\omega_{ih}}{\omega} = \frac{f}{\sqrt{2}} \quad \text{or} \quad \nu^2 = \frac{f^2}{2} = F \qquad \text{...(114)}$$

As in axial field $\nu^2 = n$, hence flatter F plays the same role in azimuthally varying fields as n does in axial fields. By using the flatter we could adjust cu in resonance due to the variation of B.

Spiral Focusing

In the region where the field lines are bowed there is a horizontal component B_h at points off the mid plane ⊥ to the spiral locus of constant phase of field. If ϕ is the spiral angle, $B_r = -B_\theta \tan\phi$ and the axial force $F_z = r\, qv_\theta\, Br = -\omega^2 zf\, NM \cos N\theta \tan\phi$. This is focusing and defocusing depending on the field sign of $\tan\phi$ and changing sign of $\cos N\theta$. Let us take account of the different path lengths in the focusing and defocusing fields. Now F_z becomes

$$F_z = \omega^2 zfNM \cos N\theta \tan\phi[1 - (f/N) \cos N\theta \tan\phi]$$

$$d^2z/d\theta^2 = zfN \cos N\theta \tan\phi - zf^2 \cos^2 N\theta \tan^2\phi \qquad \text{...(115)}$$

$$\text{or} \qquad < d^2z/d\theta^2 > = 1/2\ zf^2 \tan^2\phi$$

$$\text{Its solution is } \bar{z} = \bar{z}_m \sin(\omega_{a\upsilon}\theta) \qquad \text{...(116)}$$

where radial frequency $\omega_{av} = (f/\sqrt{2}) \tan\phi$

The larger the value of $\bar{z}$, the stronger are the in and out radial components of field, so the ion is forced into rapid axial oscillations about the path $\bar{z}$. Using $z = \bar{z} + \Delta x$ in eqn (115), and solving, we get

$$\Delta z = -\bar{z}\ (f/N) \cos N\theta \tan\phi \qquad \text{...(117)}$$

$$\text{and} \qquad z = \bar{z} - \bar{z}\ (f/N) \cos N\theta \tan\phi$$

Using this value of z in eqn. (115), and neglecting term in f^3 as f is very small (~ 1/20), we get

$$d^2\bar{z}/dt^2 = \omega^2 d^2\bar{z}/d\theta^2 = -\bar{z}\ 2\omega^2 f^2 \cos^2 N\theta \tan^2\phi$$

Using its average value as $\bar{z}\omega 2\ f^2 \tan^2\phi$, hence

$$d^2\bar{z}/dt^2 + \bar{z}\omega^2 f^2 \tan^2\phi = 0 \qquad \text{...(118)}$$

But the Thomas force is also present, hence the proper eqn of motion in any azimuthally variation of field is

$$\frac{d^2z}{dt^2} + \omega^2 \left(\frac{f^2}{2} + f^2 \tan^2 \phi \right) z = 0 \qquad ...(119)$$

Hence the axial betatron frequency

$$\omega_z^2 = \frac{1}{2} \omega^2 f^2 (1 + 2 \tan^2 \phi)$$

and $$v_z^2 = \omega_z^2/\omega^2 = \frac{1}{2} f^2 (1 + 2 \tan^2 \phi) = F(1 + 2 \tan^2 \phi)$$

The *average field index k* is defined as

$$k = \frac{d < B >/< B >}{dR/R} = \frac{dp/p}{dR/R} - 1 = \alpha - 1$$

Above method of finding v_z is an approximate. The complicated mathematics gives

$$v_x^2 = 1 + k \text{ ....}$$

$$v_z^2 = -k + \frac{N^2}{N^2 - 1} F(1 + 2 \tan^2 \phi) \qquad ...(120)$$

The first SF cyclotron to accelerate protons was operated at Delft in 1958 and it gave 12 MeV protons. The first SF cyclotron in the relativistic energy region came into operation in 1961 at UCLA (California)—protons of 50 -75 MeV range were successful accelerated. The largest of 850 MeV with eight spiral sectors is in progress at Oak Ridge.

3.11 INDIAN ACCELERATION

The accelerators in operation in India are:

(a) Cock Croft-Walton Accelerators

(a) Physics University Andhra, University Waltair, maximum proton energy 400 keV, in operation since 1970.

(b) B.A.R.C. Bombay (400 keV. 1970) used for ion-implantation,

(c) Bose Institute of Science & Technology, Calcutta (250 keV, 1959),

(d) Saha Institute of Nuclear Physics, Calcutta (250 keV, 1956),

(e) B.A.R.C, Bombay (150-250 keV, 1956, (f) Physics Deptt., A.M.U. Aligarh (150 keV, 195K).

(b) Van de Graaff Accelerators

(a) Physics Deptt, B.H.U. Varansi (400 keV, 1976),

(b) Physics Deptt, Punjabi University, Patiala (400 keV, 1977),

(c) I.I.T. Kanpur (2 MeV, 1971),

(d) B.A.R.C. Bombay (5 MeV, 1961). Accelerators at Kanpur and Bombay are vertical type, the beam is bent to 90° by electro-magnetic focusing.

(c) Fixed Energy Cyclotron

It is in operation since 1965 at Saha Institute of Nuclear Physics, Calcutta. The maximum proton energy available is 3.7 MeV.

(d) Variable Energy Cyclotron

(Weak Focusing or Classical type). It is in operation since 1975 at Physics Deptt, Panjab University Chandigarh. The maximum possible energy is 8 MeV for protons. The deuterons, He^3 and alphas are accelerated and used in research. The machine has following characteristics:

(i) maximum magnetic field 14 kilo gauss,

(ii) weight of the magnet 20 tonnes,

(iii) size of pole pieces 26",

(iv) frequency of the oscillator 10-20 MHz,

(v) dee voltage 30-40 kilo volts,

(vi) ion source-hooded arc type,

(vii) vacuum 5×10^{-6} mm of Hg,

(viii) total power consumption 100 KVA.

(e) Variable Energy Cyclotron (AVP Type)

It is in operation since 1977 at Calcutta and is capable of delivering a beam of protons of energies between 6-60 MeV, deuterons between 12-65 MeV and alpha particles between 25-130 MeV. The machine has following characteristics:

(i) Electro-magnets of weight 262 tonnes, 610 cm length, 224 cm width and 290 cm height. 3 spiral shaped sector pole tips are placed 120°, apart. Valley coils are five sets of coils, each set having three pairs. The seventeen trim coils have 34 leads.

(ii) A single RCA 6949 is used as the main oscillator tube. The frequency range is 5.5 to 16.5 MHz.

(iii) Ion source is a hot cathode P.I.G type source.

SOLVED EXAMPLE

Example 1:

A linear accelerator for the acceleration of protons to 45.3 MeV is designed so that, between any pair of accelerating gaps, the protons spend one complete radio-frequency cycle inside a drift tube. The rf-frequency used is 200 Mc/sec.,

(a) What is the length of the final drift tube?

(b) If the first drift tube is 5.35 cm long, at what K.E. are the protons injected into the linac?

(c) If the peak accelerating potential is 1.49×10^6 volts, calculate the total length of the accelerator.

Solution:

Length of the final drift tube $L_F = \frac{v_F}{f} = \frac{1}{f}\sqrt{\left(\frac{2E}{M}\right)}$

$$= \frac{(2 \times 45.3 \times 1.6 \times 10^{-13})^{1/2}}{200 \times 10^6 \times (1.6724 \times 10^{-27})^{1/2}}$$

$= 0.4654$ m. $= 47$ cm

K.E. of the injected protons $= \frac{1}{2} M v_1^2 = \frac{1}{2} M L_1^2 f^2$

$= 1/2 \times 1.6724 \times 10^{-27} \times (5.35 \times 10^{-2})^2 \times (200 \times 10^6)^2$

$= 0.60$ MeV.

$\therefore$ Increase in energy $E = 45.3 - 0.6 = 44.7$ MeV.

Energy increase in N gaps $NqV = NeV$.

$\therefore N = E/eV = 44.7 \times 10^6/1.49 \times 10^6 = 30$

Total length $L = \Sigma L_n = \frac{1}{f}\left(\frac{2qV}{M}\right)^{1/2} \int_0^N n^{1/2}\, dn$

$$= \frac{2}{3f}\left(\frac{2qV}{M}\right)^{1/2} N^{3/2}$$

$$= \frac{2(2 \times 1.6 \times 10^{-19} \times 1.49 \times 10^{6})^{1/2} \times 30^{3/2}}{9 \times 200 \times 10^{6} \times (1.6724 \times 10^{-27})^{1/2}} = \mathbf{9.4m.}$$

Example 2:

Show that the radius of curvature R of the path of a particle inside the dees of a cyclotron is proportional to the $N^{1\ 2}$, where N is the number of times the particle has been accelerated across the space between the dees.

Solution:

If there are N accelerations and the average potential difference between D's each time the ions cross the gap is V, the final energy will be

$$E = NqV = 1/2\ mv_m^{\ 2}.$$

As the kinetic energy $E = q^2B^2R^2/2M$, hence we have

$$q^2B^2R^2/2M = NqV \quad \text{or} \quad R = (2MNqV)^{1/2}/qB$$

$$\therefore \qquad \mathbf{R \propto N^{1/2}.}$$

Example 3:

Deuterons are accelerated in a fixed frequency cyclotron to a maximum dee orbit radius of 88 cm. The magnetic field, is 14000 gauss. Calculate the energy of the emerging deuteron beam and the frequency of the dee voltage. What change in magnetic flux density is necessary if doubly changed helium ions are accelerated. Given atomic masses :

$$H^2 = 2.014102\ mu.,\ He^4 = 4.002603\ mu.$$

Solution:

$$\text{Energy of the emerging deuteron} = = \frac{1}{2} B^2R^2q^2/M$$

$$= \frac{1}{2} \times \frac{(1.4)^2 \times (.88)^2 \times (1.602 \times 10^{-19})^2}{2.014102 \times 1.66 \times 10^{-27}} \text{ joules}$$

$$= 36.3 \text{ MeV}$$

Frequency of the dee voltage = Frequency of oscillations

$$= \frac{Bq}{2\pi M} = \frac{1.4 \times 1.602 \times 10^{-19}}{2 \times 3.14 \times 2.014102 \times 1.66 \times 10^{27}}$$

$$= 10.67 \text{ Mc/sec.}$$

Magnetic field required for the acceleration of $(He^4)^{++}$

$$= \frac{2\pi Mf}{q} = \frac{2 \times 3.14 \times 4.002603 \times 1.66 \times 10^{-27} \times 10.67 \times 10^6}{2 \times 1.602 \times 10^{-19}}$$

$$= 1.39 \text{ web/m}^2.$$

$\therefore$ Change in magnetic field = 1.4 – 1.39 = **0.01 tesla.**

Example 4:

In a certain betatron the maximum magnetic field was 4000 gauss, operating at 50 cycles/sec with a stable orbit diameter of 60 inches. Calculate the average energy gained per revolution and the final energy of the electrons.

Solution:

If B is the magnetic field at the end of the accelerating cycle at the orbit of radius R, the final energy

$$E = BeRc = 0.4 \times 1.602 \times 10^{-19} \times 30 \times 2.54 \times 10^{-2} \times 3 \times 10^8 \text{ joule}$$

$$= 91 \text{ MeV. (Assuming } v = c \text{ for simplicity).}$$

Total no. of revolution,

$$N = \frac{c}{4\omega R} = \frac{3 \times 10^8}{4 \times 2 \times 3.14 \times 50 \times 30 \times 2.54 \times 10^{-2}}$$

$$= 3.1 \times 10^5.$$

$\therefore$ Average energy gained per revolution

$$= 91 \times 10^6/3.1 \times 10^5 = \mathbf{294\ eV.}$$

Example 5:

A 100 MeV betatron has on orbit radius of 35 cm. in which the electrons acquire 480 volts per revolution. How far will the electrons travel in attaining full energy? The magnet i energized at 180 Hz and produces pulses of 2μ sec duration, each consisting of 3×10^9 electrons. What is the peak and the average beam current? What is the duty cycle?

Solution:

Total no. of revolutions N = $100 \times 10^6/480 = 2.083 \times 10^5$.

$\therefore$ Total distance traversed by the electron L = $2\pi RN$

$$= 2 \times 3.14 \times 0.35 \times 2.083 \times 10^{5\,=\,4.57} \times 10^5 \text{ m.}$$

As 3×10^9 electrons or 4.8×10^{-10} coulomb of charge passes in 2μ sec, hence peak value of current

$$I_p = 4.8 \times 10^{-10}/2 \times 10^{-6} \times 10^{-4} \text{ amp.}$$

In one second magnetic field reaches to its peak value 180 times, 4.8×10^{-10} coulomb of charge is flowing in every time, hence the flow of charge per sec or the average current

$$I_{av} = 4.8 \times 10^{-10} \times 180 = 8.64 \times 10^{-8} \text{ amp.}$$

As pulses are produced only for 2μ sec, and the time period of the field is 1/180 sec, hence the duty cycle

$$= \frac{dt}{T} = \frac{2 \times 10^{-6}}{1/180} = \mathbf{3.6 \times 10^{-4}.}$$

Example 6:

Deuterons are accelerated in the synchrocyclotron which has magnetic field of 15000 gauss at the centre and 14310 gauss at the periphery of the dee. Calculate the maximum frequency of the dee voltage. If the dee voltage frequency is modulated between this maximum and a minimum of 10 Mc/see., calculate the gain in energy of a deuteron.

Solution:

Maximum frequency of the dee voltage $f_0 = \omega_0/2\pi = B_0q/2\pi M_0$

$$= \frac{1.5 \times 1.602 \times 10^{-19}}{2 \times 3.14 \times 2.014102 \times 1.66 \times 10^{-27}} = 11.44 \text{ Mc/sec.}$$

The relativistic mass of the particle at the periphery

$$M = \frac{B'q}{2\pi f'} = \frac{1.431 \times 1.602 \times 10^{-19}}{2 \times 3.14 \times 10 \times 10^6} \text{ kg}$$

$$= 2.199 \text{ mu.}$$

$\therefore$ K.E. of the particle = $(M - M_0)\, c^2 = (2.199 - 2.0141)$

$$= \mathbf{0.1849 \text{ mu.} = 171.6 \text{ MeV.}}$$

Example 7:

Assume that in the 70 MeV betatron synchrotron the radius of the stable electron orbit is 28 cm. Calculate:

(a) the frequency of the applied electric field;

(b) the value of the magnetic field intensity at the orbit for this energy, and

(c) the energy loss by radiation during a single revolution of an electron.

Solution:

Relativistic equation for energy of electrons is E = BeRc.

$$\therefore\ B = \frac{E}{eRc} = \frac{70 \times 1.6 \times 10^{-13}}{1.602 \times 10^{-19} \times 0.28 \times 10^{8}} = 0.83 \text{ tesla}$$

Frequency of the applied electric field = $eBc^2/2\pi E$

$$= \frac{1.602 \times 10^{-19} \times 0.83 \times (3 \times 10^{8})^2}{2 \times 3.14 \times 70 \times 1.6 \times 10^{-13}}$$

$$= 1.706 \times 10^{8} \text{ cycles/sec.}$$

Energy radiated by an electron of energy E in one revolution

$$\Delta E = \frac{88.5 E^4}{R} = \frac{88.5\,(0.07)^4}{0.28} = \mathbf{7.5 eV.}$$

Example 8:

In the cosmotron proton synchrotron there are four quadrants of radius 9.144 m. and four connecting straight sections each of length 3.048 m. Protons are injected at 3.6 MeV. Calculate the magnetic flux density at the moment of injection, the speed of the protons, and the frequency of the accelerating voltage. The protons are accelerated to a final kinetic energy of 3 GeV, calculate the maximum magnetic flux density and the maximum frequency of the accelerating voltage.

Solution:

Energy of the injected proton, when it moves in a circular path of radius R is given by $E = B^2R^2q^2/2M$.

$$\therefore\ B = \frac{(2ME)^{1/2}}{Rq} = \frac{(2 \times 1.67 \times 10^{-27} \times 3.6 \times 1.6 \times 10^{-13})^{1/2}}{9.144 \times 1.6 \times 10^{-19}}$$

$= 0.02999$ web/m^2 300 gauss

and $$v = \frac{BqR}{M} = \frac{0.03 \times 1.6 \times 10^{-19} \times 9.144}{1.67 \times 1027}$$

$= 2.6 \times 10^7$ m/sec.

Frequency in a circular orbit

$f_c = v/2\pi R = 2.6 \times 10^7/2 \times 3.14 \times 9.144$

$= 0.453$ mega-cycles/sec.

As proton moves in four quadrants, hence frequency reduces to

$$f_0 = \frac{2\pi Rf_c}{2\pi R + 4L} = \frac{2 \times 3.14 \times 0.453 \times 10^6}{2 \times 3.14 \times 9.144 + 4 \times 3.048}$$

$= 370$ kc/sec.

For the protons of energy 3 GeV, we have to use relativistic eqn. (56)

$$B_m = \frac{3.33\,[T(T + 2M_0c^2)}{R} = \frac{3.33\,[3(3 + 2 \times 0.938)]^{1/2}}{9.141}$$

$= 1.392$ web/m^2.

Maximum frequency can be obtained by combining eqns. (58) and (59) as

$$f_0 = \frac{c^2eBR}{(2\pi R + 4L)\,(T + M_0c^2)} = \textbf{4.193 mega-cycles/sec.}$$

EXERCISES

1. The H^+ and H_2^+ ions accelerated in a Van de Graaff generator to 5 MeV are to be magnetically separated from one another. Approximately over what distance must a 10,000 gauss field be applied if the two beams are to diverge by 20^0? **(18")**
2. Estimate the total electromagnetic energy in the accelerating tube of the SLAC two miles linac during the acceleration pulse

when it is operated to give 40 BeV electrons. The diameter of the wave guide is 8.2 cm. What is the total energy in joules of the electrons in one pulse after acceleration?

3. A standard cyclotron of 120 cm. pole diameter is operated with a 10 Me oscillator:

   (a) What magnetic field is required for the acceleration of deuterons?

   (b) What will be the final deuteron energy?

   (c) With the same rf frequency, what is the maximum K. E. to which He^3 ions could be accelerated in this cyclotron?

   (d) Under the assumption that the magnetic field calculated in (a) is the maximum available, what oscillator frequency would be required to obtain the highest possible H^3 energy and what is that energy ?

   (a) 1.31 weber/m^2; (b) 14.8 MeV; (c) 39.4MeV; (d) 6.67Me, 99 MeV.

4. By what fraction would the time required to pass through a dee be increased as compared with the time required at non-relativistic speed, if the cyclotron is designed to accelerate particles to half the speed of light? **(1.16)**

5. A 98 MHz oscillator supplies an effective voltage of 50 kV to the dees of a cyclotron accelerating α-particles to a maximum energy of 28 MeV. Calculate the closest approach of adjacent turns of the spiral ion path, assuming the beam to lie between radii of 10 and 75cm. **(1.35 mm at the outer radius)**

6. In a certain betatron the maximum magnetic field was 0.4 tesla, operating at 60 Hz with a stable orbit diameter of 66 inches. Calculate the maximum K.E. of an electron, injected with energy 50 kV and the approximate total time of flight.

   (100 MeV, 4m. sec.)

7. A synchrocyclotron is being designed to accelerate protons to 700 MeV. What percentage frequency change will be required in the rf-oscillator? **(42.6% decrease)**

8. Protons are accelerated to 740 MeV in the Berkeley synchro-cyclotron which has a B = 2.3 tesla. What is the oscillator

frequency at the time a pulse of protons is injected at the centre? What is the oscillator frequency at the time the protons have reached the rim ? **(3.5×10^7, 1.95×10^7 cycles/sec.)**

9. The radius of the stable electron-orbit of a small electron synchrotron is 12 cm. and the maximum magnetic field is 9000 gauss. Calculate the frequency of the applied electric field and the maximum electron energy produced.

   (4.04×10^8cycles/sec., 32.4 MeV)

10. A synchrotron is to be designed to accelerate protons to 12 GeV kinetic energy:
    (a) Assuming a maximum field strength of 1.43 web/m^2 estimate the radius of curvature of the proton orbit.
    (b) If about 25 per cent of the protons path is spent in field free straight sections, what is the final revolution frequency?
    (c) Assuming one revolution per rf-cycle, at what kinetic energy must the protons be injected if the frequency over the entire acceleration cycle is to vary by a factor of 5? What type of device would you suggest for the injector?

    [(a) 30m., (b) 1.2 Mc/sec.]

11. Using the data given in Example 8, and the field index n = 0.6: Calculate the expected radial displacement of the protons from their correct orbit if the frequency error is +0.1 per cent
    (a) soon after injection and
    (b) when B reaches 1.4 weber/m^2.

4

Nuclear Forces

4.1 INTRODUCTION

To have a stable state of a nucleus we have to assume that some other forces act on the protons. These attractive forces exceed the forces of repulsion. Even aside from the sign of the force, electrostatic forces are much too weak to account for the main effects, gravitational forces are weaker still and offer no assistance in the problem. It is therefore necessary to postulate an entirely new type of interaction, known as nuclear force. Let us collect following information on nuclear forces from study of complex nuclei.

(a) Shape of the Potential

As the nuclear interactions do not extend to very large distances beyond the nuclear radius, and l/r^n $(r > l)$ character is not useful to solve the problem. To avoid this difficulty, the idea of cut-off in the potential is introduced. The parameters are taken to be the strength V_0 (the value of potential at the origin and the range α (the distance beyond which potential goes to zero rapidly). It is less than nuclear dimensions.

(b) Saturation Property

We know that the nuclei have the same density ($R = R_0\, A^{1/3}$) the binding energy per nucleon for nuclei with $A > 40$ is constant. These facts imply that the nuclear *force saturates*. Nucleons attract each other strongly only if they are in the same orbital state. According to the Pauli principle only two neutrons and two protons will be found in the same orbital state. Therefore, it is possible to find four nucleons strongly bound or α-particle structure, also confirmed by binding energy curve.

(c) Charge Independence

We know that the atomic mass number A is approximately equal to twice the atomic number Z for the light and intermediate nuclei. It shows that light nuclei prefer to add nucleons in n-p pairs, *i.e.,* there is a strong interaction between neutrons and protons. The neutron excess in heavy nuclei confirms that n-n force is attractive, but it is sufficiently attractive to lead to a stable di-neutron. For a good approximation one can write

n-n $\simeq$ p-p $\simeq$ n-p or the forces *as charge independent.*

We use the principles of wave mechanics on nuclear problems and then by comparison with experimental data, find a consistent description of the nuclear forces acting between two nucleons (*two body problem*). There are two general methods of investigation, the study of n-p and p-p scattering events over a wide range of energy and the study of deuteron (only bound state of two nucleons).

4.2 DEUTERON

The deuteron does possess measurable properties which might serve as a guide in the search for the correct nuclear interaction. These properties are:

1. The parity of deuteron as measured, indirectly, by studies of nuclear disintegrations and reactions for which certain rules of parity changes exist, is even.
2. The angular momentum quantum number, often called the nuclear spin, of the ground state of the deuteron determined by a number of optical, radio frequency and micro-wave methods is one. It suggests that the spins are parallel (triplet state) and the orbital angular momentum of the deuteron about their common center of mass is zero. Thus, the ground state is $^3 5$ state.
3. The binding energy of the deuteron is very small. Its experimental value is 2.225 ± 0.002 MeV. Since the energy needed to pull a nucleon out of a medium mass nucleus is about 8 MeV, we must regard the deuteron as loosely bound.

4. The extraordinary stability of the alpha particle shows that the most stable nuclei are those in which number of neutrons and photons are equal. The deuteron consists of two particles of roughly equal masses M, so that the reduced mass of the system is 1/2 M.

5. A radio frequency molecular beam method has been employed to determine the quadrupole moment of the deuteron as $Q = 0.00282 \times 10^{-28}$ m^2. This shows the departure from spherical symmetry of a charge distribution. The +ve sign indicates that this distribution is prolate rather than oblate.

   The electric quadrupole moment and the magnetic moment discrepancy can be explained if the ground state is a mixture of the triplet states $^3S_1$ and $^3D_1$ having even parity. The percentage probability of finding the deuteron in D-state is $4 \pm 2\%$. As deuteron spends most of the time in the spherically symmetrical state (S-state), we will for the moment ignore the D-state contribution to the deuteron wave function.

6. The sum of the magnetic dipole moments of the proton (2.79275 μN) and neutron (–1.91315 μN), do not exactly equal to magnetic moment of the deuteron (0.85735 μN) measured by magnetic resonance absorption method.

7. Since the neutron has no charge, the force between the neutron and proton can not be electrical. This force can not be magnetic as magnetic moments are very small. It can not be gravitational force, as the masses are very small. So we must accept the nuclear force as a new type of force. This force is short range, attractive and along the line joining the two particles (central force). Since a central force can not account for the quadrupole moment of the deuteron. As a quadrupole moment is small, the assumption will be approximately correct.

8. The force depends only on the separation of the nucleons not on the relative velocity or orientation of the nucleon spins with respect to the line. This force can be derived from a potential. Since the force is attractive, V(r) is negative and decreases with decreasing r. Since it is short range, V(r) vanishes for $r > b$, where $b \sim 3$ fermi.

The Schrodinger wave equation for the two body problem is

$$\Delta^2\phi + (2m/h^2)(E - V)\psi = 0, \qquad ...(1)$$

where m is the reduced mass, E the total energy of the system equal to the binding energy of deuteron and V the potential energy describing the forces acting between the two bodies.

In this terms of spherical polar coordinates, eqn. (1) becomes

$$\left[\frac{1}{r^2}\frac{\partial}{\partial r}\left(r^2\frac{\partial\phi}{\partial r}\right)+\frac{1}{r^2\sin\theta}\frac{\partial\theta}{\partial}\left(\sin\theta\frac{\partial\phi}{\partial\theta}\right)+\frac{1}{r^2\sin^2\theta}\frac{\partial^2\psi}{\partial^2\phi}\right]$$

$$+ (2m/\hbar^2)\,[E - V(r, \theta, \phi)]\,\psi = 0 \qquad ...(2)$$

Let V (r, θ, φ) actually depends on r only and not on θ and φ. The solution of the above equation can be written as a product of a function of r only and one of θ and φ only as $\psi(r, \theta, \phi) = \phi(r)\,\phi(\theta, \phi)$. Substituting it into equation (2) we get

$$\frac{1}{\psi(r)}\frac{d}{dr}\left(r^2\frac{d\phi(r)}{dr}\right)+\frac{2mr^2}{\hbar^2}[E - V(r)] = -\frac{1}{\psi(\theta, \phi)}$$

$$\left[\frac{1}{\sin\theta}\frac{\partial}{\partial\theta}\left(\sin\theta\frac{\partial\phi(\theta,\phi)}{\partial\theta}\right)+\frac{1}{\sin^2\theta}\frac{\partial^2\psi(\theta,\phi)}{\partial\phi^2}\right] \qquad ...(3)$$

The L.H.S. of this equation depends only on r, the R.H.S. depends only on θ and φ. For all values of the variables, each side of this must separately equal to the some constant, which comes out to be $l(l + 1)$. Thus we get

$$\frac{1}{r^2}\frac{d}{dr}\left(r^2\frac{d\psi(r)}{dr}\right)+\frac{2m}{\hbar^2}\left[E - V(r) - \frac{l(l+1)\,\hbar^2}{2mr^2}\right]\psi(r) = 0$$

$$...(4)$$

where l is the angular momentum quantum number of the system.

The last term in the bracket appears as a straight addition to the actual potential V(r) and is known as centrifugal potential.

The Schrodinger equation for ^3S state ($l = 0$) of the deuteron is

$$\frac{1}{r^2}\frac{d}{dr}\left(r^2\frac{d\psi(r)}{dr}\right)+\frac{2m}{\hbar^2}[E - V(r)]\,\psi(r) = 0 \qquad ...(5)$$

In this case reduced mass m = 1/2 M. We expect the ground state to be spherically symmetric (S-state), so that φ(r) depends only or r. Substituting ψ(r) = u(r)/r in equation (5), where u(r) is called the radial wavefunction, we get

$$\left(\frac{d^2u}{dr^2}\right) + \left(\frac{M}{\hbar^2}\right)[E - V(r)]\, u = 0 \qquad ...(6)$$

The wavefunction of the bound state of the deuteron is not markedly dependent on the exact shape of the potential V(r) between a proton and neutron provided that a potential of short range is chosen. For simplicity, we represent V(r) by a *square well* of depth V_0 and radius b, where b is the range of the nuclear force. In it V(r) has a constant negative value $-V_0$ for separations less than a certain value b and the value zero for all greater separations. The other central force potentials may be of

Gaussian well type $\quad V(r) = -V_0 e^{-(r/\alpha)^2}$

Exponential well type $V(r) = -V_0 e^{-r/\alpha}$

Yukawa well type $\quad V(r) = -\dfrac{V_0 e^{-r/\alpha}}{(r/\alpha)}$

For the ground state of the deuteron, the total energy E is negative and equal to – B, where B is the binding energy of deuteron. Thus, equation (6) can be written as

$$\frac{d^2u}{dr^2} + \frac{M}{\hbar^2}[V_0 - B]u = 0 \quad \text{for } r < b \qquad ...(7)$$

and

$$\frac{d^2u}{dr^2} + \frac{M}{\hbar^2}(-B)u = 0 \quad \text{for } r > b \qquad ...(8)$$

These equations can be written as

$$\frac{d^2u}{dr^2} + K^2u = 0, \quad r < b \qquad ...(9)$$

and

$$\frac{d^2u}{dr^2} - \alpha^2u = 0, \quad r > b \qquad ...(10)$$

where $\dfrac{K^2 = M(V_0 - B)}{\hbar^2}$

and $\dfrac{\alpha^2 = MB}{\hbar^2}$

General solutions of eqns. (9) and (10) are given as

$$u = A_1 \sin Kr + B_1 \cos Kr \quad ...(11)$$

$$u = A_2 e^{\alpha r} + B_2 e^{-\alpha r} \quad ...(12)$$

The following boundary conditions must be imposed:

1. u (r → 0) = 0, to keep wave function y finite.
2. u (r → ∞) = 0, u must not diverge faster than r as r → ∞.

To satisfy the conditions at zero and infinity, the solutions reduce to

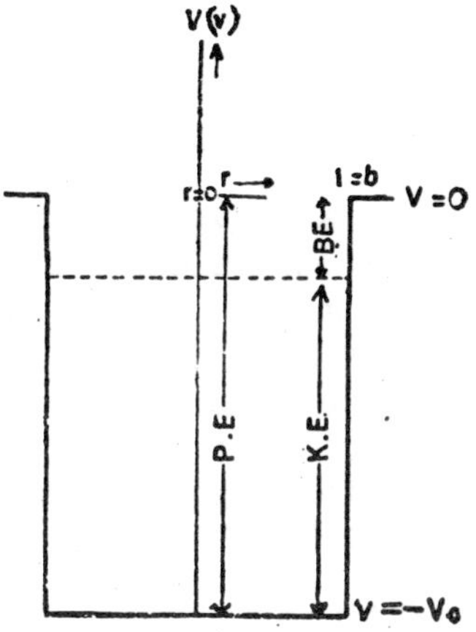

Fig. 1.1 : Square well potential of deuteron.

$$u = A_1 \sin Kr \quad \text{for } r < b \quad ...(13)$$

$$u = B_2 e^{-\alpha r} \quad \text{for } r > b \quad ...(14)$$

Since these two solutions join smoothly at r = b. Hence equating the values and first derivatives of u at r = b, we have

$$A_1 \sin Kb = B_2 e^{-\alpha b} \quad ...(15)$$

and $$A_1 K \cos Kb = -B_2 e^{-\alpha b} \quad ...(16)$$

∴ $$K \cot Kb = -\alpha \quad ...(17)$$

The constants A_1 and B_2 are obtained from the requirement that the integral of $|\psi|^2$ over all space must be equal to unity.

$$4\pi \int_0^\infty |\psi|^2 r^2 dr = 4\pi \int_0^\infty u^2 dr = 1$$

or $$4\pi \int_0^b A_1^2 \sin^2 Kr dr + 4\pi \int_0^\infty B_2^2 e^{-2\alpha r} dr = 1$$

$$A_1^2 \left[b - (1/2K) \sin 2Kb\right] + \left(\frac{B_2^2}{\alpha}\right) e^{-2\alpha b} = \frac{1}{2}\pi \quad ...(18)$$

Thus, A_1 and B_2 can be obtained from this equation with the help of eqns. (13) and (14). With the values of b and α as given above, the second term is about twice as large as the first. Hence the *nucleons in the deuteron spend only one third of the time within the range of nuclear force and thus the deuteron is loosely bound.* This can be seen from Fig. 4.2, where u(r) is plotted against r.

No excited S-states—Equation (17) can be written as

$x \cot x = -\alpha b$ ...(19)

where x = Kb. Using B = 2.225 ± 0.002 MeV and b ≤ 3f we find that α = 0.232 and αb ≤ 07. If we draw now curves y = cot x and y = –αb/x, the intersections give the roots of eqn (19). From Fig. 4.3 it is clear that roots are slightly greater than π/2, 3π/2, 5π/2.... The correct solution is $x = Kb \approx \frac{1}{2}\pi$, since if Kb were greater than π, the wave function would have a node at Kb = π and thus, u(r) and hence ψ(r) would not be the wave function of the ground state—a *contradiction of our hypothesis.* We may thus put $Kb = \frac{1}{2}\pi + \epsilon$ in eqn (19) and get

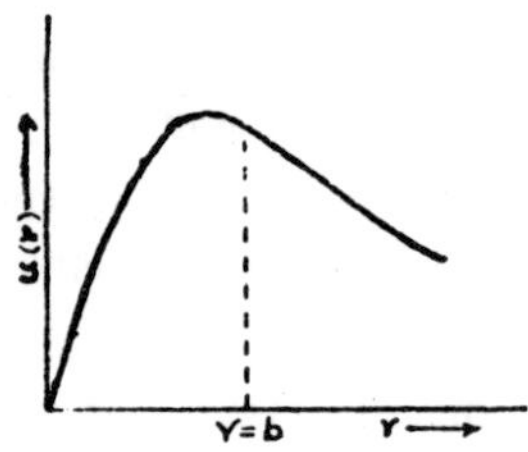

Fig. 4.2 : Ground state deuteron wave function.

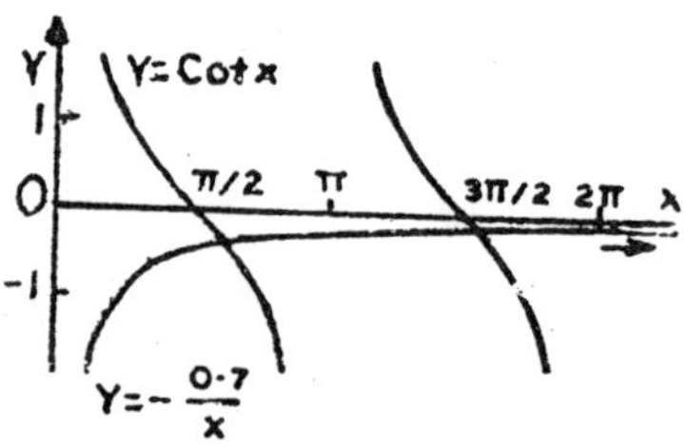

Fig. 4.3 : Solution of equation x cot x = –0.7.

$$\left(\frac{1}{2}\pi + \epsilon\right)\cot\left(\frac{1}{2}\pi + \varepsilon\right) = -\alpha b \qquad ...(20)$$

As ∈ is small, hence $\cot\left(\frac{1}{2}\pi + \epsilon\right) \simeq -\epsilon$, and $\epsilon \simeq 2\alpha b/\pi$

$$\text{Thus, } Kb = \frac{\pi}{2} + \frac{2\alpha b}{\pi}. \qquad ...(21)$$

This shows that *there can not be any excited S-states.*

Range and Depth of Potential

Using $Kb = \frac{\pi}{2}$ and again neglecting B in the expression for K [as the depth of potential is very much greater than the binding energy *i.e.*, $K^2 >> \alpha^2$ which can be obtained from eqn (21)], we get

$$\frac{MV_0b^2}{h^2} = \frac{\pi^2}{4} \quad \text{or} \quad V_0b^2 = \frac{\pi^2\hbar^2}{4M} \qquad ...(22)$$

It is the relation between range b and potential depth V_0. Actually V_0b^2 is slightly greater than $\frac{\pi^2\hbar^2}{4M}$, as Kb is slightly greater than $\pi/2$. By accepting the approximate value of range b = 2 fermi, the value of potential depth is V_0 = 36 MeV.

Another result which does not depend on the form of potential is the wavefunction outside the range of nuclear forces. In this region function u(r) decreases exponentially with r and reduces to zero at infinity. The radial distance where the amplitude decreases to 1/e of its maximum amplitude is often called the *radius* of the deuteron.

$$\therefore \text{Radius } R = l/\alpha = \hbar/\sqrt{(MB)} = 4.31 \times 10^{-5} \text{ m} \qquad ...(23)$$

It is about twice that of the range b. This explains that the *deuteron is a loosely bound system.* As b < R, hence the nuclear forces can be said to be short range.

$$\text{or} \quad \frac{d^2u_l(r)}{dr^2} + \left[K^2 - \frac{l(l+l)}{r}\right] u_l(r) = 0 \quad r \le b \qquad ...(24)$$

$$\text{and} \quad \frac{d^2u_l(r)}{dr^2} - \left[\alpha^2 + \frac{l(l+l)}{r}\right] u_l(r) = 0 \quad r > b \qquad ...(25)$$

where K^2 and α^2 are having their usual values.

The general solution of these equations involve spherical Bessel functions j_l and spherical Neumann functions n_l. As the latter approaches $-\infty$ as $r \to 0$, thus the solution of eqn (24) is

$$u_l = A\, j_l(Kr), r \le b \qquad ...(26)$$

$$\text{where} \quad j_l(Kr) \left(\frac{\pi}{2Kr}\right)^{1/2} j_{l+1/2}(Kr) \qquad ...(27)$$

The solution of eqn. (25) is

$$u_l(r) = B\ h_l(i\alpha r) = B[j_l(i\alpha r) + in_l(i\alpha r)] \qquad ...(28)$$

where $$m(i\alpha r) = (1)^{l+1}\left(\frac{\pi}{2i\alpha r}\right)^{1/2} J_{-l-1/2}(i\alpha r) \qquad ...(29)$$

Using boundary conditions that the function and its first derivative are continuous at the edge of the well, we get

$$\left[\frac{1}{u_l}\frac{du_l(r)}{dr}\right]_{inside} = \left[\frac{1}{u_l}\frac{du_l(r)}{dr}\right]_{outside,} \quad (r = b) \qquad ...(30)$$

Using the relation $\dfrac{dj_l(\rho)}{d\rho} = j_{l-1}(\rho) = \dfrac{l+1}{\rho} j_l(\rho)$, we get

$$K\left[\frac{j_{l-1}(Kb)}{j_l(Kb)} - \frac{l+1}{Kb}\right] = i\alpha\left[\frac{h_{l-1}(i\alpha b)}{h_l(i\alpha b)} - \frac{l+1}{i\alpha b}\right]$$

or $$\frac{j_{l-1}(Kb)}{j_l(kb)} = \left(\frac{\alpha}{K}\right)\left[\frac{ih_{l-1}(i\alpha b)}{h_l(i\alpha b)}\right] \qquad ...(31)$$

For $b < 1.43 \times 10^{-15}$m,$\alpha b < 1$ and since $\alpha << K$ the expression in the bracket on R.H.S is less than one, and is approximately zero. Thus

$$j_{l-1}(Kb) \approx 0 \qquad ...(32)$$

This condition holds for all angular momenta except $l = 0$. We have already discussed the case $l = 0$. For $l = 1$, we get j_0 (Kb) $\approx 0 = \sin$ (Kr). Hence Kb $\simeq \pm\pi, \pm 2\pi, \pm 3\pi$,.... Thus, the minimum well depth is

$$V_0 \simeq \frac{\pi^2\hbar^2}{Mb^2} \qquad ...(33)$$

If we choose $b = 2 \times 10^{-15}$, we get $V_0 = 144$ MeV, which is almost four times as large as the actual well depth in the ground state. Repeating this procedure for larger and larger values of l we find that a deeper and deeper well depth is required to produce a bound state. Thus, we conclude that no bound state exists for $l > 0$.

4.3 NEUTRON PROTON SCATTERING AT LOW ENERGIES

Two kinds of the reactions can be involved in neutron proton interaction: One scattering and other radiative capture. The latter has low probability and cross section for high energy neutrons, as the cross section for the competing radiative capture reaction decreases with 1/v, where v is the neutron velocity. In practice protons are bound in molecules. The chemical binding energy of the proton in a molecule is about 0.1 eV. Thus for neutron energies = > 1eV the proton can be assumed as free. *This sets a lower limit to the neutron energy*. If the neutron energy is less than 10 MeV, only the S-wave overlaps with the nuclear potential and is scattered.

In the centre of mass system, the Schrodinger equation for the two body (n-p system) problem is

$$\Delta^2\phi + \frac{M}{\hbar^2}[E - V(r)]\,\psi = 0 \qquad ...(34)$$

where,

M = Proton or neutron mass = 2 × Reduced mass of the system.

E = Incident kinetic energy in C-M system = (incident K. E. in L-co-ordinates)

and V(r) = Inter-nucleon potential energy.

At large distances from the centre of scattering the solution of this equation is expected to be of the form

$$\psi = e^{ikz} + \frac{e^{ikr}}{r} f(\theta) \qquad ...(35)$$

The term e^{ikz} represents a plane wave describing a beam of particles moving in the z-direction towards the origin (scattering centre). The second term represents the scattered wave. The complex quantity f(θ) is the scattering amplitude in the direction θ and is to be evaluated in terms of k. In the case of a spherically symmetric potential the entire arrangement is axially symmetric about the incident direction and hence does not depend on the azimuthal angle ϕ. The 1/r dependence is necessary for the conservation of particles in the outgoing wave. The volume of a spherical shell, between r and r + dr is $4\pi r^2$ dr and hence the density of particles in it or the probability of finding one particle in the spherical shell must vary with $1/r^2$ which is proportional to the square of the implitude of the scattered wave in the shell. Hence the amplitude of the scattered wave must vary with 1/r.

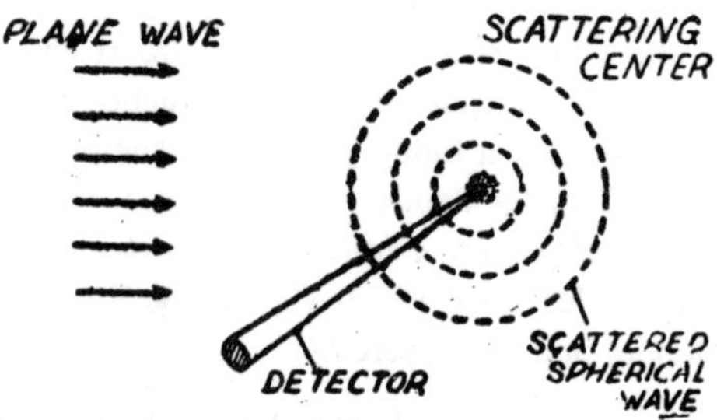

Fig. 4.4 : Scattering process.

To compute the differential scattering cross section, we must find the number of particles dN scattered in unit time by one target nucleus into a solid angle dΩ and the incident flux F. If v is the speed of an incoming particle with respect to the scatterer, then

Incoming flux of particles $F = \psi_{in}^{*}\, \psi_{in} v = v$

Similarly, dN is equal to the flux of scattered particles $\psi_{sc}^{*}\psi_{sc}v$ multiplied by the area $r^2 d\Omega$ cut out by dΩ on a spherical surface of radius r and is given by

$$dN = \psi_{sc}^{*}\psi_{sc}\, vr^2 d\Omega = |\, f(\theta)\,|^2\, v d\Omega$$

∴ The differential cross section $d\sigma = |\, f(\theta)\,|^2 v d\Omega = |\, f(\theta)\,|^2 \Omega$

or $$\sigma = \int |f(\theta)|^2\, d\Omega = 2\pi \int |f(\theta)|^2 \sin\theta d\theta \qquad ...(36)$$

First of all let us consider the wave equation (34) in the absence of a scattering centre [V(r) = 0 for all values of r].

$$\Delta^2\phi + \left(\frac{ME}{\hbar^2}\right)\psi = 0 \qquad ...(37)$$

This has the solution $\psi = e^{ikz}$ ...(38)

where $k = \dfrac{1}{\lambda} = \dfrac{\sqrt{(ME)}}{\hbar}$

Rayleigh proposed that this type of wavefunction can be expanded into a series in terms of spherical harmonic functions. Thus, eqn. (38) can be written as an infinite series

$$\psi = e^{ikz} = e^{ikr\cos\theta} = \sum_{i=0}^{\infty} R_l(r)\, Y_l,\, (\theta),$$

where l is the integer representing the number of the partial waves. It, as usual, signifies the orbital angular momentum of the system. The radial functions $R_l(r)$ are solutions of the radial part of equation (37).

$$\frac{1}{r^2}\frac{d}{dr}\left(r^2\frac{dR}{dr}\right)+\left(k^2-\frac{l(l+1)}{r^2}\right)R=0 \qquad ...(40)$$

This equation has two solutions, one is not finite at the origin and cannot represent the plane wave. The other is finite at origin and can be represented in terms of spherical Bessel functions as

$$R_l(r) = i^l \sqrt{[4\pi(2l+1)]}\, j_l(kr) \qquad ...(41)$$

The square of this gives the r-dependence of the probability density for each partial wave in expression (39). The values of first few spherical Bessel functions are given as

$$j_0(kr)=\frac{\sin kr}{kr},\ j_1(kr)=\frac{\sin kr}{(kr)^2}-\frac{\cos kr}{kr},$$

$$j_2(kr)=\left(\frac{3}{(kr)^3}-\frac{1}{kr}\right)\sin kr-\frac{3\cos kr}{(kr)^2},$$

The spherical harmonic function is given by

$$Y_{l,0}(\theta)=\frac{(2l+1)^{1/2}}{(4\pi)^{1/2}}\, Pt(\cos\theta), \qquad ...(42)$$

where $P_l(\cos\theta)$ is the Legendre polynominal of order l. The square of the spherical harmonic function gives the angular dependence of the probability density. The values of the first few Legendre polynomials are

$$P_0(\cos\theta)=1,\ P_1(\cos\theta)=\cos\theta,\ P_2(\cos\theta)=\frac{1}{2}(3\cos^2\theta-1).$$

For incident neutrons kinetic energy less than 10 MeV (in the lab-system), the only partial wave involved in scattering is the $l = 0$ or S-wave. The scattering is then spherically symmetric in the centre of mass system. The higher the l-value, the larger the impact parameter has to be for a particle with given linear momentum. In the absence of a scattering potential equation (39) can be written as

$$\psi = R_0(r)Y_{0,0}(\theta)+\sum_{l=1}^{\infty}R_l(r)=\frac{\sin kr}{kr}+\left(e^{ikz}-\frac{\sin kr}{kr}\right) \qquad ...(43)$$

The averaged value of the quantity within the brackets over all directions in space is zero. The first term corresponds to the spherically symmetric partial wave (S-wave). For S-wave scattering, only the first term is affected and can therefore be written as ψ_s in the presence of the scattering potential V(r), We can write it as $\psi_s = u(r)/r$.

Thus, as $r \to \infty$, the solution u(r) assumes the form c sin $(kr + \delta_0)$, where c is an arbitrary constant and δ_0 is some phase angle. Thus the complete wave function outside the scattering potential is

$$\psi = \frac{c \sin(kr + \delta_0)}{r} + \left(e^{ikz} - \frac{\sin kr}{kr}\right)$$

$$= e^{ikz} + \frac{c}{r}\,\frac{e^{ikz}\,e^{i\delta_0} - e^{0kr}\,e^{i\delta_0}}{2i} - \frac{1}{kr}\,\frac{e^{ikr} - e^{-ikr}}{2i}$$

$$= e^{ikr} + \frac{e^{ikr}}{r}\left(\frac{ce^{i\delta_0} - 1/k}{2i}\right) - \frac{e^{-ikr}}{r}\left(\frac{ce^{-\delta_0} - 1/k}{2i}\right)$$

This scattered wave must contain no incoming wave. Therefore, we can write the coefficient of e^{-ikr} as zero. Thus, we have

$$ce^{-i\delta_0} - \frac{1}{k} = 0 \quad \text{or} \quad c = \left(\frac{1}{k}\right) e^{i\delta_0},$$

hence $$\psi = e^{ikz} + \frac{e^{2i\delta_0}}{r}\,\frac{e^{2i\delta_0} - 1}{2ik} \qquad ...(44)$$

Comparing this with the standard solution (35), we get

$$f(\theta) = \frac{e^{2i\delta_0} - 1}{2ik} = \frac{e^{i\delta_0}}{k} \cdot \frac{e^{i\delta_0} - e^{-i\delta_0}}{2i} = \frac{e^{i\delta_0}}{k}\sin\delta_0, \qquad ...(45)$$

The total elastic scattering cross section is given as

$$\sigma_0 = 2\pi \int_0^{\pi} \frac{\sin^2\delta_0}{k^2} \sin\theta\, d\theta = \frac{4\pi}{k^2}\sin^2\delta_0$$

$$= 4\pi\,\lambda\!\!\!^{-2}\sin^2\delta_0 \qquad ...(46)$$

The analysis here is carried through only for $l = 0$ scattering. Higher orbital angular momentum waves also have to be considered at higher energies. The total cross-section can be written as a sum of partial cross-sections, one for each l-wave. The partial cross-sections are

$$\sigma_l = 4\pi \lambda\!\!\!^{-2} (2l + 1) \sin^2 \delta_l \qquad ...(47)$$

Scattering Length

For neutrons of very low energy scattered free protons, $\lambda\!\!\!^{-}$ is very large and hence k is very small. It can seen from eqn. (45) that as $k \to 0$, δ_0 must also approach zero, otherwise $f(\theta)$ would become infinite. Thus for low energy neutrons $f(\theta)$ can be written as

$$f_0 = \lim_{\delta_0 \to 0} \frac{e^{i\delta_0} \sin \delta_0}{k} = \frac{\delta_0}{k} = -a \qquad ...(48)$$

where the quantity +a is called the *scattering length* in the convention of Fermi and Marshall. Hence, for low energy neutrons

$$u(r) = c\,(kr + \delta_0) = ck\,(r - a) \qquad ...(49)$$

This is the equation of a straight line intersecting the r-axis at r = a, and is obtained by extrapolating the radial wave function u(r) from the point just beyond the range of the nuclear force. Scattering from a potential giving a bound state produces a positive a. If the potential gives only a virtual state, the slope of the inner wave function at r = b is positive and a is negative.

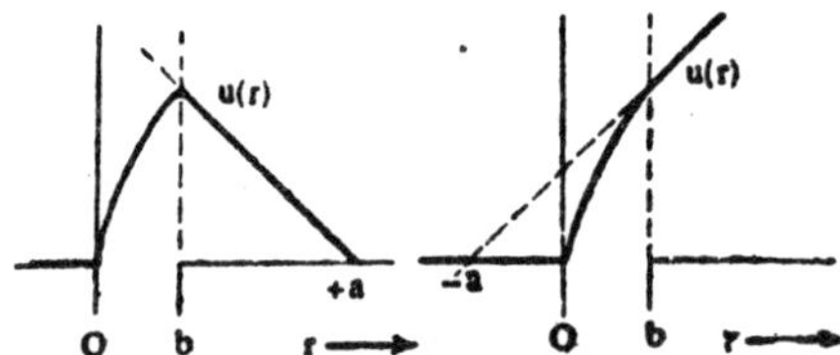

Fig. 4.5 : (left) Positive scattering length (bound state); (right) Negative scattering length (unbound state).

From eqns. (46) and (48) the zero energy scattering cross-section becomes

$$\sigma_0 = 4\pi a^2 \qquad ...(50)$$

This is identical with the scattering cross-section of an impenetrable sphere of radius a, in the limit of zero energy. The measurement of σ_0 *determines the magnitude of the scattering length a but not its sign.*

Determination of the Phase Shift δ_0

We shall now attempt to determine the phase shift δ_0 for low energy neutron-proton scattering by solving also the Schrodinger's equation in the region where the interaction between the two particles takes place. For this we make the simple assumption of a square well for the nuclear potential. Inside the well of depth V_0 and radius b the radial wave equation for particles whose total energy has the positive value E is

$$\frac{d^2u}{dr^2} + \frac{M}{\hbar^2}[E + V_0]\,u(r) = 0 \qquad ...(51)$$

Inside the square well this equation has the simple solution

$$u(r) = A \sin k_1 r, \text{ where } k_1 = \sqrt{[M(E + V_0)]}/\hbar \qquad ...(52)$$

Outside the square well, the solution can be written as

$$u(r) = B \sin (kr + \delta_0) \qquad ...(53)$$

At the edge of the rectangular well (r = b), the two solutions and their derivatives with respect to r must be continuous.

$$\therefore \qquad A \sin k_1 b = B \sin (kb + \delta_0)$$

and $\qquad k_1 A \cos k_1 b = Bk \cos (kb + \delta_0)$

Hence $\qquad k_1 \cot k_1 b = k \cot (kb + \delta_0) \qquad ...(54)$

This relationship is analogous in form to equation

$$K \cot Kb = \alpha$$

which describes the binding energy B of the deuteron in terms of the same rectangular well (V_0, A).

Here we have $K = \sqrt{[M(V_0 - B)]}/\hbar$ and $\alpha = \sqrt{[MB)}/\hbar$

For low energy neutrons ($E < V_0$), we may assume $K = k_1$ (as $V_0 > B$), hence *the wavefunction u(r) inside the well is nearly the same for the deuteron and the n-p-scattering system.* Thus, for approximation we have

$$\frac{\sqrt{(ME)}}{\hbar} \cot \left(\frac{\sqrt{(ME)}}{\hbar} b + \delta_0 \right) = \frac{\sqrt{(ME)}}{\hbar}$$

As the scattering length a is much larger than the range b of the potential, thus for very low energy neutrons kb can be neglected in comparison to δ_0.

$$\therefore \qquad \cot \delta_0 = -\sqrt{\left(\frac{B}{E}\right)} \text{ or } \sin \delta_0 = \frac{E}{E + |B|} \qquad \text{...(55)}$$

Substituting this value of $\sin^2 \delta_0$ in equation (46) we obtain the approximate value of total scattering cross-section as

$$\sigma = 4\pi\hbar^2 \frac{E}{E + |B|} = \frac{4\pi\hbar^2}{M} \cdot \frac{1}{E + |B|} \qquad \text{...(56)}$$

The Spin Dependence of Nuclear Forces

At very low energies the situation is very different Numerical substitution of B = 2.22 MeV In equation (56) gives a predicted value for zero energy neutrons (E = 0), $\sigma_0 \simeq 2.3$ barns. Which is in violent disagreement with the measured value $\sigma_0 = 20.36 \pm 0.10$ barns. This disagreement is a sign of some fundamental error in our assumptions. This point was cleared up by E. P. Wigner in 1935. He suggested that the scattering occurs not only in the triplet state (^3S), but in the singlet state (^1S) as well.

Experimentally the total angular momentum of the deuteron nucleus is unity. The spins are, therefore, correlated in the ground state of the deuteron. The situation is different in n-p scattering experiment. When unpolarized neutrons strike randomly oriented protons, their uncorrelated spins add up to unity in three fourths of the collisions and to zero in one fourth of the collisions. In other words we can say that the triplet state (S = 1) has three times the statistical weight (2S + 1) of a singlet (S = 0) state. The total n-p scattering cross-section will then be

$$\sigma = \frac{3}{4}\sigma_t + \frac{1}{4}\sigma_s,$$

where σ_t and as indicate the cross-sections in triplet and singlet states respectively.

One test of Wigner's hypothesis is by measurement of the cross-section over the range 0 to 5 MeV, where the theoretical expression for a, should hold. We can use the calculated value $\sigma_t = 2.3$ barns and the experimental information on σ_0 (zero energy cross-section) to calculate σ_s. To fit the experimental data, we must have (in barns)

$$20.36 = \frac{3}{4}(2.3) + \frac{1}{4}\sigma_s$$

$\therefore \qquad \sigma_s = 74$ barns. ...(57)

Introducing the singlet and triplet scattering lengths as and a_t, where $\sigma_s = 4 + 4\pi a_s^2$ and $\sigma_t = 4\pi a_t^2$, we obtain the rough estimates $a_t \sim 4.3$ fermi and as $a_s \sim 24.3$ fermi.

Coherent Scattering of Slow Neutrons

Equation (50) shows that we can find the magnitude, but not the sign, of the scattering length a The most important evidence is the scattering of neutrons by *ortho and para hydrogen.* An experimental comparison of the coherent scattering from ortho and para hydrogen was first suggested by Teller in 1936 to test the spin dependence of the neutron proton interaction. Not only we are able to verify that the scattering is spin dependent but also will be able to show that singlet state scattering length is negative. In para-hydrogen molecules the proton spins are anti-parallel. The molecules can rotate with energies given by

$$E_J = J\,(J + 1)\,\frac{\hbar}{2g}, \quad (J = 0, 2, 4, ...),$$

where **g** is the moment of inertia. In ortho hydrogen proton spins are parallel. The rotational states are given by J = 1, 3, 5..... If temperature is low enough practically all the para hydrogen will be in the state J = 0 and all the ortho hydrogen will be in state J = 1.

We shall now derive an expression for the scattered intensity from a molecule of ortho or para hydrogen when the incident neutron energy is so small that $\lambda\!\!\!^{-}_n$ is much greater than 0.78×10^{-10}m the distance between the atoms in hydrogen molecule. The theoretical expressions for total cross-section for the special case of neutron energy of 0.001463 eV and a gas temperature of 19.5° K, as derived by Schwinger and Teller in 1937, are

$$\sigma_{para} = 6.69\,(3a_t + a_s)^2 \qquad ...(58)$$

$$\sigma_{ortho} = 6.69\,[(3a_t + a_s)^2 \text{ or } (a_t - a_s)^2] + 1.74\,(a_t - a_s)^2 \qquad ...(59)$$

where a_t and a_s are triplet and singlet scattering lengths. The last term in σ_{ortho} was added to take into account inelastic scattering by conversion of ortho to para. This process is energetically possible but its cross-section is small. It is clear from above relations that σ_{para} and σ_{ortho} become identical if $a_t = a_s$. In this case the total nuclear forces are said to be spin independent.

According to the experimental results of Sutton (1947), $\sigma_{para} = 4.0$ barns and $\sigma_{ortha} = 12.5$ barns. This great difference between these cross-sections indicates that the n-p force is strongly spin dependent.

$|3a_t + a_s|$ is very much smaller than $|a_t \times a_s|$ when $a_s < 0$ but very much larger when $a_s > 0$. Since $\sigma_{para} < \sigma_{ortho}$, hence $|3a_t + a_s|$ is smaller than $|a_t - a_s|$ or $a_s < 0$. This –ve value, of a_s indicates that the *singlet state of the deuteron* is virtual or unbound. We can now solve equations (58) and (59) for a_t and a_s. There are four possible pairs of a' and a values. Two can be discarded because they have –Ve a_t values. A third pair can be shown to be inconsistent with n-p scattering data. The remaining solution is $a_t = 0.52 \times 10^{-14}$m and $a_s = 2.3 \times 10^{-11}$m. Therefore, the total cross-section will be

$$\sigma_0 = \frac{3}{4}\sigma_t + \frac{1}{4}\sigma_s = \frac{3}{4} \times 4\pi a_t^2 + \frac{1}{4} \times 4\pi a_s^2 = 19.8 \text{ barns},$$

which is in fair agreement with experimental value 20.36 barns.

Spin of Neutron

The large observed value of $\sigma_{ortho}/\sigma_{para}$ relates to the spin of the neutron. The ground state of the deuteron ha I = 0 and I = 1, if

$$S_r + S_p = \frac{1}{4} + \frac{1}{2} = 1 \qquad \text{(spin parallel)}$$

$$S_n + S_p = \frac{3}{2} - \frac{1}{2} = 1 \qquad \text{(spin anti-parallel).}$$

In the second case the relative statistical weights for n-p collisions with free protons would change from their values of 3/4 and 1/4 to values of 5/8 and 3/8. This will change the scattering cross-sections.

$$\sigma_{ortho}/\sigma_{para} \simeq 2.$$

This is against experimental results, hence the neutron has spin 1/2 not 3/2.

Shape Independent Effective Range Theory

While defining the scattering length $a = -\delta_0/k$ we make the use of scattering of S-wave by using a square well potential. We shall discuss here a more general method, which gives a very good approximation for phase shift δ_0 and at the same time does not require a specified form of potential V(r). It is possible to expand the quantity cot δ_0 in a simple power series of k. This will simplify the procedure of computing phase shifts. In the energy range 0 to 15 MeV, two terms in the power series suffice. The first term contains the scattering length and the second term contains a quantity called the effective range, to be defined below.

In developing the effective range theory we begin by writing Schrodinger's equation for an S-wave and positive energy for the relative motion. Let u and u_0 be the radial wavefunctions corresponding to two energies E and E_0. Then we have

$$\frac{d^2u}{dr^2} + \frac{M}{\hbar^2}[E - V]\,u = 0 \quad ...(60)$$

$$\frac{d^2u_0}{dr^2} + \frac{M}{\hbar^2}[E_0 - V]\,u_0 = 0 \quad ...(61)$$

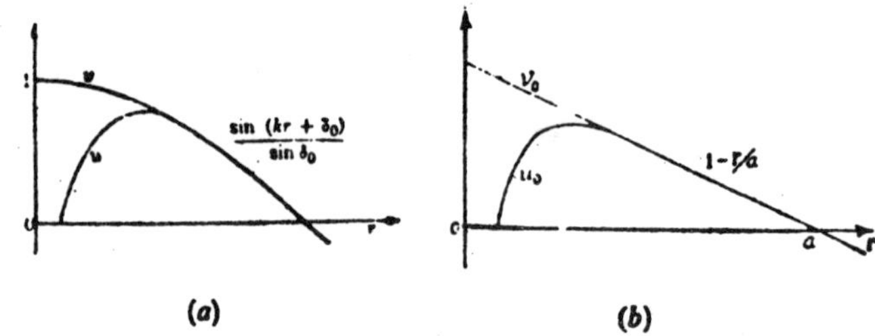

Fig. 4.6 : Radial wavefunction in effective range theory.

where V is the nuclear potential of any shape. Multiply eqn. (60) by u_0 and eqn. (61) by u, subtract and integrate from 0 to ∞. Then

$$\left[uu'_0 - u_0u'\right]_0^\infty = \left(k^2 - k_0^2\right)\int_0^\infty uu_0\,dr \quad ...(62)$$

Consider a new function v which is identical with u outside the potential well and extrapolates the behaviour of u to the origin. The functions are given as v = c sin (kr + δ). These are normalized in such a way that v → 1 at the origin. Thus

$$v = \sin(kr + \delta)/\sin\delta \quad ...(63)$$

We get a relation like eqn. (62) for the wave functions v and v_0. Thus

$$\left[vv'_0 - v_0v'\right]_0^\infty = \left(k^2 - k_0^2\right)\int_0^\infty vv_0\, dr \qquad ...(64)$$

We may now subtract eqn. (62) from eqn. (64). Since v = u and $v_0 = u_0$ outside the potential well and $u = u_0 = 0$ for r = 0, hence

$$-\left[vv'_0 - v_0v'\right]_{r=r_0} = \left(k^2 - k_0^2\right)\int_0^\infty (vv_0 - uu_0)\, dr \qquad ...(65)$$

As relation (63) gives

$$v' = k \cos(kr + \delta)/\sin\delta$$

$$\therefore \qquad (v')_{r=0}\ k \cot\delta \text{ and } (v)_{r=0} = 1$$

Similarly $(v'_0)_{r=0} = k_0 \cot\delta_0$ and $(v_0)_{r=0} = 1$

Substituting these values in equation (65), we have

$$-k_0 \cot\delta_0 + k \cot\delta = (k^2 - k_0^2)\int_0^\infty (vv_0 - uu_0)\, dr \qquad ...(66)$$

Consider now the case when $k_0 = 0$ (the case of zero energy). By definition of scattering length $k_0 \cot\delta_0 = -1/a$ when $k_0 \to 0$. Thus eqn. (66) becomes

$$k \cot\delta = -1/a + \frac{1}{2} k^2 \rho(0, E) \qquad ...(67)$$

where $\frac{1}{2}\rho\,(0, E) = \int_0^\infty (vv_0 - uu_0)\, dr$

Since u and v differ only inside the range of nuclear forces, therefore, the integrand vanishes outside the nuclear force range. Inside the range, the wavefunctions are almost independent of the energy, because the potential energy is much larger than the energy of the system E. We can thus write

$$v \simeq v_0 \text{ and } u \simeq u_0$$

and $$\frac{1}{2}\rho\,(0, 0) = \int_0^\infty \left(v_0^2 - u_0^2\right) dr = \frac{1}{2} r_0 \qquad ...(68)$$

The quantity r_0 is constant and is known as *effective range* of nuclear potential. It is one of the average properties of the potential to be determined from the scattering. The factor $\frac{1}{2}$ is included in the definition to make r_0 approximately equal to the radius. This justifies the name "effective range" for r_0. Thus eqn (67) becomes

$$k \cot \delta = -1/a + \frac{1}{2} r_0 k^2. \qquad ...(69)$$

$$\therefore \text{ Cross-section } \sigma = \frac{4\pi a^2}{\left\{\left[1 - \frac{1}{2} a r_0 k^2\right]^2 + a^2 k^2\right\}} \qquad ...(70)$$

Whatever the shape of the potential, the scattering can be described by the two parameters a and r_0. The experimental values of δ and k serve to determine a and r_0. Any reasonable potential shape can be made to give matching values of a and r_0 by suitable choice of its depth V_0 and range b. Hence the two experimentally determined lengths a and r_0 do not determine the shape of the nuclear potential, but if the shape is chosen then a and r_0 fix the depth and range. For this reason above relation is known as the *shape independent approximation.* Triplet effective range r_0 is obtained from a and has value $1.70 \pm 0.03 \times 10^{-15}$m. Singlet effective range r_0 is obtained from a_s and has the value $2.5 \pm 0.2 \times 10^{-15}$m.

4.4 PROTON-PROTON SCATTERING AT LOW ENERGIES

The p-p scattering is the only way to get direct evidence on proton-proton force. There are two essential differences between p-p and n-p scattering. First, the p-p scattering is caused not only by the nuclear forces but also by the Coulomb force. Second, the scattering and scattered particles are identical and obey the Pauli exclusion principle and, therefore, the wavefunction describing the two protons must change sign on the interchange of the two particles. Hence a symmetric space wavefunction (S, D states, etc.) can only be associated with an anti-symmetric (singlet) spin wavefunction, whereas a symmetric (triplet) spin wavefunction is required for an anti-symmetric space wavefunction (P, F states, etc.). For incident protons of energy below 10 MeV, only the S state interaction is of any importance in the scattering, since protons in higher orbital angular momentum states stay apart from each other beyond the range of the nuclear force.

Experimentally the study of p-p scattering, is capable of much higher accuracy than n-p scattering, for the following reasons:

1. Protons are easily available over a wide range of energies.
2. Protons can be made monoenergetic.

3. Protons can be produced in well collimated beam.
4. Protons can be easily detected by their ionizing properties.
5. Protons undergo Coulomb scattering simultaneously with nuclear scattering. This increases the sensitivity in case one of the scattering probabilities is small and also gives sign of the phase shifts resulting from nuclear scattering.
6. The proton combination obeys Fermi Statistics. This simplifies the analysis of proton-proton scattering.

We are now interested in obtaining a theoretical expression for the differential elastic scattering cross-section of protons by protons. The theory is more complicated than that of n-p scattering because the Coulomb potential appreciably distorts the incident wave eve at finite distances. The radial wave equation for the proton-proton system is

$$\frac{d^2u}{dr^2} + \frac{2m}{\hbar^2}\left[E - V(r) - \frac{e^2}{4\pi\varepsilon_0 r} - \frac{l(l+1)\hbar^2}{2mr^2}\right]u = 0$$

where V(r) is the square well nuclear potential. For low energies, the nuclear potential will affect only the $l = 0$ partial wave, but the Coulomb potential produces higher l-value scattering. Rutherford first investigated the scattering by the Coulomb field from the classical stand point. The Rutherford cross-section for Coulomb scattering of particle of charge Z_1e by a particle of charge Z_2e in the centre of mass system is

$$d\sigma = \left[\frac{(Z_1e)^2\,(Z_2e)^2}{4(4\pi\epsilon_0)^2\,m^2v^4}\operatorname{cosec}^4\frac{\theta}{2}\right]2\pi\sin\theta\,d\theta, \qquad ...(72)$$

where v is the velocity of the incident particle, m is the reduced mass of the system and θ is the angle of scattering in the centre of mass system. For p-p scattering $Z_1 = Z_2 = 1$, $m = M/2$, $\theta = 2\theta_1$ and $\sin\theta\,d\theta = 4\sin\theta_1\cos\theta_1\,d\theta_1$ (where θ_1 is in laboratory system). Thus, in laboratory system, the cross-section becomes

$$d\sigma = \frac{e^4}{(4\pi\epsilon_0)^2\,E_0^2}\left[\frac{1}{\sin^4\theta_1} + \frac{1}{\cos^4\theta_1}\right]\cos\theta_1\,2\pi\sin\theta_1\,d\theta_1, \qquad ...(73)$$

where $E_0 = \frac{1}{2}Mv^2$ (kinetic energy of incident proton in the laboratory system). The term containing $\cos^4\theta_1$ is added because each proton at angle θ_1 in the L-system is accompanied by a recoil proton potential is repulsive or attractive, as δ_0 is positive corresponding to attractive potentials and negative to repulsive potentials For large energies the last term becomes the most important because of the v_2 coefficient.

The differential cross section is a complicated function of the energy E_0, the angle of scattering θ_1 and the phase shift δ_0. Differential cross-section for p-p elastic scattering for 2.4 MeV protons is shown in Fig. 4.7. At small scattering angles the scattering is essentially pure Coulomb scattering. The nuclear scattering predominates in the central region of angles, near a C.M. angle of 90° (lab angle of 45°). At forward and backward angles, the dips are caused by interference between nuclear and Coulomb scattering.

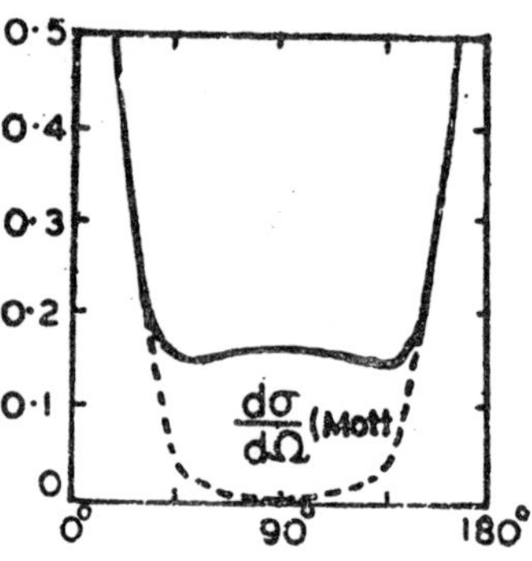

Fig. 4.7 : Differential cross-section for p-p scattering.

Experiments on p-p Scattering

The experimental data are analyzed further along lines similar to those used in n-p scattering but are complicated considerably by the presence of the Coulomb force. In the interpretation of the p-p scattering results, the phase shifts δ_0 are the common meeting ground of experiment and theory. Jackson and Blatt showed that the observed variation of δ_0 with incident proton energy can be matched by any of the conventional shapes of potential well, with suitable choices of depth and range. A shape independent theory, therefore, his a natural attractiveness. Such a theory, for proton interaction, contains as parameters, the effective range r_0 and scattering length a_p. The best fit to the data compiled by Jackson and Blatt is obtained with

$$r_0 = 2.65\ (1 \pm 0.03) \times 10^{-15}\ \text{m and } a_p = 7.7\ (1 \pm 0.07) \times 10{-}^{15}\ \text{m.}$$

The fact that a_p is negative indicates that the S^1 state for the p-p system cannot be bound. Therefore the di-proton or He^2 can not exist, The exclusion principle prevents any ^3S state, which would be the analog of the stable deuteron. The absolute value of a_p requires further interpretation because it includes Coulomb as well as specifically nuclear effects. The latter can be isolated by treating the Coulomb force, with in the range of the nuclear force, as a small perturbation. Thus a/i is increased to the value equal to

$$a_{eq} = -17 \times 10^{-15} \text{ m.}$$

We know that for neutron-proton estate $a_s = -24 \times 10^{-15}$m and $r_{0s} = 2.5 \times 10^{-15}$m. The effective ranges do not differ significantly in view of their large experimental uncertainties. The scattering lengths are different significantly. Such a difference might arise from the magnetic moment interaction, as the interinsic magnetic dipole moments of the neutron and proton are of opposite sign. Schwinger has evaluated a net attractive magnetic interaction in the ^{3}[np) force and a net attractive magnetic interaction in the ^{1}(pp) force, at an angle $(90 - \theta_1)$ and it is not possible to distinguish between the two protons.

Equation (73) does not agree with experiment even at low energies where nuclear scattering is negligible. Mott introduced a third term from wave mechanical consideration. This places symmetry requirements on the wave-function and represents the interference between the wave functions of protons. Thus, eqn. (73) becomes

$$d\sigma = \frac{e^4}{(4\pi \epsilon_0)^2 E_0^2}\left[\frac{1}{\sin^4\theta_1} + \frac{1}{\cos^4\theta_1} - \frac{\cos\{e^2/4\pi \epsilon_0 \hbar v)\log\tan^2\theta_1}{\sin^2\theta_1\cos^2\theta_1}\right]$$
$$\cos\theta_1\ 2\pi\sin\theta_1 d\theta_1$$

The negative sign is because the protons obey Fermi Statistics. Second and third terms drop out for unlike charged particles. For proton, energies above 1 MeV, $\frac{e^2}{4\pi \epsilon_0 \hbar v} > \frac{1}{7}$ and the cosine term is nearly unity, unless θ_1 is close to zero or $\pi/2$. Thus away from $\theta_1 = 0$ or $\pi/2$, for $E_0 > 1$ MeV, the Mott term is simplified and above equation reduces to

$$d\sigma \frac{e^4}{(4\pi \epsilon_0)^2 E_0^2}\left[\frac{1}{\sin^4\theta_1} + \frac{1}{\cos^4\theta_1}\ \frac{1}{\sin^2\theta_1\cos^2\theta_1}\right]$$
$$\cos\theta_1\ 2\pi\sin\theta_1 d\theta_1 \qquad ...(74)$$

The main difference between proton and neutron seems to be of electric charge and the nuclear force apparently does not arise from charge. It is reasonable to assume that the nuclear potential between two protons has the some characteristics as that between proton and neutron. Therefore, in p-p scattering at low energies it is expected that only the $l = 0$ scattering processes will be affected by the nuclear potential, just as in n-p scattering. When the effects of S wave-nuclear scattering are included we obtain scattering term. For S wave-nuclear scattering the complete differential scattering cross-section in laboratory system is given as

$$\frac{d\sigma}{d\Omega}=\frac{e^4}{(4\pi \in_0)^2 E_0^2}\left[\frac{1}{\sin^4\theta_1}+\frac{1}{\cos^4\theta_1}-\frac{1}{\sin^2\theta_1\cos^2\theta_1}-\frac{8\pi \in_0 \hbar v \sin\delta}{e^2}\right.$$

$$\left\{\frac{\cos\left[\delta_0+\frac{e^2}{4\pi \in_0 \hbar v}\log\ \sin^2 1/2\theta\right.}{\sin^2(\theta/2)}+\frac{\cos\left[\delta_0+\frac{e^2}{4\pi \in_0 \hbar v}\log\cos^2 1/2\theta^2\right.}{\cos^2(\theta/2)}\right\}$$

$$\left.+\left(\frac{8\pi\varepsilon_0 \hbar v}{e^2}\right)^2 \sin^2\delta_0\right]\cos\theta_1 \qquad ...(75)$$

In this equation 1st, 2nd and 3rd terms are due to Coulomb repulsion, the 5th term due to nuclear scattering and the fourth term due to interference between the Coulomb and nuclear scattering effects. This relation reduces to equation (74) for a pure Coulomb field when nuclear phase shifts δ_0 is zero *i.e.*, when there is n nuclear scattering. The fourth term is useful in determining δ_0 when δ_0 is small, because it is linear in $\sin^2\delta_0$ and it is large compared to the last term which involves $\sin^2\delta_0$. The linearity in δ_0 of the interference term also permits determination of whether the nuclear. *This equality of the parameters of the (pp) and (np) singlet interactions gave rise to the hypothesis of the charge independence of the nuclear forces.* This hypothesis states that the forces between two nucleons in the same spin and orbital angular momentum state are independent of the charge of the two nucleons.

4.5 SIMILARITY BETWEEN (nn) AND (pp) FORCES

(i) Similar to diprotons, di-neutrons are not found in a stable stationary state. The scattering length for the (nn) interaction appears to have the same (negative) sign as that for the (pp) interaction.

(ii) The study of lightest pair of mirror nuclei, H^3 and He^3 provides us the direct evidence on the equivalence of $^1(nn)$ and $^1(pp)$ forces. The possible interactions between two particles in these nuclei can be written as

$$H^3 \rightarrow {}^3(np) + {}^1(np) + {}^1(nn),\ He3 \rightarrow {}^3(np) + {}^1(np) + {}^1(pp)$$

These are all S-state interactions because orbital angular momentum is zero for each nucleus. The observed binding energy difference is 0.76 MeV. This corresponds to the Coulomb energy of two discreate protons

$$\left[\frac{3}{5}\left(\frac{e^2}{4\pi \epsilon_0 R}\right) z(z-1) = 0.75 \text{ MeV}\right].$$

This proves that (nn) = (pp) force.

(iii) Inelastic scattering of deuteron by neutrons has given quantitative support to the similarity of (nn) and (pp) force.

(iv) The differential cross-sections at ~ 10 MeV for the H^3 (d, n) He^4 and He^3 (d, p) He^4 reactions show very similar asymmetries. These observations support the equality of (nn) and (pp) force.

(v) Strong confirmation for equality of n-n and p-p forces comes also from the steady of heavier mirror nuclei, such as ${}_{16}S^{31}$ and ${}_{51}P^{31}$.

4.6 NON-CENTRAL FORCES

(a) Experimental Evidence for the Existence of Non-Central Forces

A serious discrepancy was discovered between theory and experiment, performed by Kellogg in 1939. The experimental results showed a fine structure in the radio frequency magnetic resonance spectrum of deuterium, but it was different from the theoretical prediction. The quadrupole moment established by this method was 2.82×10^{-31} m^2. It was assumed that n-p force was a central force. The ground state of deuteron must be a S-state whose quadrupole moment should be zero and the magnetic moment should he the algebraic sum of the magnetic moments μ_p, and μ_n. The existence of quadrupole moments in heavier nuclei can be explained by assuming the force between neutron and proton as partly a central force and partly noncentral (*or tensor*) force. From the small deviation of the deuteron wavefunction from spherical symmetry, we can calculate that the non-central part of the force is also very small compared to fie central force part. The tensor force resembles a magnetic interaction between two dipoles. It results in a lower potential energy for n and p when the line joining them is | | to the spin direction than when it is ⊥.

(b) General Form of the Non-Central Force

Let us now see whether we can construct a potential which depends not only on the relative position vector of the particles r, but also on their spin orientations $\vec{\sigma}_1$ and $\vec{\sigma}_2$ and which can account for the quadrupole moment. This potential must be invariant under rotations and reflections of the co-ordinate system which we use to describe the relative motion of the particles and hence must be a scalar.

Wigner in 1941 has showed that, if the interactions are assumed to be invariant with respect to displacement, rotation and inversion of the observer's co-ordinate system and independent of the particle velocities, the general potential can be written as

$$V(r) = V_1(r) + V_2(r)\, \vec{\sigma}_1 . \vec{\sigma}_2 + V_3(r) S_{12} \qquad ...(76)$$

where the potentials V may depend on the orbital momentum of the two particle system as well as on the charge of the particles. The reasons of this limited choice are:

1. The vector r changes to –r under reflection and thus mutt occur in even powers only.
2. Derivatives of r cannot occur since forces are velocity independent.
3. Since $\vec{\sigma}_1$ and $\vec{\sigma}_2$ are not invariant against rotation and $\vec{\sigma}_1 \cdot \vec{\sigma}_2$ is invariant.
4. $\vec{\sigma}\ \vec{r}$ is invariant against rotation, but not against inversion, since $\vec{r} \to -\vec{r}$ and $\vec{\sigma} \to \vec{\sigma}$ on inversion. Because of this, only even power of $(\vec{\sigma} \cdot \vec{r})$ may occur such as $(\vec{\sigma}_1 \cdot \vec{r})(\vec{\sigma}_2 \cdot \vec{r})$.

The first term is corresponding to central force. The second term corresponding to a spin dependent but central force. The value of $\sigma_1 . \sigma_2$ for a triplet state and –3 for a single state. The last term is corresponding to the interaction potential which depends not only on the separation between neutron and proton but also on the angle which their spins make with the line joining the two particles. The non-central character of the interaction is contained in the tensor operator.

$$S_{12} = 3(\vec{\sigma}_1 \,.\, r)\,(\vec{\sigma}_2 \,.\, r)/r^2 - \vec{\sigma}_1 . \vec{\sigma}_2 \qquad ...(77)$$

The first term gives the dependence of the interaction on spin angles. The second term has been subtracted so that the average of S_{12} over all directions r is zero.

(c) Properties of the Non-Central Force

Since S_{12} is a scalar, the tenser force potential is a scalar. It can be shown that whin $V_3(r)$ is finite, the total angular momentum **I** and the parity remain good quantum numbers but that the orbital angular momentum **L** and the spin angular momentum S are not constants of motion. The magnitude of the spin angular momentum S^2 is still a constant of motion for the two body systems.

The tensor operator S_{12} is defined in such a way that its average over all directions is zero. In the singlet spin state there is no preferred direction for the spin orientation hence total spin angular momentum

$S = \frac{1}{2}(\vec{\sigma_1} + \vec{\sigma_2}) = 0$. Thus, we get

$$S_{12} = -3(\vec{\sigma}_1 \,.\, \vec{r})\,(\vec{\sigma}_2 \,.\, r)/r^2 - \vec{\sigma}_1 . \vec{\sigma}_2$$

This results that the tensor force is zero in the singlet state and all previous results for singlet states are unaffected entirely.

We also know that a triplet states of a given total angular momentum **I** can be written as a linear superposition of triplet states with orbital angular momenta L = I – 1, I + 1. Since parity of a wave function is $(-1)^L$, hence the state with L = I has opposite parity to the two states with L = I – 1 and L = I + 1. Hence the I = 1 states of even parity are mixtures of $^3S_1$ and $^3D_1$ and the I = 1 states of odd parity are pure $^3P_1$.

(d) The Ground State of the Deuteron

Magnetic Moment

We know that the small discrepancy between the sum of the magnetic moment of proton and neutron and the measured value for the deuteron can be interpreted as a contribution of the orbital motion of the proton in the D-state in the deuteron ground state. If nuclear forces are supposed to be central forces, the difference will be zero. This contribution can appear only with the non-central forces. The operator describing the magnetic moment of deuteron is

$$\vec{\mu} = \vec{\mu}_p \times \vec{\sigma}_p + \vec{\mu}_n \times \vec{\sigma}_n + \vec{L}r \qquad ...(78)$$

where $\vec{\mu}_n$ and $\vec{\mu}_p$ are the magnetic moments of neutron and proton measured in nuclear magnetons, $\vec{\sigma}_n$ and $\vec{\sigma}_p$ their unitary spin operators and $\mathbf{L}_p$ is the orbital angular momentum of the proton. The uncharged neutron cannot contribute any magnetic moment by orbital motion alone. In the centre of mass system the orbital angular momentum of the proton is half of the combined orbital angular momentum **L**. Thus the above equation can be written as:

$$\mu = (\mu_n + \mu_p).\frac{1}{2}(\sigma_n + \sigma_p) + \frac{1}{2}(\mu_n - \mu_p)(\sigma_n - \sigma_p) + \frac{1}{2}L.$$

Since the operator $(\sigma_n - \sigma_p)$ in the second term vanishes for a triplet state, and I = L + S, hence the magnetic moment operator becomes

$$\vec{\mu} = (\mu_n + \mu_p)\,\mathbf{I} - (\mu_n + \mu_p - \frac{1}{2})\mathbf{L} \qquad ....(79)$$

The observed value of μ is the expectation value of this expression in the state with $\mathbf{I}_z = \mathbf{L}$. We, therefore, can replace L by L_z.

$$\therefore \qquad \mathbf{L} \to \mathbf{L}_z = \frac{\mathbf{LI}}{\mathbf{I}^2}I_z \frac{I(I+1)+l(L+1)-S(S+1)}{2I(I+1)}I_z \qquad ...(80)$$

For, the deuteron, I(I + 1) = S(S + 1) = 2 and in a mixture of S and D states, $[L(L+1)]_{av} = 0 \times p_S + 6 \times p_D = 6p_D$. Here p_D and p_S represent the D-and S-state probabilities respectively. For the deuteron I = 1, the value of I_z in the state with $I_z = I$ is unity.

$$\therefore \qquad \mu \text{ of deuteron} = \mu_n + \mu_p - \frac{3}{2}(\mu_n + \mu_p - \frac{1}{2})p_D \qquad ...(81)$$

The last term gives a direct measure of the D-state probability. The experimental results imply a D-state probability of about 4%. The above relation does not give accurate results. There are various other causes which can give corrections, especially of relativistic effects. Hence the measured magnetic moment gives only a rough estimate of the D-state probability. One may expect that p_D is really between 2 and 8%.

Quadrupole Moment

The quadrupole moment Q was defined as the average value of $(3z^2 - r^2)$ in the state with m = I. When evaluating the quadrupole moment of the deuteron we must remember that only the proton contributes to the quadrupole moment and its distance from the centre of gravity is half of the proton neutron separation r.

$$\therefore \qquad Q = \frac{1}{4}(3z^2 - r^2) = \frac{1}{4}r^2\,(3\cos^2\theta - 1) \qquad ...(82)$$

Since the quadrupole moment is estimated from the S and D waves beyond the potential well, hence the expectation value of this operator Q is given by

$$\therefore \quad (\psi, Q\psi) = (\psi s, Q\psi s) = (\psi_D, Q\psi_D) + 2\,(\psi s, Q\psi_D) \qquad ...(83)$$

The first term is zero because the S state is spherically symmetrical and cannot have a quadrupole moment. The second term is a pure D-state term and is smaller than the cross term (as $p_S >> p_D$), hence only the cross term contributes. The measured quadrupole moment Q is, therefore, be written as

$$Q = \frac{1}{\sqrt{(50)}}\int_0^\infty r^2 u(r)\omega(r)\,dr \qquad ...(84)$$

where the constant comes from the spin sum and angular integration u(r) is the deuteron ground state S-wave function outside the range of nuclear forces and can be written as $u(r) = Ns\, e^{-kr}$.ω(r) is the deuteron D-wave function at distances beyond the range of the specific potentials and can be written as

$$\omega(r) = N_D e^{-kr}\,(1 + 3/kr + 3/k^2r^2).$$

Here N_S and N_D are normalization constants and $k = \sqrt{(MB)}/\hbar^2$.

The rough estimate of Ns can be obtained by neglecting the small D-state probability compared to unity by substituting

$$p_s = \int_0^\infty u^2 dr = 1 \qquad N_s = \sqrt{(2k)} \qquad ...(85)$$

Since the weighing factor r^2 flavours the contribution of the outside wave function, hence we can estimate the quadrupole moment by substituting the values of u(r) and ω(r) into the equation (84). The result will be

$$Q = \frac{N_S N_D}{\sqrt{8k^2}} \qquad ...(86)$$

This relation can be used for a good estimate of N_D. If value of N_S is taken from eqn. (85), we have

$$Q = \frac{N_D}{2k^{5/2}} \quad \text{or} \quad N_D = 2Qk^{5/2} \qquad ...(87)$$

Thus, we see that the function ω(r) outside the range of the forces is determined completely by the quadrupole moment. This result implies that the D-state probability p_D depends strongly on the tensor force range RT and increases rapidly as R_T is made shorter. The integral of ω^2 from the radius R_T on out is given by

$$\int_{R_T}^{\infty} \omega^2 \, dr = \frac{N_{D^2}}{R_{T^3} k^4} \qquad ...(88)$$

To take into account the contribution from $r < R_T$ we can roughly double this and thus get the physically important quantity

$$P_D = \int_0^{\infty} \omega^2 \, dr = 2\int_{R_T}^{\infty} \omega^2 \, dr = \frac{6N_{D^2}}{R_{T^3} k^4} = \frac{24Q^2 k}{R_{T^3}} \qquad ...(89)$$

This equation implies that the tensor force cannot have an arbitrarily small range otherwise the ground state would become a predominantly D state rather than predominantly S state. The experimental value of Q is + 2.73(e × 10^{-31}m^2).

A rough measurement of p_D is obtained from the deuteron magnetic moment (eqn. 81). This value of p_D leads to a fair estimate of R_T, since it is in the cube of R_T which occurs in eqn. (89). This gives that the tenser force range R_T falls nest 3 × 10^{-15} m, which is almost independent of the range of the central force and is slightly larger than the range of central forces.

(e) Neutron-Proton Scattering Below 10 MeV

In principle the analysis of the scattering is very much altered by non-central forces. The separation of the various partial waves is no longer complete, since the orbital angular momentum *l* is not a constant of the motion. At energies below 10 MeV, it is advantageous to decompose the wave function into spherical harmonics by the use of the spin angular function. The spin S can have values 0 and 1 corresponding to the singlet and triplet states. The S-scattering in the singlet state (S = 0) is unchanged by the presence of tensor forces since they do not act in that state. Therefore, we have to discuss only the triplet scattering. We know that the tensor force couples the $^3D_1$ state with the $^3S_1$ state. Thus both states contribute to the scattering even at low energies.

If the ingoing wave is a pure $^3S_1$ wave, the outgoing wave is a mixture of S and D waves owing to the action of the tensor force during the impact. Similarly, an ingoing pure $^3D_1$ wave emerges as a mixture outgoing wave. The waves, which have the same mixture ratio of S to D states before and after the scattering event, are called *eigenstates for the scattering*. There are two such eigenstates: One is called the α-wave in which S wave is predominant. Other is called the β-wave in which D wave is predominant. Since we are dealing with the same angular momentum state $^3S_1$ and $^3D_1$ as the ground state of the deuteron, the asymptotic form of perturbed u(r) and ω(r) can be written in the form

$$u(r) = a \sin (kr + \delta)$$

$$\omega(r) = b \sin (kr + \delta - \pi),$$

where the angle δ represents the phase shift compared to the asymptotic form of the unperturbed plane wave. The solution u, ω of the wave equation is an eigenstate of the scattering provided that the phase shift δ is the same for both u(r) and ω(r). There are two (α and β) solutions satisfying this condition, with two different phase shifts (δ_α and δ_β) and two different values of the ratio b/a which determine ω(r)/u(r).

If we neglect the phase shifts of all the waves except the α-wave, the differential scattering cross-section in C-M system becomes

$$\frac{d\sigma}{\delta\Omega} = \frac{\sin^2 \delta_\alpha}{k^2}\left[\left(1+\frac{\tan \in}{\sqrt{8}}\right)\sin^2 (2 \in) P_2(\cos\theta)\right] \quad ...(90)$$

where P_2 (cos θ) is the second Legendre polynomial and $\varepsilon = \tan^{-1} b_\alpha/a_\alpha$, *i.e.*, the asymptotic ratio of a to u, can be determined by the quadrupole moment as

$$\in \simeq \tan \in \simeq b_\alpha/a_\alpha \simeq \sqrt{(2)}Qk^2. \quad ...(91)$$

If we substitute the numerical value of Q in this eqn. and evaluate the angle dependent term in eqn. (90), it gives that the coefficient of P_2(cos θ) in the angular distribution is less than 0.01 even at 10 MeV. Hence, this tensor effect in the scattering is much too small because in this energy region the experimental errors for angular distribution measurements are at present of the order of 10%. P-wave scattering can be expected to contribute terms. The triplet parameters a_t and r_{0i} are not much affected by the tenser forces, if the potentials have similar shape.

4.7 SATURATION OF NUCLEAR FORCES

Thus, far we have discussed mainly the character of the forces acting between pairs of nucleons. When we come to consider systems comprising more than two particles, we must expect some complications to enter. We shall deal with certain quantitative aspects of nuclear forces which become important when not only the two, body problem but also the properties of other nuclei are studied. There are two important facts: the density of nucleons is roughly equal for all nuclei *(saturation of density)*, and the binding energy per nucleon is roughly equal for all nuclei (*saturation of binding energy*).

In the case of two body problem, following assumptions have been made:

(a) The nuclear force acts between the pair of nucleons and does not influenced by the presence of neighbouring nucleons.

(b) The nuclear force between the nucleons is velocity independent, attractive, independent of the type of the nucleons and central.

These assumptions do not correspond to reality. The basic force law would have to be modified, if we add nucleon to a nucleus. The force between two nucleons is attractive for distances $r > d$, and is repulsive otherwise. Hence d is of the order of the distance between nearest neighbours in the nuclei.

(A) Exchange Forces

In 1932, Heisenberg proposed, in order to explain the saturation of nuclear forces, that nuclear forces were *Exchange Forces* which would depend explicitly on the symmetry of the wave function. At the time of Heisenberg's idea of exchange forces, mesons were known and it was known that π-meson was being exchanged between nucleons, by any of the following processes:

$$\begin{array}{llll} n, p & ; \; p, n & ; \; p, p & ; \; n, n \\ p + \pi^-, p & ; \; n + \pi^+, n & ; \; p + \pi^o, p & ; \; n + \pi^o, n \\ p, \pi^-, p & ; \; n, \pi^+ + n & ; \; p, p + \pi^o & ; \; n, n + \pi^o \\ p, n & ; \; n, p & ; \; p, p & ; \; n, n, \end{array}$$

The exchange of a pion is thus equivaler t to charge exchange. We can think of the nucleons as exchanging their space and spin co-ordinates. The wave equation of the two body system for an ordinary central force is

$$[(\hbar^2 M)\nabla^2 + E]\Psi(r_1, r_2, \sigma_1, \sigma_2) = V(r)\Psi(r_2, r_1, \sigma_1, \sigma_2). \qquad ...(92)$$

The force is known as no *exchange, ordinary* or *Wigner force*. The interaction does not cause any exchange. Exchange forces are classified as:

1. *Majorana Forces*

The Majorana interaction is that in which it is assumed that two particles attract one another if the wave function describing the entire system does not change sign when the special co-ordinates of the two particles are interchanged and that they repel if the wave function changes sign, *i.e.*,

$$[(\hbar^2 M)\nabla^2 + E]\psi(r_1, r_2, \sigma_1, \sigma_2) = V(r)\psi(r_2, r_1, \sigma_1, \sigma_2). \qquad ...(93)$$

This interaction reflects the coordinates and replaces r by –r in the wave-function. Hence eqn. (93) may be written as

$$[(\hbar^2/M)\nabla^2 + E]\psi(r) = (-1)^l\, V(r)\psi(r). \qquad ...(94)$$

This eqn. indicates that the force is always attractive for states of even *l*(S, D, G,...) and always repulsive for states of odd *l*.

2. *Bartlett Forces*

This involves the exchange of the spin hut not the position coordinates of the two interacting nucleons. For such an interaction, the Schrodinger eqn is

$$[(\hbar^2 M)\nabla^2 + E]\psi(r_1, r_2, \sigma_1, \sigma_2) = V(r)\psi(r_1, r_2, \sigma_2, \sigma_1). \qquad ...(95)$$

The wave function of two particles is symmetric if the total spin S = 1 and anti-symmetric if S = 0. Thus eqn. (95) gives

$$[(\hbar^2/M)\nabla^2 + E]\psi(r) = (-1)^{s+1}\, V(r)\psi(r). \qquad ...(96)$$

This relation is equivalent to an ordinary potential which changes sign between S = 0 and S = 1. The nuclear force can not be totally of the Bartlett type because it is clear from neutron proton scattering data that both the ^{3}S and ^{1}S potentials are attractive.

3. *Heisenberg Forces*

In the type of interaction there is an exchange of both the position and the spin co-ordinates of the two nucleons. For such an inter Action, the Schrodinger eqn. is

$$[(\hbar^2 M)\nabla^2 + E]\psi(r_1, r_2, \sigma_1, \sigma_2) = V(r)\psi(r_2, r_1, \sigma_2, \sigma_1). \quad ...(97)$$

Since the wave function of two particles is symmetric for (l + S) even and anti-symmetric for (l + S) odd, hence eqn (97) may be written as

$$[(\hbar^2 M)\nabla^2 + E]\ \psi\ (r) = (-1)^{l+s+1}\ V(r)\psi(r) \quad ...(98)$$

This relation indicates that sign of ordinary potential is positive if (l + S) is odd and is negative if (l + S) is even. This gives that the force is attractive for even l triplet states and odd l singlet states, but is repulsive in odd l triplet and even l singlet states.

The three types of exchange operators P^M, P^B and P^H are used to construct these three types of exchanged forces. The reversal of sign between ^3S and ^1S states indicates that the nuclear force cannot be wholly of the Heisenberg type. The difference between the n – p interactions in these states can be explained by assuming that the interaction is roughly 25 per cent Heisenberg or Bartlett and 75 per cent Wigner or Majorana.

The three types of exchange operators can be related as

$$P^H = P^M P^B \text{ and } (P^M)^2 = (P^B)^2 = (P^H)^2 = 1. \quad ...(99)$$

This shows that each operator has only two eigenstates +1 and –1.

The Majorana exchange operator is +1 for the states of even l and –1 for the states in odd l. The Bartlett exchange operator gives +1 in triplet states and –1 in singlet states, independent of l. In the various states of the two particle systems, the exchange operators have the values given below:

Operator	Even parity states		Odd parity states	
	Triplet	Singlet	Triplet	Singlet
P^H	1	–1	–1	1
P^M	1	1	–1	–1
P^B	1	–1	1	–1

The most general potential of the exchange type is written in the form

$$V_W(r) + V_M(r)P_M + V_B(r)P^B + V_H(r)P^H \quad ...(100)$$

This potential also has tensor operator S_{12} term for mixtures of Wigner and Majorana forces.

(B) Isotopic Spin Formalism

In this formalism we are considering the proton and the neutron as different quantum states of the same particle, the nucleon. The total wave-function is written as a product of the space part, a spin part and an isospin part. The nucleus must obey Fermi-statistics in order to the consistent with the ordinary theory. Thus, the total wave-function for the two or more particles

$$\psi = \psi(\text{space})\ \psi\ (\text{spin})\ \psi\ (\text{isotopic spin}) \qquad ...(101)$$

must be anti-symmetric with respect to interchange of all co-ordinates of two nucleons. In the ground state of the deuteron, for example, ψ (space) is symmetric, as it a mixture of an S-state and a D-state, ψ (spin) is symmetric (the two spins are parallel), so that the ψ (isotopic spin) must be anti-symmetric and thus T = 0 (the two isotopic spins are oppositely oriented); The lowest state of deuteron in which the two nucleon spins are opposed, ψ (spin) is then antisymmetric, is the lowest one in which T = 1.

The concept of the T multiplet has been applied to β-decay, γ-decay and to nuclear reactions. *The success of these applications supplies additional support for the hypothesis of the charge independence of nuclear forces.*

In order to confirm with isospin conservation, the Hamiltonian describing the interaction between two nucleons must be rotationally invariant in isospace. Thus it must contain scalar quantities formed with the isospins $\vec{\tau}_1$ and $\vec{\tau}_2$. The product of operators $\vec{\tau}_1 \cdot \vec{\tau}_2$ gives + 1 when applied to isotopic spin triplet states and –3 when applied to isotopic spin singlet states. We can use $\vec{\tau}_1 \cdot \vec{\tau}_2$ to define an isotopic spin operator P^r analogous to spin operator σ.

$$P^r = \frac{1}{2}(1 + \vec{\tau}_1 \cdot \vec{\tau}_2) \qquad ...(102)$$

The operator gives + 1 when applied to the (symmetric) isotopic spin triplet states, –1 when applied to the (anti-symmetric) isotopic spin singlet states. Hence, it is equivalent to simple exchange of the isotopic spin co-ordinates η_1 and η_2 of the two particles. Thus, we can write

$$P^r\psi(r_1, \zeta_1, \eta_1; r_2, \zeta_2, \eta_2) = \psi(r_1, \zeta_1, \eta_2; r_2, \zeta_2, \eta_1) \qquad ...(103)$$

where r_1 denotes the position of the one particle and ζ_1 its spin direction.

We have required complete anti-symmetry under the full exchange of all the co-ordinates of the two particles. Since Heisenberg exchange operator P^H exchanges position mechanical spin and the isotopic spin operator P^r exchanges the isotopic spin co-ordinates, hence we can write.

$$P^H P^r \psi = -\psi \qquad ...(104)$$

This is a condition on ψ not an operator entity. We multiply equation (104) by P' on both sides and have

$$P^H \psi = -P^r \psi \qquad ...(105)$$

because equation (103) shows that $(P^r)^2 = 1$.

It is clear from equation (105) that we can replace the Heisenberg exchange operator by the isotopic spin operator $-P^r$. As the Bartlett exchange operator ^{PB} can be written as

$$P^B = \frac{1}{2}(1 + \vec{\sigma}_1 . \vec{\sigma}_2) \qquad ...(106)$$

Thus, we can replace the Majorana exchange operator by the combination

$$P^M = \frac{1}{4}(1 + \vec{\sigma}_1 . \vec{\sigma}_2)(1 + \vec{\tau}_1 \cdot \vec{\tau}_2) \qquad ...(107)$$

Equations (105), (106) and (107) show the expressions for the three linearly independent exchange operators in terms of the mechanical and isotopic spin operators. From the point of view of the isotopic spin formalism it is more convenient to use three other linearly independent operators $(\vec{\sigma}_1 . \vec{\sigma}_2)(\vec{\tau}_1 \cdot \vec{\tau}_2)$ and $(\vec{\sigma}_1 . \vec{\sigma}_2)(\vec{\tau}_1 \cdot \vec{\tau}_2)$.The values of these products in states of the two particle system are listed below:

Spin Product	Even Parity States Triplet	Even Parity States Singlet	Odd Parity States Triplet	Odd Parity States Singlet
$\vec{\sigma}_1 . \vec{\sigma}_2$	1	–3	1	–3
$\vec{\tau}_1 \cdot \vec{\tau}_2$	–3	1	1	–3
$(\vec{\sigma}_1 . \vec{\sigma}_2)(\vec{\tau}_1 \cdot \vec{\tau}_2)$	–3	–3	1	9

8.8 HIGH ENERGY n-p AND p-p SCATTERING

High energy scattering experiment can be used to investigate details of the potential shape and to learn about the two nucleon forces in states other than the S-state. The development of the high energy accelerators has made it possible to extend the energy range of n-p and p-p scattering to energies at which the de Broglie wavelength of the relative motion is considerably smaller than the range of nuclear forces. At high energy, the presence of higher angular momentum states results in a large number of parameters, making the data much more difficult to interpret. In the words of Blatt and Weisskopf, *the data available are strange and unexpected.* The analysis is still more complicated for neutrons since a mono-energetic neutron beam is not available. The cross-section, experimentally observed is an average over a rather large energy region and may consequently fail to show rapid energy dependence. The high energy p-p scattering data are easier to interpret than the n-p scattering data, because the Pauli exclusion principle cuts the number of phase shifts in half and the experimental cross-section can be obtained with mono-energetic protons.

1. n-p Scattering Above 10 MeV

Incident neutrons upto 20MeV laboratory kinetic energy are scattered by protons isotropically. Deviations from spherically symmetric scattering are clearly measurable at energies of 27 MeV. At still higher energies

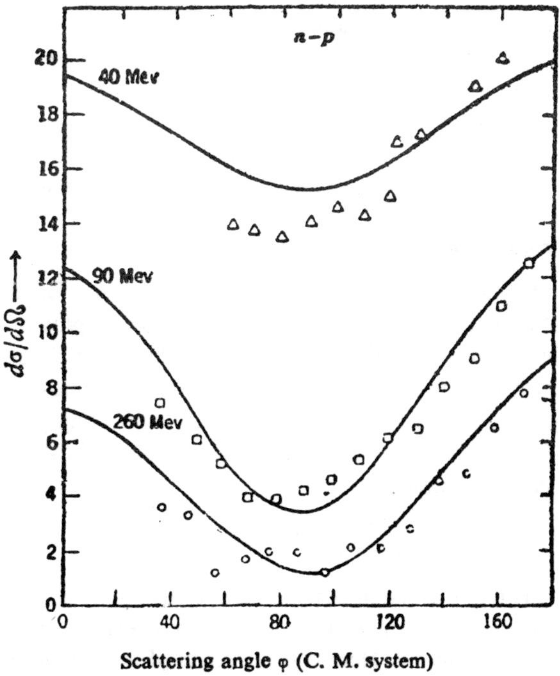

Fig. 4.8. Angular distribution of neutrons scattered by protons.

the angular distribution of the scattered neutrons takes on a characteristic valley shape. The differential cross-section for neutrons of 40, 90 and 260 MeV lab kinetic energy, as a function of neutron scattering angle φ(C.M. system) is shown in Fig. 4.8. The most striking fact about these distributions is the forward ($\phi = 0$) and backward ($\phi = 180°$) maxima. The peak at 0° can be understood in terms of absorption and diffraction of neutrons in the scattering medium. The backward peaking at 180° is the result of the exchange force which interchanges neutron and proton position co-ordinates, the Majorana.

Serber Force

The observed neutron angular distribution has been explained by assuming a half and half mixture of ordinary and exchange force. This mixture is known as a Serber force and is represented by the operator $1/2(1 + P^M)$. The Majorana exchange operator P^M (has a value +1 for even values of l and –1 for odd values of l. Thus, the Serber force is zero for odd states (P, F,...), and hence cannot account for the saturation. In contrast with n-p scattering below 10MeV, the effective range for high energy scattering is dependent on the assumed shape of the potential well. Christian and Hart have used several potential shapes and found that the distribution near 90° was flat with pure central force. The addition of some tensor force makes the theoretical curve less flat near 90°.

Born Approximation

The influence of exchange force and of ordinary forces on high energy scattering can be understood qualitatively by employing the Born approximation, inspite of the fact that the nucleon interactions are not weak. Born Approximation can be expected to give some indication of the true behaviour at least of the less strongly interacting partial waves of higher l. The scattering amplitude as defined in eqn. (35) is given in the Born approximation by

$$f(\theta) = -\frac{M}{4\pi\hbar^2}\int dr\ V(r)e^{i}q.r \qquad ...(108)$$

where V(r) is the interaction potential and q the momentum transferred in scattering, having a value ($k_{final} - k_{initial}$). For elastic collisions with $E_f = E_i$, $|q| = 2k \sin \theta/2$. Here θ is the angle of deflection in the C.M. system and $2\hbar^2 k^2/M = E_i$. Calculation of the high energy cross-section involves the shape of the potential, and no simple general treatment of the problem can be given except at such high energies that the Born approximation is applicable. The approximation is not qualitatively wrong for energies well above the depth of the nuclear well (say 30 MeV or so) and is not satisfactory for quantitative work at any energy.

If the potential V(r) vanishes outside the range R, the Born approximation predicts isotropic scattering, more or less independent of energy if qR < 1. The oscillations of exp. (iq.r.) cause the scattering to fall off rapidly if qR > 1. Near q = 0 (forward scattering) the cross section should stay rather large even as the energy increases. For q = 2k (backward scattering) the cross-section will fall rapidly with increasing energy. The general behaviour of the above eqn. is shown in Fig. 4 9. Eqn. (108) indicates that neutrons scattered at energies in the 100 MeV range mostly

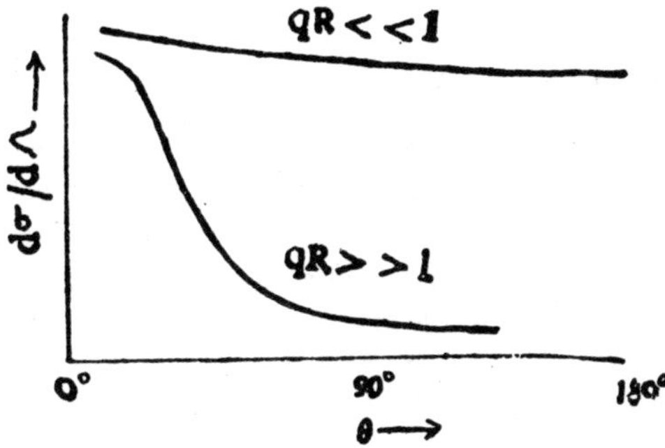

Fig. 4.9 : Born approximation result for n-p scattering.

go forward and recoil protons mostly have low energies in the neighbourhood of 10 MeV. But experimental results showed that the majority of the emergent particles were fast protons moving with nearly the energy and the direction of the incident neutrons. This can be explained if the neutron and proton change roles, or is exactly the consequence of Majorana force.

As the S wave scattering amplitude is independent of the exchange character of the nuclear force, it can be shown that, for the same radial dependence of the force and to the extent that the Born approximation tor scattering in states of $l \neq 0$ is valied. A pure Majorana force given the same scattering in the direction θ as a Wigner force does in the direction $\pi - \theta$.

In other words we can say that the cross-section shows a strong maximum in the forward direction for ordinary forces and in the backward direction for Majorana exchange force. Since the recoiling proton goes forward in the latter case, the exchange potential can be said to cause an exchange of the electric charge from the proton to the neutron which then continues on its way as proton.

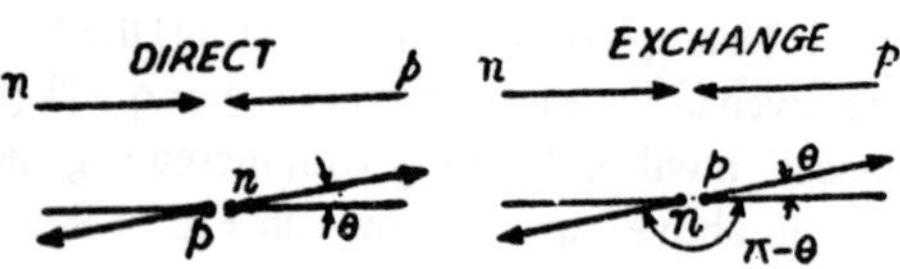

Fig. 4.10 : Direct and exchange n-p scattering.

2. p-p Scattering Above 10 MeV

Just as in the case of n-p scattering, it is expected that more information concerning the shape and exchange character of the potential can be obtained from high energy p-p scattering. The Coulomb potential plays a smaller and smaller part in the scattering as the energy is increased and a direct comparison of the n-p and p-p interactions should be possible at very high energies. In the p-p scattering the identity of the two protons reduces considerably the number of parameters necessary to describe the scattering. As we have already noted, two protons in a singlet state have even parity while those in triplet states have odd parity. If the Serber exchange, suggested by the high energy neutron-proton data, is assumed to act between protons, no scattering occurs in states of odd parity. This result indicates that there would be no triplet scattering of one proton by another. An estimate of high energy scattering can be again made by the Born approximation. If the Coulomb scattering is ignored the differential cross-section calculated in this way for any potential with *Serber Exchange* is just twice that for the n-p scattering calculated in the same approximation with the same potential. A central potential with the range for singlet and triplet states leads, in Born, approximation, to the same relative angular distribution in both states. Therefore, the p-p angular distribution calculated in this case is the same as in the n-p.

The experimental differential cross-sections for high energy p-p scattering are shown in Fig. 4.11. At low angles, cross-section is large because it is dominated by Coulomb scattering. For angles above this small angle region, the cross-section is nearly independent of angle (or *isotropic*). The accuracy is not high at all energies, but around 150 MeV the departure from isotropy is about 3%. An isotropic angular distribution generally implies pure S-wave scattering. This is highly unlikely at incident proton energies of 100 MeV or more, as the total cross-section is almost twice as large as the maximum possible for pure S-wave scattering. At one particular energy, it might happen that interference between P and D wave scattering results in a nearly isotropic scattering intensity, but this would not occur over a range of energies. Another puzzling result is the energy dependence of the total cross-section. The total cross-section is nearly independent of energy for energies from 150 to 400 MeV, and has a value 3.4 ± 0.4 millibarns/steradian. Above 400 MeV the observed cross-section begins to rise significantly as the creation of pions becomes important.

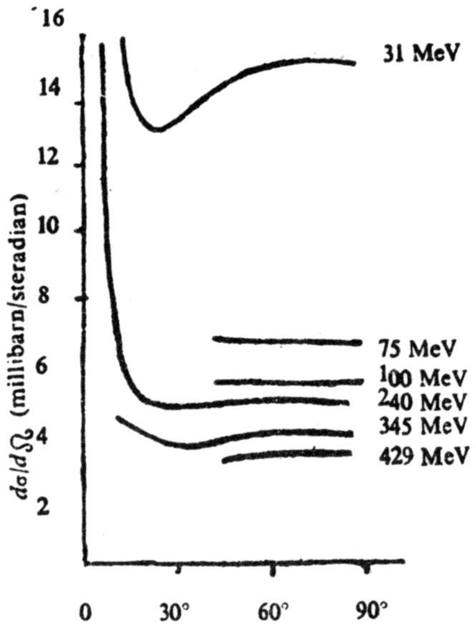

Fig. 4.11 : Experimental differential cross-section for high energy p-p scattering.

Nuclear Core

An interesting explanation of these strange results was suggested by Jastrow. For any conventional potential the S-phase decreases to zero with increasing energy and approximate isotropy is not possible to obtain when phases other then S-phases are involved. It is possible if the S-wave phase shift is assumed-negative which corresponds to a repulsive potential. Since at high energies the S-wave is strongly influenced by the behaviour of the potential at very short distances hence we require a potential that attractive at larger distances but has a strongly repulsive core. The S-wave will have a phase shift which will depend on this central repulsive core and become repulsive over all the high enough energies. The higher l-values are kept away from the core by centrifugal forces. The core radius is of the order of 0.5 fermi. Another useful feature of the repulsive core hypothesis is that it helps to explain the saturation of nuclear forces.

At high energies in n-p and p-p interactions, tensor forces are important. In 1958. Marshak has introduced another term V(r) L.S into the two body potential, where **L** is the orbital and S the spin, angular momentum of the system. Recently a very promising technique involving the scattering of polarized protons has been developed for the more detailed informations about the nucleon-nucleon interaction.

4.9 POLARIZATION, HIGH-ENERGY NUCLEAR SCATTERING

The spin orientations of particles in a beam from an accelerator are normally randomly distributed. If some how the spins are oriented in some preferred direction, the beam is then said to be polarized in that direction. If the number of particles in the beam with spin component parallel to this preferred direction is N_+ and the number with anti-parallel spin component is N_-, then the polarization P, often called the polarizing power, is defined as

$$P = (N_+ - N_-)/(N_+ + N_-). \qquad ...(109)$$

Nucleons may be polarized by elastic scattering on a spinless centre provided that the interaction producing the scattering is spin dependent. We may scatter a beam of protons on a target and select the beam scattered under angle θ_1. By scattering this beam a second time on a target under angle θ_2, we find that the scattered intensity depends only on θ_2, but also on angle ϕ, between the two planes of scattered and can be written as

$$I = A(\theta_2) + B(\theta_2) \cos \phi \qquad ...(110)$$

Let n be the number of particles in the initial unpolarised beam. We can assume that in this beam there are 1/2n particles with spins of each sign (+ or –). After scattering by target T_1 to the left, the number of particles reaching the target T_2 with polarization P_1 and spin up is $1/2nf_1(1 + P_1)$ and with spin down is $1/2nf_1(1 - P_1)$, where f_1 is th fraction of all particles reaching T_2 Similarly those particles scattered to the right by T_1. $1/2nf_1(1 - P_1)$ will reach T_2 with spin up and $1/2nf_1(1 + P_1)$ will reach T_2 with spin down. Let P_2 and f_2 be respectively to polarization with spin up and the fraction of all particles scattered by the target T_2. The number scattered twice to the left with spin up = $1/2nf_1f_2(1 + P_1)(1 + P_2)$ and with spin down = $1/2nf_1f_2(1 - P_1)(1 - P_2)$. Hence, the number of particles scattered twice to the left (LL) is $1/2nf_1f_2(2 + 2P_1P_2)$. The number of particles scattered twice, first to the left by T_1 and then to the right by T_2, with spin up is $1/2nf_1f_2(1 + P_1)(1 - P_2)$. Hence the number of particles scattered twice, first to the left and then to the right (LR) = $1/2nf_1f_2(2 - 2P_1P_2)$.

$$\therefore \text{ Left right symmetry } \in = \frac{(LL)-(LR)}{(LL)+(LR)} = P_1P_2 \qquad ...(111)$$

If the second scattering is identical to the first, then $P_1 = P_2 = P$ and $\in = P^2$. Hence the magnitude, not the sign, of P can be easily obtained from actual experiment. The sign may be determined by studying the interference of nuclear scattering with Coulomb scattering.

Low energy nucleon scattering cannot show any polarization, because only the S-wave contributes. In the energy range above 100 MeV, the scattering of protons and neutrons from heavier nuclei has been observed to polarize the nucleon beam. These scattering experiments provide valuable information concerning the spin dependence of nuclear forces. With a purely central interaction potential and an unpolarized target nucleus, the scattered particles would not show any left-right asymmetry, even with a polarized incident beam. This asymmetry can be explained, if the interaction has a tensor character with the spin-orbit coupling. The asymmetry is larger if orbital angular momentum l is large Thus, the polarization experiments become important at high energy, where high values of l contribute to the phase shift.

4.10. MESON THEORY OF NUCLEAR FORCES

Any attractive force between two particles is regarded as the exchange of an attractive property. The exchanged attractive property between two protons is an electron for an $(H_2)^+$ molecule and is the surrounding electric field for the Coulomb force between two charges. We may introduce similarly a new nuclear field surrounding each nucleon. The application of quantum mechanics to the electromagnetic field surrounding a charged particle leads to the conclusion that the electrical force is exerted by the transfer of a photon, which is referred to as *the field particle of the electromagnetic field,* from one charged body to another. Yukawa (1935) thought that *the strong interaction between nucleons might be accounted for in a similar manner by postulating an appropriate field panicle of rest mass different from zero. This virtual particle has been given the name meson.* To show that the range of force is related to the mass of exchanged particle, assume that the π^o-meson is contained virtually in a proton. A temporary dissociation would be allowable if it does not take a time longer than Δt, given by uncertainty principle

$$\Delta t \approx \frac{\hbar}{\Delta E} \qquad ...(112)$$

Here $\Delta E = (M_2 + m_\pi)\, c^2 - M_p c^2 = m_\pi c^2$

If this virtual particle travels with the velocity of light, as might be expected for a field particle, then the greatest distance the meson could travel in this time, also known as *range of the pion exchange force,*

$$R \simeq c\Delta t \approx \frac{\hbar}{m_\pi c} \qquad ...(113)$$

Based on known nuclear dimensions, Yukawa assumed that the range of the nucleon force and of the field particle would be about 2×10^{-15} m and this led to a value of m_π roughly 200 times the mass of an electron. About two years after Yukawa's theory, particles of mass about 200 m, were discovered in cosmic radiation. These virtual particles are now known as mesons which were thought to be Yukawa field particles for more than ten years. Several different mesons with different charges and rest masses have been discovered. It was found that the Yukawa particle was not the μ-meson but its parent, the shorter lived μ-meson. The pion, which occurs in positive, negative and neutral forms has a mass about 270 times that of the electron and interacts very rapidly with matter. It has other properties spin and parity, which qualify it to be the field particle for nucleon forces.

The attraction between any two nucleons could arise from the transfer of a π^o-meson from one nucleon to the other. The force between a proton and a neutron could result from the transfer of a π^+-meson from the former to the latter or of a π^--meson in the opposite direction. To interpret the nucleon-nucleon scattering data in terms of a potential function, let us compare the meson theory with the quantum theory of electromagnetic interactions. To obtain a pion wave-equation we express the total energy E of a pion in terms of the pion rest mass energy $m_\pi c_2$ and momentum p as

$$E^2 = c^2p^2 + m\pi^2c^2 \quad ...(114)$$

The energy E and momentum component p are represented by the operators

$$E = \frac{i\hbar\partial}{\partial t},\ p_z = -\frac{i\hbar\partial}{\partial x},$$

$$p_v = -\frac{i\hbar\partial}{\partial y} \text{ and } p_z = -\frac{i\hbar\partial}{\partial z}$$

Introducing a pion wavefunction ϕ (a scalar), we obtain the Klein Gordon equation for a free particle of spin 0.

$$-\frac{\hbar^2\partial^2\phi}{\partial t^2} = -c^2\hbar^2\nabla^2\phi + m\pi^2c^4\phi$$

$$\nabla^2\phi - \frac{1}{c^2}\frac{\partial^2\phi}{\partial t^2} - \frac{m\pi^2c^2}{\pi\hbar^2}\phi = 0 \quad ...(115)$$

From $m\pi = 0$, it reduces to the well known wave-equation

$$\left(\nabla^2\phi - \frac{1}{c^2}\frac{\partial^2}{\partial t^2}\right)\phi = 0 \quad ...(116)$$

for the quanta of the electromagnetic field. The simplest type of electromagnetic field is the electrostatic field, where $\partial\phi/\partial t = 0$. The corresponding static pion field is the analogue of Laplace's equation in the absence of electric charges. The presence of charge requires the analogue of Poisson's equation $\nabla^2\partial\phi = ep/\in_0$. where e is the magnitude of the electronic charge and p is the electron particle density. In our present case we have nucleon charges interacting with the pion field. For a static pion potential due to a single point nucleon charge g at the origin, we have an equation

$$\nabla^2\phi - \left(\frac{m\pi^2c^2}{\hbar^2}\right)\phi = 4\pi g\delta(r) \quad ...(117)$$

The solution of this eqn. which vanishes at infinity is:

$$\phi(r) = -\int -\frac{e^{-r}|r-r'|}{|r-r'|}g\delta(r')\delta\tau \quad ...(118)$$

Here $\mu \approx m\pi c/\hbar$ and $\delta\tau'$ is the volume element. If we have a point source of strength g at r_1, then we have,

$$\phi(r) = -g\frac{e^{-\mu}|r-r_1|}{|r-r_1|}$$

$\therefore$ Potential energy of a source g at is in this field is given by $V = g\phi(r_2)$. If the potential ϕ is due to another source g at r_1, then the interaction energy between two sources

$$V = -g^2\frac{e^{-\mu}|r_1-r_2|}{|r_1-r_2|} = -\frac{g^2e^{-\mu r}}{r} \quad ...(119)$$

Comparing with the relation for Coulomb potential, we see that this nuclear potential decreases more rapidly with distance from the source than the Coulomb potential by an exponential factor. At separations small compared to μ^{-1}, this potential energy varies with r just as the Coulomb energy loss. Yukawa noted that this interaction energy may account for file fact that nuclear forces act only over a short range and that the range

$$R = \frac{1}{\mu} = \frac{\hbar}{m\pi c} \quad ...(120)$$

Substitution of the value of $m\pi$ as 270 m_e, gives R = 1.4 fermi.

Since in this theory the nuclear particle does not change its charge, we find that according to the theory n-n, n-p and p-p forces are equal. However, the theory does not explain the exchange nature of nuclear forces, which is established from high energy scattering experiments and is able to explain saturation of nuclear forces. The theory in its simple form cannot explain the spin dependence or the presence of non-central forces. The theory is modified by several theoretical physicists taking into account:

1. the tensoral character of the meson field wave functions,
2. the isobaric-spin character of the meson field wave functions,
3. the intrinsic nature of the source field coupling, and
4. the strength of the source-coupling constant.

The interaction potential between two nucleons so calculated gives approximately $g^2/\hbar c$ = 17. It may be compared with the fine structure constant $e^2/4\pi\epsilon_0\hbar c = 1/137$. The large dimensionless coupling constant $g^2/\hbar c$ suggests that more than one meson is transferred simultaneously between two nucleons. For small distances, we have strong repulsive core arising from the δ function. This core prevents the nucleons from coming close together. When two nucleons collide at very high energy (~BeV), the nucleons can penetrate or disturb each other's core. Thus we see that only the exchange of a relatively small number of mesons can have an appreciable effect on the nuclear force at low and medium energies. On the one hand, we see that the meson theory is qualitatively correct, but on the other hand, not a single quantity has been. calculated and measured which confirms quantitative correctness.

SOLVED EXAMPLES

Example 1:

Recall that b~λ/4'~π/2K for the ground level of the deuteron. From this, show that the radius of the deuteron, in the rectangular well model is approximately given by

$$R = \frac{2bV_0^{1/2}}{\pi B^{1/2}}.$$

Solution:

From eqns. (9) and (10), we have

$$K^2 = \frac{M}{\hbar^2}(V_0 - B) \text{ and } \alpha^2 = \frac{MB}{\hbar^2}$$

$$\therefore \quad \frac{K}{\alpha} = \frac{\sqrt{[(V_0 - B)}}{B}$$

Since $b = \dfrac{\pi}{2K}$ (given) and $R = \dfrac{1}{\alpha}$, hence

$$\frac{\pi/2b}{1/R}\sqrt{\left(\frac{V_0 - B}{B}\right)} \text{ or } R\,\frac{2b}{\pi}\left(\frac{V_0 - B}{B}\right).$$

As $B < V_0$, we can write for an approximation $V_0 - B \simeq V_0$.

$$\therefore \quad \mathbf{R} = \frac{\mathbf{2bV_0^{1/2}}}{\mathbf{\pi B^{1/2}}}.$$

Example 2:

Show that for a square well of depth V_0 and range b, the scattering length a for a spinless neutron is given by the relation

Solution:

K cot Kb = $(b - a)^{-1}$, where $K = (MV_0)^{1/2}/\hbar$.

For a square well of depth V_0 and radius r_0, we have

$$k_1 \cot k_1 b = k \cot (kb + \delta_0)$$

where $$k_1 = \frac{\sqrt{M[(E + V_0)]}}{\hbar} \quad \text{and} \quad k = \frac{\sqrt{(ME)}}{\hbar}$$

Since $V_0 > E$, hence we can write $k_1 = \frac{\sqrt{[MV_0]}}{\hbar} = K$. As the phase shift $\delta_0 = m - ak$ (eqn. 48).

$$\therefore K \cot Kb = k \cot (kb - ak) = k \cot k(b - a)$$

$$= \frac{\mathbf{k}}{\mathbf{\tan k\,(b - a)}} = \frac{\mathbf{k}}{\mathbf{k(b - a)}} = \frac{\mathbf{1}}{\mathbf{b - a}}.$$

Example 3:

Calculate scattering lengths a_t a, d, a_s. Given that σ_{para} = 4.19 and σ_{ortho} = 128 barns.

Solution:

The theoretical expressions for the total cross-sections are

$$\sigma_{para} = 7.69\,(3a_t + a_s)^2$$

$$\sigma_{ortho} = 6.69\,[(3a_t + a_s)^2 + 2(a_t - a_s)^2] + 1.74\,(a_t - a_s).$$

Substituting values of σ_{para} and σ_{ortho} in above relations and solving these equations, we get following four pairs:

$$\begin{matrix} a_t = \\ a_s = \end{matrix} \begin{vmatrix} 0.91 \\ -1.95 \end{vmatrix} \begin{vmatrix} 0.52 \\ -2.34 \end{vmatrix} \begin{vmatrix} -9.52 \\ -3.38 \end{vmatrix} \begin{vmatrix} -0.91 \\ -3.77 \end{vmatrix} \times 10^{-14}\,\text{m}$$

Last two pairs can be discarded immediately because they have negative a_t values. The first pair can also be discarded on the basis that it is inconsistent with the n-p scattering data. Thus, the remaining solution is

$$\mathbf{a_t = 0.52 \times 10^{-14}\ m,\ a_s = 2.34 \times 10^{-14}\ m.}$$

Example 4:

Calculate the total cross-section for n-p scattering at neutron energy 2MeV (lab). Given at = 5.38F, a_s = –23.7F, r_{0l} = 1.70F and r_{0s} = 2.40F.

Solution:

Total cross-section $\sigma = \frac{3}{4}\sigma_l + + \frac{1}{4}\sigma_s$.

Using equation (70), this equation can be written as

$$\sigma = \frac{3\pi a_{t^2}}{\left[1-\frac{1}{2}a_t\, r_{at}\, k^2\right]^2 + a_{t^2}k^2} + \frac{\pi a_{s^2}}{\left[1-\frac{1}{2}a_s\, r_{as}\, k^2\right]^2 + a_{s^2}k^2}$$

In C.M. system $E_{C.M.} = \frac{1}{2}E_{Lab} = 1MeV$

$$\therefore \qquad k^2 = \frac{ME}{\hbar^2} = \frac{1.6748\times10^{-27}\times1.6\times10^{-13}}{(1.0549\times10^{-34})^2}$$

Substituting numerical values, we have

σ = **1.83 × 80^{-28} + 1.074 × 10^{-28} = 2.904 barns.**

EXERCISES

1. Use the well parameters $^3a = {}^1a$ = 2.8 F. $^3l_0$ = 21MeV and $^1V_0$ = 12MeV to compute the scattering cross-section for 2MeV (lab) neutrons. **(3.0 barns)**
2. Use the parameters a_t = 5.38F, r_{0l} = 1.70 F and as = –23.7 F to determine r_0s from the experimental observation that σ = 1.69 barns at energy 4.75 MeV (lab). **(2.4F)**
3. Worthington and others measured a p-p deferential scattering cross section of 0.1114 barns/steradian (C.M. system) for a proton energy of 4.203MeV (lab) and a scattering angle of 60° (C.M. system). Calculate the phase shift. **(46°)**
4. Prove that $(\vec{\sigma}_1 \cdot \vec{\sigma}_2)^2 = 3 - 2\vec{\sigma}_1 \cdot \vec{\sigma}_2$ and that $(\vec{\sigma}_1 \cdot \vec{\sigma}_2)^2 = 0$.
5. Show that p-p scattering can take place only in ^1S, ^2P. ^1D....states, that can you say about the total isospin of the n-p system in these states ?